THE IMPACT OF TECHNOLOGICAL CHANGE ON EMPLOYMENT AND ECONOMIC GROWTH

THE IMPACT OF TECHNOLOGICAL CHANGE ON EMPLOYMENT AND ECONOMIC GROWTH

Papers Commissioned by the
Panel on Technology and Employment
Edited by
Richard M. Cyert
and
David C. Mowery

Committee on Science, Engineering, and Public Policy

National Academy of Sciences
National Academy of Engineering
Institute of Medicine

BALLINGER PUBLISHING COMPANY
Cambridge, Massachusetts
A Subsidiary of Harper & Row, Publishers, Inc.

1988

NOTICE: The National Academy of Sciences was established in 1863 by Act of Congress as a private, nonprofit, self-governing membership corporation for the furtherance of science and technology for the general welfare. The terms of its charter require the National Academy of Sciences to advise the federal government upon request within its fields of competence. Under this corporate charter, the National Academy of Engineering and the Institute of Medicine were established in 1964 and 1970, respectively.

The Committee on Science, Engineering, and Public Policy is a joint committee of the National Academy of Sciences, the National Academy of Engineering, and the Institute of Medicine. It includes members of the councils of all three bodies.

This publication was prepared by the Panel on Technology and Employment of the Committee on Science, Engineering, and Public Policy. The statements, findings, conclusions, and recommendations are those of the authors and do not necessarily reflect the views of the Economic Development Administration or other sponsors.

International Standard Book Number: 0-88730-290-4

Library of Congress Catalog Card Number: 88-11970

Printed in the United States of America

Library of Congress Cataloging-in-Publication Data

The impact of technological change on employment and economic growth / edited by Richard M. Cyert and David C. Mowery : papers commissioned by the Panel on Technology and Employment [of the] National Academy of Sciences, National Academy of Engineering, [and] Institute of Medicine.
    p. cm.
    Includes index.
    ISBN 0-88730-290-4
    1. Labor supply—United States—Effect of technological innovations on. 2. United States—Economic policy—1981-
I. Cyert, Richard Michael, 1921-   . II. Mowery, David C.
III. Committee on Science, Engineering, and Public Policy (U.S.). Panel on Technology and Employment.
HD6331.2.U5147   1988
331'.0973—dc19
                                                                 88-11970
                                                                      CIP

PANEL ON TECHNOLOGY AND EMPLOYMENT

RICHARD M. CYERT *(Chairman)*, President, Carnegie-Mellon University

MORTON BAHR, President, Communications Workers of America

DAVID CASS, Director, Center for Analytic Research in Economics and Social Science, University of Pennsylvania

ALONZO A. CRIM, Superintendent, Atlanta Public Schools

DOUGLAS A. FRASER, Past President, United Auto Workers; Professor of Labor Studies, Wayne State University

RICHARD B. FREEMAN, Professor of Economics, Harvard University

SAMUEL H. FULLER, Vice President, Research and Architecture, Digital Equipment Corporation

JUDITH M. GUERON, President, Manpower Demonstration Research Corporation

ANNE O. KRUEGER, Professor of Economics, Duke University

LAWRENCE LEWIN, President, Lewin and Associates, Inc.

JAMES N. MORGAN, Professor of Economics and Research Scientist, Institute for Social Research, University of Michigan

THOMAS J. MURRIN, President, Energy and Advanced Technology Group, Westinghouse Electric Corporation

ELEANOR HOLMES NORTON, Professor of Law, Georgetown University Law Center

D. RAJ REDDY, Director, Robotics Institute, and Professor of Computer Science, Carnegie-Mellon University

NATHAN ROSENBERG, Professor of Economics, Stanford University

WILLIAM W. SCRANTON, III, Lieutenant Governor, Commonwealth of Pennsylvania, 1979–1987

G. RUSSELL SUTHERLAND, Vice President, Engineering, Deere & Company

MARTA TIENDA, Professor of Rural Sociology, University of Wisconsin

LOUISE TILLY, Chair, Committee on Historical Studies, Graduate Faculty, New School for Social Research

AMY D. WOHL, President, Wohl Associates

STAFF

DAVID C. MOWERY, Study Director

DENNIS HOULIHAN, Assistant to the Director

NINA HALM, Administrative Assistant

SARA COLLINS, Research Assistant

COMMITTEE ON SCIENCE, ENGINEERING, AND PUBLIC POLICY

GILBERT S. OMENN *(Chairman)*, Dean, School of Public Health and Community Medicine, University of Washington, Seattle

H. NORMAN ABRAMSON, Executive Vice President, Southwest Research Institute

FLOYD E. BLOOM,* Director and Member, Division of Pre-Clinical Neuroscience and Endocrinology, Scripps Clinic and Research Foundation

W. DALE COMPTON, Senior Fellow, National Academy of Engineering

EMILIO Q. DADDARIO, Esq., Wilkes, Artis, Hendrick, and Lane

GERALD P. DINNEEN, Vice President, Science and Technology, Honeywell, Inc.

ALFRED P. FISHMAN, William Maul Measey Professor of Medicine and Director, Cardiovascular-Pulmonary Division, University of Pennsylvania School of Medicine

RALPH E. GOMORY, Senior Vice President and Chief Scientist, IBM Corporation

*Term expired February 1987.

CONTENTS

ix

LIST OF FIGURES

LIST OF TABLES

ACKNOWLEDGMENTS

The work of the Panel on Technology and Employment was supported with both public and private funds. Within the federal government, support was provided by the U.S. Department of Labor (the Assistant Secretary for Policy), the U.S. Department of Commerce (the Economic Development Administration), and the U.S. Army Recruiting Command. The following private organizations provided support for the study: the AT&T Foundation, the American Federation of Labor and Congress of Industrial Organizations, Citicorp, the Computer and Business Equipment Manufacturers Association, the General Motors Foundation, IBM Corporation, and the Xerox Foundation.

The panel also received support from the Thomas L. Casey Fund of the National Academy of Sciences and the National Research Council (NRC) Fund. The NRC Fund, a pool of private, discretionary, nonfederal funds, is used to support a program of Academy-initiated studies of national issues in which science and technology figure prominently. The fund consists of contributions from a consortium of private foundations, including the Carnegie Corporation of New York, the Charles E. Culpeper Foundation, the William and Flora Hewlett Foundation, the John D. and Catherine T. MacArthur Foundation, the Andrew W. Mellon Foundation, the Rockefeller Foundation, and the Alfred P. Sloan Foundation; the Academy Industry Program, which seeks annual contributions from companies that are concerned with the health of U.S. science and technology and with public policy issues that have science and technology content; and the National Academy of Sciences and the National Academy of Engineering endowments.

We would also like to express our gratitude to Dr. Leonard Rapping, the panel's study director from June 1985 through March 1986, who was responsible for commissioning a number of the papers published in this volume. Thanks also are due the panel's professional staff: Dennis Houlihan, Nina Halm, and Sara Collins. Finally, we are indebted to the members of the Panel on Technology and Employment, who identified topics for research and reviewed the results of that research.

Richard M. Cyert
Chairman

David C. Mowery
Study Director

INTRODUCTION

Richard M. Cyert and David C. Mowery

The Panel on Technology and Employment was organized by the Committee on Science, Engineering, and Public Policy (COSEPUP), a joint committee of the National Academy of Sciences, the National Academy of Engineering, and the Institute of Medicine, and first met in September 1985. The panel included members from corporate management and organized labor, academic experts on labor markets and worker adjustment, educational leaders, and former federal and state government officials. The report of the panel, *Technology and Employment: Innovation and Growth in the U.S. Economy,* was released in June 1987.[1]

COSEPUP's charge to the panel asked for an assessment of a diverse array of issues related to the employment-related effects of technological change, ranging from the effects of new technologies on firm size to their implications for the distribution of income and employment in the future U.S. economy. In considering these issues, the panel commissioned a number of research papers (a complete listing may be found in the Appendix), asking scholars to survey areas of importance for its inquiry. In most cases, experts in the field were asked to survey an extensive literature and assess its implications for issues of concern to the panel. In other cases, authors undertook original empirical or theoretical analyses of these issues. Many of the papers were presented at meetings of the panel and discussed by the panel. Conclusions of these papers, however, are those of the author(s), and do not necessarily represent the findings or conclusions of the Panel on Technology and Employment.

This volume contains a selection of the papers commissioned by the panel. Limited resources prevented the publication of all of the papers; those not contained in this book can be obtained through the Publication on Demand program of the National Academy Press.[2] The selection of a subset of the papers for publication in this volume—a decision that was very difficult—was based on the relevance of the topics to the panel's intellectual and policy concerns, and does not represent a judgment on the quality of the papers not included in this collection.

The published papers are grouped into four broad areas, dealing with the dynamics of employment growth, decline, and sectoral shifts in the U.S. economy; the effects of technological change on skill requirements and the distribution of earnings and incomes; sectoral patterns of adoption of new technologies; and selected policy issues.

THE EMPLOYMENT AND LABOR MARKET EFFECTS OF TECHNOLOGICAL AND OTHER SOURCES OF ECONOMIC CHANGE

The first section contains three chapters that analyze recent trends in worker displacement, job creation and loss, and the intersectoral flows of jobs and workers within the U.S. economy. In recent years, permanent job losses among experienced workers, worker displacement have contributed to concerns over technological unemployment. Limited anecdotal and statistical data suggest that worker displacement expanded during the early 1980s, particularly when the 1981–82 recession is compared with prior cyclical downturns.

Michael Podgursky utilizes data from the special survey of displaced workers that was included in the Current Population Survey in 1984 and 1986 to provide a statistical description of the characteristics of displaced workers, an assessment of changes between 1984 and 1986 in the rate of worker displacement, and an analysis of the duration of unemployment following displacement.

Podgursky extends the work of Flaim and Sehgal (1985) in several ways, incorporating data from the 1986 survey of displaced workers and developing a statistical model of the determinants of unemployment duration following displacement. Unfortunately, the displaced worker survey data do not provide information on the causes of displacement, making it virtually impossible to assess the role of technological change in the displacement of these workers and preventing any rigorous assessment of the characteristics of technologically displaced workers.

Jonathan Leonard's chapter on labor market dynamics presents a different perspective on the factors affecting aggregate unemployment trends. Leonard uses detailed data on employment and unemployment trends in a single state to develop a statistical description of the flows of newly created and vanishing jobs that underpin aggregate fluctuations in unemployment. Leonard's statistical analysis of the labor market emphasizes the high rates of job creation and destruction within the economy that occur throughout the business cycle—in an average year, nearly 11 percent of jobs vanish, while one-seventh of all jobs are newly created. These large flows of newly created and vanishing jobs, the causes of which are not analyzed in Leonard's chapter, mean that modest changes in the difference between the rates of job creation and destruction can have a substantial impact on aggregate unemployment levels. Leonard's analysis suggests that the U.S. labor market is very dynamic, characterized by large flows of workers between jobs and rapid change in the mix of available openings. This dynamic character of the U.S. labor market should aid adjustment to technological change, inasmuch as high rates of new job creation expand openings for workers displaced from their former positions. Workers must have the skills and information to make the transition between jobs, however, in order for the economy to adjust to technological or other forms of economic change.

Both Leonard and Podgursky discuss trends in employment and unemployment that reflect the operation of many factors that extend well beyond technological change, rather than focusing specifically on the labor market and employment effects of technological change. Robert Costrell's chapter discusses the employment-related effects of technological change, employing a theoretical analysis of the effects of technological change on wages and the distribution of employment and analyzing data on recent trends in wages and employment. Theoretical predictions of the employment and wage effects of technological change depend on a number of offsetting factors, including the extent to which new technologies reduce labor requirements per unit of output or increase the productivity of the capital with which labor is combined to produce output. Costrell's analysis of recent trends in wages and employment concludes that workers moving from manufacturing to the nonmanufacturing sector in recent years (since 1980) have incurred greater wage losses than during any previous portion of the postwar period. The slow growth in real wages and labor productivity that has characterized the U.S. economy since 1973 remains largely unexplained in Costrell's analysis. Costrell argues, however, that the post-1979 recovery of labor productivity growth in manufacturing, especially within durables manufacturing, may have contributed to increases in worker displacement.

THE EFFECTS OF TECHNOLOGICAL CHANGE ON SKILLS AND THE DISTRIBUTION OF EARNINGS AND INCOMES

The next section contains three chapters discussing two controversial aspects of technology's effects on employment—the effects of technological change on the skill requirements of employment and the effects of technological change on the distribution of earnings and incomes within the U.S. economy. Kenneth Spenner surveys the large literature on the skill impacts of technological change. According to Spenner, the conclusions of this research are subject to such enormous uncertainties that policymakers concerned with training and education are well advised to avoid large resource commitments to any specific vision of the detailed occupational structure and skill requirements of the future U.S. economy. Spenner argues that the definitions and measures of employment-related skills on which the study of skill impacts is based are flawed and do not command universal agreement. The data available to measure these skill impacts with any precision are unreliable. Moreover, the ways in which new technologies are implemented affect the skills required to operate them, further obscuring the relationship between technological change and skill requirements.

Many of Spenner's arguments concerning the difficulties in predicting and measuring the skill impacts of new technologies are supported by the next chapter, a study by Martin Binkin of the U.S. military's experience in forecasting and adapting to the changing skill requirements of new weapons systems. As one of the largest employers in the nation and a leading adopter of advanced technologies, the U.S. military has devoted considerable resources to the prediction of the skill impacts of new technologies, while attempting to reduce the skill requirements of advanced new weapons systems. Binkin argues that both of these efforts have been remarkably unsuccessful—not only has the military establishment failed to forecast skill and manpower requirements accurately, but new generations of weapons systems appear to impose unrealistic demands on the skills of enlisted military personnel for field maintenance and operation. Even within an environment in which the design and introduction of new technological systems and the training of personnel to operate these systems are largely controlled by a single organization, the skill impacts of new technologies have created severe difficulties for policymakers.

Recent work on the "vanishing middle class" (such as Kuttner 1983) argues that changes in the structure of the U.S. economy, including technological change, have produced growing polarization in the distribution of incomes within the United States. More recent discussions of "deindustralization" and the "two-tiered workforce" have suggested that technological and other factors have contributed to increased inequality in the earnings of individual workers. Participants in this debate have employed inconsistent data and definitions of such concepts as the distribution of earnings and incomes, contributing to sharply divergent conclusions. McKinley Blackburn and David Bloom survey this debate in their chapter analyzing the factors affecting changes in the distribution of household incomes and individual earnings within the U.S. economy during the past two decades. Blackburn and Bloom agree with other analysts that the distribution of household incomes within the United States has become more unequal during this period, but attribute little if any of this increased inequality to technological change. Instead, these scholars emphasize the role of changes in the structure of households within this economy, especially the distributional implications of increased female labor force participation, in explaining growing inequality.

Changes in the distribution of individual earnings, the area in which any distributional effects of technological change should be particularly visible, have been less substantial, are more uncertain in direction, and appear to reflect nontechnological factors.

SECTORAL PATTERNS OF TECHNOLOGY ADOPTION

The three chapters by Kenneth Flamm, Walter Oi, and Larry Hirschhorn adopt an approach to the analysis of the employment-related effects of technological change that contrasts with that of the previous chapters, focusing on the adoption of specific technologies or the effects of new technologies in specific sectors. Flamm's chapter considers the reasons for slow adoption of this important manufacturing technology within the U.S. economy. Purely economic factors, such as labor costs or interest rates, do not explain the relatively slow (by comparison with other industrial nations) adoption of robotics in the United States, according to Flamm. Although a complete explanation for lagging adoption undoubtedly incorporates some of these purely economic variables, Flamm stresses the role of managerial factors, as did the report of the Panel on Technology and Employment (1987). The tendency for U.S. firms to lag behind Japanese firms in the level of utilization and rate of adoption of robotics appears to be influenced by differences in management goals and strategies between the two nations.

Walter Oi's chapter examines the employment effects of technological change in retail trade and reveals the complexity of the channels though which technological change affects employment. Oi concludes that many of the most important technological influences on employment in the retail trades affected transportation and housing, enabling households to increase their inventories of foodstuffs and other purchased items, reducing the frequency of shopping trips' and contributing to reductions in the wages and skill requirements associated with retail trades employment. Hirschhorn discusses a topic that has only just begun to receive scholarly attention, the adoption and effects of new information technologies in the service sector. Data from a recent survey suggest, somewhat surprisingly, that relatively small firms have adopted small-scale computing technologies rapidly. Hirschhorn's detailed discussion of the employment-related effects of the adoption of advanced information technologies in a large retailing firm and a commercial bank demonstrate once again the limitations of technological determinism in explaining the effects of new technologies on either the skills of employees or the organization of work.

TRADE, TAX, AND DIFFUSION POLICY ISSUES

The final three chapters in this volume consider selected policy issues. Although the Panel on Technology and Employment did not discuss international trade policy in its report, the interaction between technological change and international trade is one of the central themes of that report. The characteristics of technological change and technologically dynamic industries create difficult challenges to the structure and administration of U.S. trade policy. C. Michael Aho discusses the implications for trade policy of new technologies and the policies increasingly adopted by both the U.S. and foreign governments for the promotion of these technologies. The postwar focus of U.S. trade policy on gradual liberalization of world trade through tariff reduction is being eroded by declining domestic political support and changes in the global economic environment. The challenges identified by Aho do not admit of ready answers but will affect both trade policy and the posture of the U.S. government toward the promotion of domestic technological development and international technological cooperation.

The federal tax system historically has been a central instrument of U.S. social and economic policy. In recent years, the R&D tax credit has been employed to support the development of new technologies. Joseph Cordes surveys the evidence on the effects of the R&D tax credit on investment in R&D and discusses the tax treatment of R&D activities in other industrial nations. Cordes also examines the influence of tax policy on the adoption of new technology, in a discussion of the investment tax credit and other revenue instruments for supporting more rapid diffusion. Cordes's chapter is not encouraging to advocates of additional tax incentives for supporting either the generation or the adoption of new technologies. Although the evidence is mixed, the costs of additional tax incentives for R&D investment appear to be substantial, relative to their impact. Direct expenditures may be a more cost-effective mechanism than tax expenditures for the support of R&D.

Technology diffusion is a central issue in the report of the Panel on Technology and Employment because new technologies can only affect employment and other variables once they are adopted within the workplace. David Mowery's chapter reviews the extensive theoretical and empirical literature on the adoption of new manufacturing technologies. Rather than "exploding" through the U.S. economy, Mowery finds that new computer-based manufacturing technologies are not being adopted much more rapidly than previous process innovations. Moreover, rates of adoption of these innovations within the U.S. economy appear to be lagging behind those of other industrial nations, a point made as well in Flamm's chapter. Mowery argues that within the U.S. economy of the near future, open as it is to international competition, employment losses resulting from relatively slow adoption of new technologies within manufacturing are likely to substantially exceed those caused by rapid adoption. The chapter briefly reviews a range of possible policies to support more rapid adoption of these technologies.

CONCLUSION

Based on these and other commissioned papers, as well as staff research and its deliberations, the Panel on Technology and Employment concluded that the rapid generation and adoption of new technologies are essential to maintaining and expanding U.S. employment and wages. Although individuals will face painful and costly adjustments as a result of technological change, society overall gains from technological change. The essential issue for policy therefore is the design of mechanisms that can assist those individuals experiencing adverse economic consequences as a result of technological change.

The papers commissioned by the panel also identify a large number of areas in which current data or knowledge prevent a full understanding of the employment-related effects of technological change. These areas are discussed in greater detail in the panel's report. The data on the rate of adoption of new technologies within the U.S. economy, on rates of growth in output, employment, international trade, and productivity in the nonmanufacturing sector, and on the firm-level employment and skills effects of new technologies, for example, are extraordinarily weak. Research on the effectiveness and design of programs for worker adjustment assistance, an essential component of policies to promote technological change, has received relatively little funding in recent years, leaving policymakers with little basis for the development of effective programs. The determinants of R&D investment and the management of the process of technological change within the firm also remain very poorly understood.

Although the Panel on Technology and Employment considered the characteristics and likely future development of specific technologies in its deliberations on the employment-related effects of new technologies, neither the report nor these commissioned papers attempt to forecast future technological developments in detail. This decision reflects the severe uncertainties that plague any exercise in technological forecasting. In addition, the effects of new technologies on employment have less to do with the specific characteristics of the technologies than the social processes of innovation, diffusion, and investment that determine where and how rapidly they are developed and adopted.

Many of the areas in which further research is needed for the development of effective policies for the generation and adoption of new technologies, policies that can contribute to the international competitiveness of this nation, lie within the social sciences. Research in the physical sciences is indispensable to the development of the scientific and engineering knowledge that supports innovation. Management of the innovation and adoption of new technologies by both public- and private-sector institutions is an area in which this nation has not performed uniformly well in recent years, however, and represents an important avenue for further research that must draw on the social sciences. We hope that these and the other research papers commissioned by the Panel on Technology and Employment will contribute to this research task.

NOTES

1. Available from the National Academy Press, 2101 Constitution Ave., N.W., Washington, D.C. 20418.
2. Copies of these papers are available through the Publication on Demand program of the National Academy Press. Orders and inquiries should be addressed to the attention of Stephen Zubal. There is a charge for reproduction of the papers.

REFERENCES

Flaim, P. O., and E. Sehgal. 1985. "Displaced Workers of 1979–83: How Well Have They Fared?" *Monthly Labor Review* 108 (June): 3–16.

Kuttner, R. 1983. "The Declining Middle." *Atlantic* 252: 60–72.

Panel on Technology and Employment. 1987. *Technology and Employment: Innovation and Growth in the U.S. Economy,* edited by R. M. Cyert and D. C. Mowery. Washington, D.C.: National Academy Press.

THE IMPACT OF TECHNOLOGICAL CHANGE ON EMPLOYMENT AND ECONOMIC GROWTH

THE EMPLOYMENT AND LABOR MARKET EFFECTS OF TECHNOLOGICAL AND OTHER SOURCES OF ECONOMIC CHANGE

1 JOB DISPLACEMENT AND LABOR MARKET ADJUSTMENT
Evidence from the Displaced Worker Surveys

Michael Podgursky

HAS THE RATE OF STRUCTURAL UNEMPLOYMENT INCREASED?

Increased import penetration, new automation technologies, and structural changes in industry have raised public concern about the problem of plant shutdowns and displaced workers in the United States. This concern is manifested in legislation such as Title III of the Job Training Partnership Act and the dozens of state and local adjustment programs it has spawned. It also is reflected in the renewed research interest in the problem of job displacement and structural unemployment.

Like the discussion of the structural unemployment problem in the early 1960s (see Gilpatrick 1966), the current discussion takes place against a background of an upward drift in the unemployment rate from one economic expansion to the next during the 1970s and 1980s (Figure 1–1). This secular rise has occurred for all major subgroups of the labor force, including groups such as married men (spouse present), who are least prone to voluntary job turnover. In the unemployment debate of the early 1960s structuralist explanations focused on the changes in the composition of labor demand and rising education and skill requirements brought about by new technologies. Explanations of the high "natural" or

Figure 1–1. Selected Unemployment Rates, 1948–85.

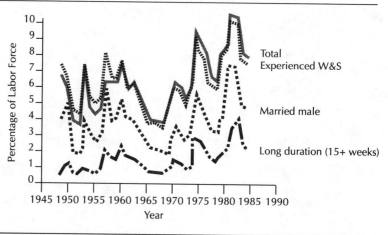

Total: civilian labor force
Experienced W&S: experienced wage and salary workers
Married male: married males, spouse present
Long duration: long duration unemployment (15+ weeks) as
percentage of civilian labor force

"nonaccelerating rate of inflation" level of the unemployment rate during the 1970s, however, tended to focus on the supply side of the labor market and emphasize the high rate of voluntary turnover among various groups of workers in the labor force (Feldstein 1983; Marston 1976).

Perhaps the most important supply-side factor at work during the last two decades was the rapid absolute and relative growth of the youth labor force due to the maturation of the post-war baby boom cohorts. It was argued that the high relative rate of unemployment among this group derived not from a shortage of jobs but rather from high rates of voluntary job turnover. This instability was blamed in part on the minimum wage, which prevented employers from producing an adequate supply of "good" jobs offering training and opportunities for advancement. Even in the absence of the minimum wage, however, the underlying frictional rate of unemployment would inevitably rise as the economy absorbed this burgeoning supply of new job entrants. Similar arguments were raised regarding women. The rising participation rate of women—particularly married women—was also considered a factor in the rising natural rate of unemployment. Finally, the effect of the unemployment insurance system in extending the duration of job search and encouraging job instability was also thought to play a role in producing a high natural rate of unemployment (Feldstein 1983).

Yet all the supply-side forces that produced a rise in the natural rate of unemployment during the 1970s should be acting to lower the natural rate in the mid-1980s. First, the share of youth in the labor force is falling rapidly and is predicted to continue to do so through the 1990s. Moreover, the wage flexibility in the youth labor market has surely been enhanced. The real value of the minimum wage has fallen sharply. In dollars of current purchasing power, the minimum wage has fallen from approximately $4.75 per hour in 1969 to its current level of $3.35 per hour. Finally, the accelerated shift from goods-producing to service-producing industries has all but eliminated any wage stickiness associated with collective bargaining in the private-sector component of the youth labor market.

Although the labor force participation rate and the labor force share of women continues to rise, it does so at a deaccelerating rate. In addition, although the level of job turnover of women may be greater, their average rate of unemployment in any month is now typically lower than that of men. Finally, whatever the quantitative effect of the unemployment insurance system on jobless duration, the share of the unemployed who are receiving unemployment insurance has fallen sharply in the 1980s relative to the 1970s and is now less than one-third of the unemployed (Burtless 1983).

Against this background, it is hardly surprising that attention has begun to focus on the level, structure, and instability of labor demand as a source of the rising natural rate of unemployment (Leonard 1987; Lillien 1982). This chapter attempts to shed some light on this problem by examining data on the extent and consequences of permanent job displacement among experienced workers using data from special "Displaced Worker Surveys" conducted by the U.S. Bureau of Labor Statistics in January 1984 and January 1986. A brief review of the literature on job displacement describes the January 1984 and 1986 "Displaced Worker Surveys" and examines regional, industrial, and occupational patterns of job displacement. This review is followed by an examination of aspects of the postdisplacement experience of workers—particularly their duration of joblessness, where they find new jobs, and the loss of earnings and benefits suffered as a result of displacement. The analysis in these two sections is based on previously unpublished tabulations from the January 1984 and January 1986 "Displaced Worker Surveys," as well as results already reported (Podgursky and Swaim 1987a, 1987b, 1987c, 1986a, 1986b). A final section summarizes the basic findings and discusses implications for labor market policy.

THE EXTENT OF JOB DISPLACEMENT

Until recently data was lacking to answer many basic questions regarding the displacement problem. How many workers lose jobs due to plant shutdowns or job displacement? Where do they find new jobs? At what wages? This section begins with a brief overview of the literature and data sources on plant shutdowns and displacement. This is followed by a description of the "Displaced Worker Survey"—a special survey developed by the U.S. Bureau of Labor Statistics to answer some of the question posed above. The section concludes with summary statistics on the level and patterns of job displacement from the "Displaced Worker Surveys."

Studies of Job Displacement

The mainstay of the literature on job displacement and plant shutdowns has been case studies of particular plant shutdowns, groups of workers, or communities (Craypo and Davisson 1983; Folbre, Leighton, and Roderick 1984; Gordon, Schervish, and Bluestone 1985; Grodus, Jarley, and Ferman 1981; Levy, Jondrow, and Jacobson 1985). Covering many different locales and time periods, these case studies have generally found large short-term losses for displaced workers and considerable adjustment difficulties for certain subgroups of workers. These studies also provide valuable qualitative insights into the adjustment problems faced by displaced workers and practical guidance for the development of labor market policy. By their very nature, however, case studies cannot tell us about the national extent of the displacement problem. Moreover, because the cases are not randomly selected, the possibility exists that their results may not be representative of general experience.

Given the inherent limitations of the case study approach, efforts have been made to develop microdata files covering a broader universe of individuals or businesses. Here it is useful to distinguish microdata files in which the unit of observation is an individual worker from those in which the unit of observation is an establishment. A number of recent studies using establishment-level data suggest that the net job growth observed in the labor market results from the simultaneous destruction and creation of a large number of jobs. The result is a considerable turbulence or "churning" in labor markets and a high risk of job displacement for workers currently employed. One of the earliest and most widely cited studies making this point is Bluestone and Harrison (1982) who use Dun & Bradstreet establishment data to estimate national and regional rates of job "births" and "deaths." They find evidence of widespread job displacement (also see Harris 1984). Preliminary results of a recent study by the GAO (Gainer 1986), however, suggest that the reporting errors in the Dun & Bradstreet file produce an inflated estimate of displacements. After correcting these errors, the GAO researchers still find a high rate of displacements, although lower than that reported in earlier studies. Similar findings are reported in studies by Leonard (1985, 1987), who estimates econometric models of employment growth and decline in business establishments on two different longitudinal establishment surveys. These studies shed a great deal of light on the job creation process—and in so doing discount certain exaggerated claims about job creation by small business.

Another establishment-level data base holding potential for future research on job displacement is under development by the Bureau of Labor Statistics. Using state unemployment claims data, the Permanent Mass Layoff and Plant Closing (PMLPC) program will identify establishments with mass layoffs or plant closings and track workers through the UI system. Thus far only limited data for seven states have been released (U.S. Department of Labor 1987), but the BLS estimates that data will be available for most states within two years.

Although analysis of establishment-level data can potentially tell us a great deal about the process of job creation and destruction in the economy, it tells us very little about the workers whose jobs are destroyed. For this we must turn to individual or worker microdata files. Numerous worker or household microdata files are available, but each has its limitations in examining the experience of displaced workers because none were explicitly designed for this purpose.[1] The creation of special retraining and adjustment programs for "dislocated" workers under Title III of the Job Training Partnership Act led to several attempts to estimate the population of potentially eligible workers using data on long-term unemployment in declining industries, occupations, and regions from the *Current Population Survey* (Bendick and Devine 1981; U.S. Congressional Budget Office 1982). Other studies use similar criteria to indirectly identify displaced workers in Social Security or UI data files to estimate earnings losses and recovery after displacement (Crosslin, Hanna, and Stevens 1984; Jacobson 1978, 1984; Lillien 1982; Sabelhaus and Bednarzik 1985). Finally, several studies have examined earnings losses and unemployment duration for relatively small samples of workers reporting permanent layoffs during the 1960s and 1970s in the Panel Study of Income Dynamics and the National Longitudinal Survey, two longitudinal data files (Ehrenberg and Oaxaca 1976; Hamermesh 1987; Ruhm 1986; Shapiro and Sandell 1985).

What has been lacking as a resource for this type of research is a large nationally representative microdata base that directly identifies workers who lose jobs due to plant shutdowns or are otherwise displaced. A logical place to identify such workers is the *Current Population Survey*, the same household survey instrument used to identify employed and unemployed workers in the labor force. In response to the growing public concern with this problem and the difficulties in gleaning information on displacement from existing data sources, the Bureau of Labor Statistics added a special "Displaced Worker Survey" to the basic *Current Population Survey* in January 1984 and again in January 1986. The remainder of this section examines data on the extent and patterns of job displacement from these new surveys.

The "Displaced Worker Survey"

In January 1984 and 1986, all respondents from the roughly 60,000 households in the *Current Population Survey (CPS)* were asked whether they or any adult member of their household (ages twenty and older) had "lost or left a job since 1979 because of a plant closing, an employer going out of business, a layoff from which . . . was not recalled or other similar reasons." An affirmative response triggered a series of supplemental questions concerning the nature of the job lost and postdisplacement labor market experience. These supplemental questions, which comprise the "Displaced Worker Survey" (DWS), augment the extensive demographic and labor force data in the basic monthly *CPS* survey. Following the Bureau of Labor Statistics, a *displaced worker* is defined as one who lost a job due to a plant shutdown, business failure, or relocation or as a worker whose employer remained in operation but who was permanently laid off due to slack work or whose job was otherwise eliminated.[2]

The analysis in this chapter limits the sample to workers between the ages of twenty and sixty-one who were displaced from full-time nonagricultural wage and salary jobs. The reasons for selecting this sample are discussed in the next section, which examines postdisplacement experience. At this point it is sufficient to note that we have tried select a group that has demonstrated a strong attachment to the workforce and that is generally the focus of policy interest. With this operational definition in mind, we now examine the survey data in Table 1–1, which presents displacements by year for various groups.

Note that these time-series arise from the retrospective data on the two surveys—that is, workers in January 1984 and January 1986 were asked the year in which they lost their job. This analysis focuses on the labor market experience of *permanently* displaced workers and not workers who are cyclically displaced and thus may return to their former employer when business improves. In order to minimize cyclical unemployment, the analysis is limited to workers who have been away from their former job for at least one year. Thus for the 1984 DWS we sample workers displaced between 1979 and 1982, and for the 1986 DWS, workers displaced between 1981 and 1984. In spite of the economic expansion under way at the time of the January 1984 and 1986 surveys, these workers have not been recalled to their old jobs in at least one year, and for some in five years.[3]

With these considerations in mind, we examine the last column of Table 1–1, which presents average displacements during 1979–82 and 1981–84 for the two surveys. Both of these surveys

Table 1-1. Total Displacement: The January 1984 and January 1986 Displaced Worker Surveys (thousands of workers).[a]

Group	1979	1980	1981	1982	1983	1984	1985	Average 1979–82	Average 1981–84
January 1984 displaced worker survey									
Total	951	1,256	1,738	2,440	2,460	—	—	1,596	—
Blue-collar	572	782	1,050	1,454	1,431	—	—	965	—
White-collar and service	379	474	688	987	1,029	—	—	632	—
Older (40–55)	199	255	355	460	551	—	—	317	—
Plant shutdowns	542	640	703	958	878	—	—	711	—
Blue-collar	315	391	385	510	464	—	—	400	—
White-collar and service	227	249	318	448	415	—	—	311	—
Older (40–55)	142	138	161	226	231	—	—	167	—
January 1986 displaced worker survey									
Total	—	—	1,265	1,676	1,510	1,694	2,409	—	1,536
Blue-collar	—	—	776	999	828	915	1,384	—	879
White-collar and service	—	—	490	677	679	779	1,025	—	656
Older (40–55)	—	—	289	399	306	359	565	—	338
Plant shutdowns	—	—	727	841	790	860	856	—	805
Blue-collar	—	—	448	507	460	462	433	—	469
White-collar and service	—	—	279	334	330	398	423	—	335
Older (40–55)	—	—	182	229	180	167	219	—	190

a. Workers displaced from full-time nonagricultural wage and salary jobs, ages twenty to sixty-one in the year of displacement. Tabulations exclude a small number of workers who were formerly employed as private household workers or who did not identify their former occupation. Agriculture includes the following 1980 Census industries: Agricultural Production (10, 11); Agricultural Services (20); and Horticultural Services (21).

indicate a considerable rate of displacements—and a rate that has not greatly diminished with the recovery following the 1981–82 recession. Plant shutdowns account for roughly one-half (45 and 52 percent) of total displacements in the two surveys.

The statistics in Table 1–1 also suggest that dropping workers laid off in the year prior to the survey is a prudent procedure because some workers who report themselves displaced but whose employer remains in operation seem to get recalled to their old jobs. For displacements in the year immediately prior to the two survey dates (1983 and 1985), the ratio of plant shutdown to total displacements is much lower—36 percent in both surveys. This suggests that some of these recently laid off workers whose employer remains in operation are not permanently terminated and will be recalled. Further evidence on this point is found when displacements in the overlapping years 1981 and 1982 are examined. Total displacements are lower in the January 1984 than the January 1986 survey, yet the proportionate discrepancy between the two surveys is much lower for plant shutdowns than for total displacements. Indeed, if we combine displacements for these two years and compare the totals for the two surveys, nonplant shutdown displacements account for 92 percent of the discrepancy. Once again, this suggests that some laid off workers whose employer remains in operation eventually drift back to their old jobs, albeit after a very long absence. Workers whose plants shut down do not.[4]

In sum, the data from the two surveys show that the level of displacements did rise during the 1981–82 recession but that the annual level of displacements remained quite high during the subsequent expansion. This suggests that these displacements stem primarily from structural rather than cyclical forces.

In Table 1–2 we examine regional patterns of job displacement. The first column presents the distribution of the full-time adult workforce across the four Census regions, and columns (2) through (5) present displacement data from the two surveys. Columns (6) through (10) in the bottom panel present regional data on payroll employment and displacements for manufacturing industries.[5]

In the upper panel, the figures in parentheses give each region's share of the experienced adult full-time workforce and displacements. The Midwest, for example, accounted for 24.2 percent of the workforce but 30.6 and 27.6 percent of total displacements in

Table 1-2. Displacement by Region.[a]

All Nonagricultural Industries	(1) Experienced Full-Time Adult Workforce 1983 (000)	Workers Displaced			
		1979–82		1981–84	
		(2) Total (000)	(3) Plant Shutdown (000)	(4) Total (000)	(5) Plant Shutdown (000)
Total United States	73,932 (100.0%)	6,812 (100.0%)	3,037 (100.0%)	6,364 (100.0%)	3,335 (100.0%)
Northeast	16,088 (21.8)	1,134 (16.6)	545 (17.9)	1,069 (16.8)	604 (18.1)
Midwest	17,923 (24.2)	2,086 (30.6)	813 (26.8)	1,755 (27.6)	860 (25.8)
South	25,561 (34.6)	2,222 (32.6)	1,024 (33.7)	2,225 (35.0)	1,198 (35.9)
West	14,360 (19.4)	1,370 (20.1)	655 (21.6)	1,315 (20.7)	673 (20.1)

Manufacturing	(6) Average Payroll Employment (000)	Workers Displaced			
		1979–82		1981–84	
		(7) Total (000)	(8) Plant Shutdown (000)	(9) Total (000)	(10) Plant Shutdown (000)
Total United States	90,652 (100)	3,277 (100.0)	1,415 (100.0)	3,040 (100.0)	1,614 (100.0)
Northeast	20,492 (22.6)	601 (18.3)	277 (19.6)	583 (19.2)	342 (21.2)
Midwest	23,229 (25.6)	1,141 (34.8)	423 (29.9)	917 (30.2)	446 (27.6)
South	17,581 (19.4)	1,030 (31.4)	474 (33.5)	1,064 (35.0)	591 (36.6)
West	29,350 (32.4)	505 (15.4)	241 (17.0)	476 (15.7)	235 (14.6)

a. Column (1): Workers ages twenty to sixty-one who worked primarily full time in nonagricultural industries (excluding private household workers) in 1983. Unpublished BLS tabulations. Columns (2)–(4) and (7)–(10): Workers ages twenty to sixty-one displaced from full-time nonagricultural wage and salary jobs. Totals exclude a small number of workers in forestry and fisheries industries and private household workers. Computed from BLS microdata tapes. Column (6): Average annual payroll employment, 1979–83 (*Employment and Earnings* May 1982: 106–17, May 1985: 118–35).

the two surveys, which suggests that the workers in the Midwest faced a risk of displacement somewhat above the national average. Surprisingly, sunbelt states in the West and South also made a contribution to displacements that tended to exceed their share of payrolls, particularly if we examine manufacturing displacements in the lower panel. The states that unambiguously fared best were in the Northeast. Although the distribution of displacements across regions is not uniform, they are not highly concentrated in any particular region. On the contrary, job displacements are pervasive in the labor markets of all the regions.

Table 1–3 presents data on displacements and employment by industry. As in the previous table, column (1) presents the distribution of the full-time adult workforce. Columns (2) and (3) present displacements from the two surveys. Comparing the share figures in parenthesis we find that goods-producing industries accounted for over 60 percent of total displacements, with durable manufacturing making the largest contributions (32.4 and 32.1 percent) of any industry group. A comparison of the share figures in column (1) shows that the goods-producing industries clearly account for a very disproportionate share of total displacements. Having noted this imbalance, we should also note that service-producing industries still account for a major fraction of total displacements. Of course, displacements in goods-producing industries may also indirectly contribute to displacements in the service industries. Other studies have noted the multiplier or ripple effect produced by the opening or closing of a manufacturing plant in a local community. In examining patterns of displacement across states, we also find that states with above-average rates of manufacturing displacements also have above-average rates of nonmanufacturing displacements.

Table 1–4 reports the distribution of displaced workers by occupational group. As in the other tables, the first column presents the occupational distribution of the full-time adult workforce, and the next columns the occupational distributions in the two surveys. A comparison of the percentage distributions for the two groups shows clearly the concentration of displaced workers in blue-collar occupations. While blue-collar workers represent approximately one-third of the full-time adult nonagricultural workforce, they made up roughly 60 percent of workers displaced from full-time jobs in the two surveys (60.5 and 56.7 percent). Among blue-collar

Table 1–3. Displacement by Industry.[a]

		Workers Displaced	
	(1)	(2)	(3)
	Experienced Full-Time Adult Workforce		
Industry	(1983)	1979–82	1981–84
Total (000) (%)	73,811 (100.0%)	6,773 (100.0%)	6,351 (100.0%)
Goods-producing	24,893 (33.7)	4,173 (61.6)	3,868 (60.9)
Mining	887 (1.2)	193 (2.9)	271 (4.3)
Construction	4,939 (6.7)	702 (10.4)	557 (8.8)
Durable manufacturing	11,055 (15.0)	2,197 (32.4)	2,042 (32.1)
Nondurable manufacturing	8,012 (10.9)	1,081 (16.0)	998 (15.7)
Service-producing	48,918 (66.3)	2,600 (38.4)	2,384 (39.3)
Transportation, Commuting, and public utilities	6,352 (8.6)	470 (6.9)	480 (7.6)
Wholesale trade	3,388 (4.6)	365 (5.4)	328 (5.2)
Retail trade	9,090 (12.3)	635 (9.4)	530 (8.3)
Finance, insurance, and real Estate	5,121 (6.9)	153 (2.3)	144 (2.3)
Business and repair services	3,316 (4.5)	370 (5.5)	449 (7.1)
Personal, entertainment, and recreation services	2,033 (2.8)	166 (2.4)	151 (2.4)
Professional services	14,995 (20.3)	300 (4.4)	325 (5.2)
Public administration	4,623 (6.2)	141 (2.1)	77 (1.2)

a. Column (1): Workers ages twenty to sixty-one who worked primarily full time in nonagricultural industries (excluding private household workers) in 1983. Unpublished BLS tabulations. Columns (2)–(4): Workers ages twenty to sixty-one displaced from full-time nonagricultural wage and salary jobs. Totals exclude a small number of workers in forestry and fisheries industries and private household workers. Computed from BLS microdata tapes.

Table 1–4.   Displacement by Occupation.[a]

| | | Workers Displaced | |
| | (1) Experienced Full-Time Adult Workforce | (2) | (3) |
Occupation	(1983)	1979–82	1981–84
Total (000) (%)	73,331 (100.0%)	6,730 (100.0%)	6,295 (100.0%)
Blue-collar	24,713 (33.7)	4,072 (60.5)	3,567 (56.7)
Precision production, craft, and repair	10,496 (14.3)	1,452 (21.6)	1,323 (21.0)
Machine operators, assemblers, and inspectors	7,375 (10.1)	1,618 (24.0)	1,408 (22.4)
Transportation operative	3,641 (5.0)	409 (6.1)	349 (5.5)
Handlers, cleaners, and laborers	3,201 (4.4)	593 (8.8)	487 (7.7)
White-collar and service	48,618 (66.3)	2,658 (39.4)	2,728 (43.3)
Managers and administrators	8,681 (11.8)	547 (8.1)	614 (9.8)
Professional	10,050 (13.7)	338 (5.0)	412 (6.5)
Technical	2,769 (3.8)	170 (2.5)	188 (3.0)
Administrative support	13,022 (17.8)	660 (9.8)	638 (10.1)
Sales	6,687 (9.1)	622 (9.2)	532 (8.5)
Service	7,409 (10.1)	321 (4.8)	344 (5.5)

a. Column (1): Workers ages twenty to sixty-one who worked primarily full time in nonagricultural industries (excluding private household workers) in 1983. Unpublished BLS tabulations. Columns (2)–(4): Workers ages twenty to sixty-one displaced from full-time nonagricultural wage and salary jobs. Totals exclude a small number of workers in forestry and fisheries industries and private household workers. Computed from BLS microdata tapes.

workers, the imbalance is particularly notable among machine operators and assemblers, who comprise 10.1 percent of the workforce but over 20 percent of displacements (24.0 and 22.4 percent). Although blue-collar workers make up a disproportionate share of the displaced, white-collar and service workers also make a substantial—and growing—contribution (39.4 and 43.3 percent).

Blue-collar workers are clearly more likely to be displaced, but in what other respects do displaced and nondisplaced workers differ? Table 1–5 provides some answers by comparing demographic characteristics of workers who were displaced with workers who were not. The columns labeled (1) report the mean characteristics by industry and broad occupational group of full-time wage and salary workers employed in January 1984 who reported that they were not displaced during the previous five years. The columns labeled (2) and (3) report the same statistics for workers displaced from full-time jobs in these industries in the two surveys. The comparison shows that displaced workers tend to be several years younger than their nondisplaced counterparts. Among blue-collar workers, women and blacks generally tend to be overrepresented among the displaced. The discrepancy is particularly apparent for women and may derive in part from their lower relative seniority. Among both blue-collar and white-collar and service workers, displaced women had approximately one year less seniority than displaced men.

What forces are causing these displacements? We have already noted that the sample has been constructed in such a way as to minimize cyclical factors. Although we can discount cyclical factors, the ultimate structural causes of the job loss are difficult to determine from a household survey. Respondents to a household survey may reasonably be expected to know about wages and labor market experience following displacement, but they can hardly be expected to know what economic forces led to the loss of their jobs or the jobs of other household members. The heavy concentration of displacements in manufacturing industries—approximately one-half of both DWS samples—certainly suggests that international competition and perhaps technological change may have played a role. Two recent studies shed some light on this matter. A Commerce Department study (Young, Lawson, and Duncan 1986) estimates that 2.4 million manufacturing jobs were lost due to a decline in net exports (that is, exports minus imports) between

Table 1-5. Demographic Characteristics of Displaced Workers in Selected Industries.

	Blue-Collar			White-Collar and Service		
	(1) Employed Nondisplaced	(2) Displaced 1979–82	(3) 1981–84	(1) Employed Nondisplaced	(2) Displaced 1979–82	(3) 1981–84
Industry						
Mining						
female (%)	1.7	4.8	6.3	—	—	—
black (%)	5.4	3.0	1.6	—	—	—
Education	11.3	11.5	11.5	—	—	—
Age	37.9	32.3	33.7	—	—	—
Construction						
female (%)	.8	6.0	1.9	—	—	—
black (%)	8.0	13.0	10.9	—	—	—
Education	11.3	11.6	11.7	—	—	—
Age	35.9	33.5	34.4	—	—	—
Manufacturing						
female (%)	29.4	32.6	33.1	33.8	42.8	39.1
black (%)	12.7	15.5	13.3	5.8	7.5	7.7
Education	11.2	11.5	11.5	13.9	13.4	13.5
Age	38.7	34.6	35.9	39.5	35.9	36.3

Transportation, Commuting, and Public Utilities

female (%)	4.6	9.1	7.0	42.3	47.7	44.3
black (%)	12.5	9.0	11.7	12.4	7.9	8.2
Education	11.8	11.9	11.8	13.3	13.0	13.6
Age	39.5	34.1	36.9	39.1	36.4	34.5

Wholesale and Retail Trade

female (%)	11.5	20.0	31.3	46.6	47.4	48.3
black (%)	10.5	10.3	15.9	6.4	6.7	8.1
Education	11.6	11.5	11.4	12.9	12.7	12.9
Age	35.2	33.5	35.5	36.9	36.3	35.6

Services

female (%)	16.2	23.3	27.2	62.9	58.1	55.3
black (%)	12.0	14.1	15.5	11.5	9.6	11.2
Education	11.3	11.7	11.9	14.2	13.6	13.6
Age	38.1	31.9	32.8	38.5	34.0	34.1

Column (1): Wage and salary workers ages twenty to sixty-one employed in full-time nonagricultural wage and salary jobs in January 1984 who did not experience job displacement over the previous five years.

Columns (2) and (3): Workers ages twenty to sixty-one displaced from full-time nonagricultural wage and salary jobs (excluding private household workers).

1977 and 1984. It should be noted, however, that such estimates of "lost jobs" do not necessarily imply worker displacement because some of the decline may be accomplished through voluntary job turnover and attrition. On the other hand, if firms that remain in operation respond to intensified foreign competition by domestic plant relocation or automation, further displacements may result. As to the effect of technological change, the GAO establishment survey mentioned in the previous section found that 16 percent of firms reporting plant shutdowns or major layoffs cited automation as a significant factor in their decision. Nonetheless, automation ranked well below product demand, increased competition, exchange rates, and government regulation among factors cited by survey respondents (U.S. Congressional Budget Office 1982).

In sum, a household survey such as the "Displaced Worker Survey" provides only limited information as to the structural causes of displacement. Its major value lay not in telling why workers were displaced, but in helping to gauge the overall incidence of this phenomenon in the labor force and of equal importance, how workers adapt—or flounder—following displacement.

LABOR MARKET EXPERIENCE FOLLOWING DISPLACEMENT

How do workers fare after displacement? How long does it take them to find new jobs? How does compensation on their new jobs compare with their former jobs? This section answers some of these questions by focusing on the postdisplacement experience of a group of workers from the DWS surveys. Before examining their experience, it would be helpful to explain how the sample is defined. First, workers ages sixty-two and older are excluded because these workers will generally be eligible for Social Security retirement payments (and possibly private pensions as well). They thus face a different set of constraints or choices regarding the labor market than do younger workers. Second, we also limit attention to workers who were displaced from full-time nonagricultural wage and salary jobs. We focus on workers losing full-time jobs for reasons of necessity and choice. The "Displaced Worker Survey" provides information on only usual weekly earnings and full-time/part-time status of the worker's former job. By limiting our sample to full-time workers we thus control in a crude manner for hours of work on the old job. Since full-time workers have made a greater commitment to the labor market and, in general, make a larger contribution to the typical household's income than do part-time workers, we judge displacement of such workers to be of the greatest policy interest. Finally, as indicated in the previous section, we include in our sample only workers who have been displaced for a year or more—that is, who have had *at least* one full year to adapt to the loss of the former job by the time of the January 1984 and January 1986 survey dates.

Having thus restricted our sample we are left with 4.2 and 3.6 million blue-collar workers (approximately 75 percent male) and 2.7 million white-collar and service workers (approximately 50 percent male) in the two surveys. The labor force status of these workers in January 1984 and January 1986 is shown in Table 5–6 below.

Although most of the workers in our sample found new jobs by the survey date the unemployment rate among blue-collar workers (25.2 and 15.0 percent) and white-collar and service workers (13.8 and 9.4 percent) remained quite high. By comparison, the unseasonally adjusted unemployment rate for blue-collar workers was 12.2 percent in January 1984 and 10.6 percent in January 1986.[6] Unemployment rates in the same two months for white-collar and service workers were 5.6 and 4.7 percent. The unemployed displaced workers in our sample accounted for approximately 12 percent of all unemployment and 15 percent of adult unemployment in January 1984. Similar displacement shares for January 1986 unemployment are 8 and 10 percent.

The Duration of Joblessness Following Displacement

How long did it take for these workers to find new jobs? In the "Displaced Worker Surveys" interviewers were to ascertain the number of weeks workers where without work and "available" for work following displacement. Several caveats are in order before we examine tabulations on this variable. First, unlike the more familiar measure of unemployment duration in the monthly *CPS*, the DWS does not ascertain whether the worker was engaged in

Table 1-6. Current Labor Force Status of Displaced Workers.[a]

Labor Force Status	Blue-Collar (000)			White-Collar and Service (000)		
	Total	Male	Female	Total	Male	Female
January 1984 (1979–82)						
Total (% of total)	4,194	3,173	1,021	2,657	1,338	1,319
	(100.0%)	(100.0%)	(100.0%)	(100.0%)	(100.0%)	(100.0%)
Employed	2,802	2,246	556	1,991	1,109	882
	(66.8)	(70.7)	(54.5)	(74.9)	(82.9)	(66.9)
Unemployed	945	721	224	319	160	159
	(22.5)	(22.7)	(21.9)	(12.0)	(12.0)	(12.1)
Not in labor force	447	206	241	347	69	278
	(10.7)	(6.5)	(23.6)	(13.1)	(5.2)	(21.1)
Unemployment rate (%)[b]	25.2	24.3	28.7	13.8	12.6	15.2
January 1986 (1981–84)						
Total (% of total)	3,635	2,725	910	2,729	1,444	1,285
	(100.0)	(100.0)	(100.0)	(100.0)	(100.0)	(100.0)
Employed	2,738	2,160	578	2,182	1,272	910
	(75.3)	(79.3)	(63.5)	(80.0)	(88.1)	(70.8)
Unemployed	482	391	91	227	101	126
	(13.3)	(14.3)	(10.0)	(8.3)	(7.0)	(9.8)
Not in labor force	415	174	241	320	71	249
	(11.4)	(6.4)	(26.5)	(11.7)	(4.9)	(19.4)
Unemployment rate (%)[b]	15.0	15.3	13.6	9.4	7.4	12.2

a. Workers ages twenty to sixty-one displaced from full-time nonagricultural wage and salary jobs.
b. Unemployment rate = 100 × Unemployed / (Employed + Unemployed).

an active job search during this period. Thus, a spell of joblessness may include a period of suspended job search and labor force withdrawal, particularly for workers who experience long spells without work. For this reason, this variable is consistently referred to as the duration of *joblessness* rather than the duration of *unemployment* in this report.

Second, it is important to keep in mind that the question on jobless duration is retrospective (as are most of the questions on the "Displaced Worker Survey") and thus subject to possible biases in the respondent's recall of past events. Evidence on possible biases in recall of unemployment spells from the one-year retrospective data on unemployment duration regularly collected in the March *CPS* shows that respondents consistently understate actual unemployment experience (Horvath 1982). This suggests that any recall bias in the "Displaced Worker Survey" on average produces an underestimate of jobless duration.

With these caveats in mind we examine the distribution of spell lengths of joblessness in Table 1–7. As noted above, the sample is limited to workers displaced for at least one year and for most of the sample, for two or more years. This long gap between the date of displacement and the DWS interview means that most workers have completed their initial spells of joblessness subsequent to displacement. Table 1–7 shows the very wide range of experience following displacement. In the 1984 DWS, 37.3 percent of the male blue-collar and 54.8 percent of male white-collar and service workers experienced fourteen or fewer weeks of joblessness. At the same time, 30.2 and 14.8 percent of the same two groups (representing a combined total of 1.2 million workers) reported spells of joblessness exceeding one year. Although workers in the 1986 DWS fared somewhat better, the same wide range of experience prevails. These skewed distributions show that the majority of workers secured reemployment relatively quickly but a sizable minority seemed to have considerable difficulties securing permanent new jobs and are thus potentially "dislocated" or "structurally unemployed" (recognizing, of course, that many of these workers may have dropped out of the labor force and were no longer counted as unemployed during part of their jobless spell).

Are displaced workers out of work longer than an average unemployed worker? The evidence in Table 1–7 suggests that they are. Although we have emphasized the difference between our

Table 1–7. Duration of Joblessness Following Displacement (percentage).[a]

Jobless Weeks	Blue-Collar			White-Collar and Service		
	Total	Male	Female	Total	Male	Female
January 1984 (1979–82)						
Total	100.0%	100.0%	100.0%	100.0%	100.0%	100.0%
0–14	35.2	37.3	28.7	49.8	54.8	44.7
15–26	13.6	13.7	13.4	13.2	13.6	12.9
27–52	19.3	18.9	20.6	16.5	16.8	16.3
52+	31.9	30.2	37.3	20.4	14.8	26.1
Median weeks	29.6	26.3	40.0	14.5	12.0	20.0
January 1986 (1981–84)						
Total	100.0	100.0	100.0	100.0	100.0	100.0
0–14	40.1	44.5	27.0	53.0	57.3	48.2
15–26	14.3	14.0	15.1	16.8	17.5	16.1
27–52	20.1	19.2	22.7	15.5	15.5	15.4
52+	25.5	22.3	35.2	14.7	9.7	20.2
Median weeks	24.9	19.7	47.7	12.4	10.5	16.0

a. Workers ages twenty to sixty-one displaced from full-time nonagricultural wage and salary jobs.

measure of joblessness and standard unemployment duration measures, it is nonetheless interesting to contrast these findings with studies of unemployment duration, which consistently find that the average completed spell of unemployment is relatively short (Bowers 1980; Sider 1985). Bowers (1980), for example, estimates that just 6.0 percent of all unemployment spells in 1979 were longer than six months in length. In Table 1–7 we see that the equivalent shares for workers displaced from 1979 through 1982 range from a low of 31.6 percent for male white-collar and service workers up to 57.9 percent for blue-collar females. Because the contribution of any group to the average unemployment rate depends not only on the relative size of the group but also on their average duration of joblessness, the above comparison suggests that the contribution of displaced workers to overall unemployment is much greater than their share of unemployed workers.[7]

What factors are associated with long-term joblessness? The statistics in Table 1–8 provide some clues. Workers with long spells tend to be somewhat older than workers with short spells, but the age gap between the two is not very large. Indeed, for all four groups, the long-term jobless worker is on average only in his or her midthirties. Local labor market conditions also matter. Long-term unemployed workers tend to be found in labor markets with above-average rates of unemployment. Workers with long spells of joblessness also tend to have lower educational attainment, but the gap is not large. Finally, the share of blacks tends to rise very sharply with jobless duration.

Multivariate analysis of the jobless duration data is consistent with the relationships shown in Table 1–8. Table 1–9 presents multivariate estimates of the effect of selected variables on the median duration of joblessness. These estimates are derived from coefficient estimates of a standard "survivor" model fit to the distribution of jobless spell lengths (Podgursky and Swaim 1987a) and indicate the effect in weeks of a one-unit change in the independent variable on expected median jobless duration for an average worker in the sample. Each additional year of age, for example, increases median jobless duration by .314 weeks for blue-collar males. For dichotomous (0–1) variables such as race, the coefficient indicates the difference in median jobless spell duration for average black and white workers. A black blue-collar worker, for example, has a median jobless spell 30.1 weeks longer than a white blue-collar worker with the same characteristics (such as age, education, area unemployment). The very large coefficients on the race variable are particularly striking in this table and are consistent with the strong relationship seen in Table 1–8. Black workers seem to have much greater reemployment difficulties than do whites.

Where Are the New Jobs?

Where do displaced workers find new jobs? The industry transition statistics in Table 1–10 provide information on reemployment patterns by old and new industries for the 1984 DWS sample

Table 1-8. Selected Mean Demographic and Economic Characteristics of Displaced Workers by Weeks of Joblessness Following Displacement, 1984 Survey.[a]

Weeks of Joblessness	Male					Female				
	%	Area Unemployment Age (years)	Rate (%)	Black (%)	Education (years)	%	Area Unemployment Age (years)	Rate (%)	Black (%)	Education (years)
Blue-collar										
0–14	44.5	30.0	7.8	9.7	11.8	27.0	30.7	7.7	8.3	12.3
15–26	14.0	31.0	8.3	5.8	11.8	15.1	30.0	7.7	19.6	11.6
27–52	19.2	31.3	9.0	13.5	11.7	22.7	33.9	8.1	16.9	11.0
53+	22.3	33.6	9.4	18.5	11.3	35.2	34.2	8.7	26.2	11.1
Total	100.0	31.5	8.6	12.5	11.6	100.0	32.6	8.1	18.2	11.5
Median weeks	26.3	—	—	—	—	40.0	—	—	—	—
White-collar and service										
0–14	57.3	33.2	7.6	5.4	14.1	48.2	30.8	7.7	6.0	13.2
15–26	17.5	33.2	8.0	3.5	13.8	16.1	31.0	7.7	8.6	12.8
27–52	15.5	34.8	8.9	7.9	13.4	15.4	32.4	8.8	11.5	12.5
53+	9.7	35.8	8.7	14.5	13.3	20.2	35.0	8.2	20.1	12.2
Total	100.0	33.9	8.0	7.1	13.8	100.0	32.2	8.0	10.7	12.8
Median weeks	12.0	—	—	—	—	20.0	—	—	—	—

a. Workers ages twenty to sixty-one displaced from full-time nonagricultural wage and salary jobs, 1979–82.

Table 1–9. The Effect of Selected Worker and Labor Market Characteristics on Median Weeks of Joblessness Following Displacement.[a]

Variable	Blue-Collar		White-Collar and Service	
	Male	Female	Male	Female
Age (years)	.314	−.465	.129	.326
Education (years)	−2.13	−7.51	−.98	−2.24
Tenure on former job (years)	.537	1.62	.332	1.000
Industry unionization rate (percent)	.364	.653	.115	.030
Area unemployment rate (%)	3.71	9.28	1.10	2.44
Race (black = 1)*	30.1	52.7	26.2	38.3
Plant shutdown* (yes = 1)	−7.67	−14.54	−4.29	−9.52

a. The change in the predicted median spell of joblessness resulting from a one unit change in the independent variable for an average worker in the sample. Variables with an asterisk (*) are dichotomous (0–1) variables. In these cases the reported effect is the difference between workers with the characteristic (1) and those without (0). Other independent variables are household headship, former wage, occupation dummies, expectation of layoff dummy, and year of displacement dummy. These results are based on the estimated means and coefficients for the Weibull survival model reported in Podgursky and Swaim (1987a).

(similar patterns are found in the 1986 DWS sample). In constructing this table, workers were included only if they were employed at the time of the January 1984 survey. Each row of the table represents an industry from which workers were displaced. The columns of the table represent industries in which workers were employed in January 1984.

The large percentages down the main diagonal of the table show that a sizable fraction of displaced workers ultimately returned to the same broad industrial sector from which they were displaced. For example, 28.5 percent of workers displaced from durable manufacturing were reemployed in durable manufacturing in January 1984. Among workers who did not return to the same industry, many ultimately found work in service industries or wholesale and retail trade. Indeed, these two sectors absorbed *31 percent* more displaced workers than they produced. Another common transition (particularly for blue-collar males) was construction. As we

Table 1–10. Where Displaced Workers Find New Jobs: Transition Rates between Industries (current industry as a percentage of total reemployed), 1984 Survey.[a]

Former Industry	Mining	Construction	Durable Manufacturing	Nondurable Manufacturing	Transportation	Commuting and Public Utilities	Wholesale and Retail Trade	Services	Public Administration
Mining	20.6%	14.6%	3.9%	12.8%	4.3%	2.4%	14.6%	20.3%	6.5%
Construction	.3	47.3	3.6	8.1	3.4	1.3	13.6	23.7	3.7
Durable manufacturing	.4	4.3	28.5	11.8	3.8	1.2	18.5	28.3	3.2
Nondurable manufacturing	1.0	9.1	7.9	35.9	4.3	1.8	15.1	21.8	3.0
Transportation	.6	7.5	4.7	7.8	36.6	2.9	14.9	19.3	5.6
Commuting and public utilities	2.0	22.0	9.3	2.7	1.0	27.5	17.1	11.6	6.8
Wholesale and retail trade	.3	6.9	5.2	8.3	2.7	2.9	44.4	26.6	2.4
Services	.6	7.0	4.7	6.7	2.0	3.6	17.6	55.4	2.4
Public administration	0	3.8	3.1	6.1	3.7	2.1	11.3	47.2	22.7

a. Workers ages twenty to sixty-one displaced from full-time nonagricultural wage and salary jobs between 1979–82 who were reemployed in January 1984. Reemployment totals exclude a small number of workers reemployed in forestry and fisheries.

will see in the next section, these transition patterns play an important role in explaining earnings losses following displacement. Not surprisingly, workers who remain in their old industry on average fare better than workers who move to a new one. This is particularly true for workers displaced from goods-producing industries.

Reemployment Earnings

When displaced workers do secure new jobs, earnings on the new job on average compare favorably with those on the former job, but once again, the range of experience is very wide. The best predictor of success is the duration of joblessness: Workers who return to work quickly have the smallest losses, while those who experience long spells of joblessness return to jobs paying much less than their former rates of pay.

Table 1–11 reports the ratio of current to former earnings for our four sex and occupation groups. Prior earnings are adjusted to take account of the growth of nominal earnings between the year of displacement and the survey date using the Employment Cost Index for wages and salaries from *Employment and Earnings* (for details see Podgursky and Swaim 1987c). For conciseness, we will hereafter refer to this as the "earnings ratio." The median earnings ratio for reemployed blue-collar workers is .86 and .88 and for white-collar and service workers .92 and .95 in the two surveys. The bottom panel of the table shows that for workers who return to full-time jobs, the median earnings ratio is roughly 2 to 4 percentage points higher.

The relatively modest losses for the median worker, however, mask a very wide dispersion. Among blue-collar males in the 1984 DWS, for example, 31.3 percent have an earnings ratio of 1.00 or higher, indicating no loss in earnings. At the same time, however, 38.3 percent have an earnings ratio of less than .75 (that is, losses in excess of 25 percent), and 17.6 percent earn less than one-half their former weekly pay.

A comparison of the top and bottom panels of the table for 1979–82 displacements shows that some of the large losses can be explained by short hours. Nonetheless, 34.4 percent of blue-collar males returning to full-time jobs report an earnings ratio of less

Table 1-11. Ratio of Current to Trend-Adjusted Earnings for Workers Employed in January 1986 (percentage).[a]

	Blue-Collar			White-Collar and Service		
	Total	Male	Female	Total	Male	Female
1979–82 displacements						
Percent employed	66.8%	70.7%	54.5%	74.9%	82.9%	66.9%
Ratio current to former earnings:						
Total	100.0	100.0	100.0	100.0	100.0	100.0
1.00+ (no loss)	32.0	31.3	34.3	38.9	43.1	34.2
.750–.999	30.1	30.3	29.3	29.4	29.2	29.7
.500–.749	20.5	20.7	19.9	17.0	17.3	16.6
Less than .5	17.4	17.6	16.5	14.7	10.4	19.5
Median ratio	.86	.86	.86	.92	.94	.89
Percentage reemployed full time	56.7	61.1	42.9	63.6	77.3	50.3
Ratio current to former earnings full-time reemployed:						
Total	100.0	100.0	100.0	100.0	100.0	100.0
1.00+ (no loss)	35.1	33.9	39.9	43.4	44.6	41.7
.750–.999	32.1	31.7	33.6	32.3	30.0	35.4
.500–.749	20.5	21.1	18.3	16.4	17.3	15.1
Less than .5	12.2	13.3	8.1	7.8	8.1	7.7
Median ratio	.88	.88	.91	.95	.97	.94

1981–84 displacements

Percentage employed	75.3	79.3	63.5	80.0	88.1	70.8
Ratio of current to former earnings:						
Total	100.0	100.0	100.0	100.0	100.0	100.0
1.00+ (no loss)	36.2	36.6	35.0	43.5	42.4	45.0
.750–.999	27.0	28.6	21.2	27.2	29.1	24.7
.500–.749	21.3	21.1	22.3	17.1	17.3	16.9
Less than .5	15.4	13.7	21.5	12.2	11.2	13.4
Median ratio	.88	.89	.84	.94	.94	.94
Percent reemployed full time	66.4	72.2	49.0	70.2	80.7	58.4
Ratio current to former earnings full-time reemployed						
Total	100.0	100.0	100.0	100.0	100.0	100.0
1.00+ (No Loss)	39.5	38.5	43.1	46.6	43.8	50.8
.750–.999	29.0	29.9	25.8	28.4	29.6	26.5
.500–.749	20.5	20.5	20.8	16.9	17.7	15.7
Less than .5	10.9	11.1	10.3	8.1	8.8	7.0
Median ratio	.92	.92	.94	.97	.95	1.00

a. Workers ages twenty to sixty-one displaced from full-time nonagricultural wage and salary jobs.

than .75, with 13.3 percent earning less than one-half their former weekly pay. A similarly wide dispersion is found in the 1986 DWS.

One important factor in explaining earnings loss is whether the worker was able to find a new job in the same general industry from which he or she was displaced. That is to say, workers who were on the main diagonal in Table 1–10 in general fared much better than workers reemployed in new industries. The data in Table 1–12 illustrates this point for displaced manufacturing workers. Displaced manufacturing workers who were reemployed in January 1984 had a median ratio of current to trend-adjusted former earnings of .83. For the 42.7 percent who returned to manufacturing jobs, however, the ratio was .88. Workers reemployed in

Table 1–12. Earnings Losses for Displaced Manufacturing Workers, by Industry of Current Employment.[a]

	1979–82 Displacements		1981–84 Displacements	
	(1)	*(2)*	*(3)*	*(4)*
Industry of Current Employment	*Number of Workers (000)*	*Ratio of Current to Former Earnings*	*Number of Workers (000)*	*Ratio of Current to Former Earnings*
All industries	2,233	.83	2,265	.86
(%)	(100.0%)		(100.0%)	
Manufacturing	954	.88	1,018	.94
	(42.7)		(44.9)	
Services	534	.67	596	.72
	(23.9)		(21.9)	
Retail trade	251	.66	281	.65
	(11.2)		(12.4)	
Construction	169	.73	197	.85
	(7.6)		(8.7)	
Wholesale	111	.69	90	.81
	(5.0)		(4.0)	
Transportation	93	.78	74	.98
	(4.2)		(3.3)	

a. Workers ages twenty to sixty-one displaced from full-time wage and salary jobs in manufacturing who were reemployed at the time of the survey.

services or retail trade fared considerably worse, with earnings ratios of .66 and .67, respectively. A similar pattern holds for the 1986 DWS data in columns (3) and (4).

Multivariate analysis of factors affecting earnings loss indicates that workers who made above-average earnings on their former job tended to have larger proportionate earnings losses. In simpler terms this means that workers who were relatively higher on the predisplacement earnings ladder had further to fall. Similarly, workers with more seniority generally had larger losses, as did workers with less education. In addition, a slack local labor market made recovery more difficult—higher unemployment rates produced larger earnings losses. Finally, workers who returned to the same three-digit occupation or industry usually fared significantly better than workers who did not.[8]

The most powerful predictor of earnings loss, unfortunately, is weeks of joblessness. This means that workers with the largest short-term losses also suffer the largest long-term losses. Table 1–13 shows the strong relationship between the duration of joblessness and earnings loss. Again taking blue-collar males as our

Table 1–13. Median Ratio of Current to Trend-Adjusted Former Earnings by Weeks of Joblessness.[a]

Weeks of Joblessness	Blue-Collar			White-Collar and Service		
	Total	Male	Female	Total	Male	Female
1979–82 displacements						
0–14	.94	.94	.95	.98	1.00	.95
15–26	.85	.85	.85	.93	.93	.92
27–52	.79	.79	.80	.78	.76	.80
53+	.63	.60	.69	.58	.68	.48
1981–84 displacements						
0–14	.95	.95	.95	1.00	.99	1.01
15–26	.91	.92	.84	.93	.94	.93
27–52	.80	.76	.82	.82	.81	.82
53+	.69	.71	.58	.72	.73	.66

a. Workers ages twenty to sixty-one displaced from full-time nonagricultural wage and salary jobs who were reemployed at the time of the survey.

benchmark, males who find new jobs quickly have January 1984 earnings nearly equal to those on their former job (that is, a .94 ratio). For workers experiencing a spell of joblessness of one year or more, however, the earnings ratio falls to .60. This strong negative relationship between jobless duration and the earnings ratio holds for the other three groups as well. It also appears in the multivariate studies described above.

Health Insurance Loss

Displaced workers also face a high risk of group health insurance loss. For most Americans who are covered by private medical insurance, coverage is a fringe benefit of their job or the job of another family member. When the job is terminated, however, this benefit quickly terminates as well—usually between thirty to sixty days after layoff. Concern with this problem has led to federal legislation allowing workers to participate in their former group health insurance plan for a limited time at their own expense. Two states (Massachusetts and Connecticut) now require employers to extend health insurance coverage in certain partial and total plant closures. Further federal and state legislation in this area has been proposed (Podgursky and Swaim 1987b).

How serious is this problem? Data from the two DWS surveys show that many displaced workers face a high risk of group health insurance loss for a prolonged period following displacement. Most of the workers in our samples were covered by a group health insurance plan on their old job. The first row of Table 1–14 shows that coverage rates on the former job averaged approximately 70 percent for blue-collar and white-collar and service workers in both surveys. The next two lines report coverage rates at the time of the DWS survey for workers who had health insurance coverage on their old job: line (2) for all such workers; and line (3) for the reemployed. The fact that these coverage rates are well below 100 percent indicates that many displaced workers do lose health insurance coverage as a result of displacement, even when they eventually secure new jobs. Cross-tabulations reported in Podgursky and Swaim (1987b) show that the risk of health insurance loss is strongly associated with reemployment earnings: Workers with the greatest loss in earnings are also more likely to lose group health insurance coverage.

Table 1-14. Job Displacement and Health Insurance Loss (percentage).[a]

	Blue-Collar			White-Collar and Service		
	Total	Male	Female	Total	Male	Female
1979–81 displacements						
(1) Covered by group health insurance on former job	74.0%	75.2%	70.4%	69.1%	77.3%	61.2%
Covered by group health insurance on former job and covered group health insurance January 1984:						
(2) Total	59.4	59.8	58.0	73.1	73.6	72.5
(3) Reemployed	71.9	72.2	71.0	80.5	80.6	80.5
1981–84 displacements						
(1) Covered by group health insurance on former job	72.0	74.0	66.3	66.9	71.9	61.2
Covered by group health insurance on former job and covered group health insurance January 1986:						
(2) Total	66.6	68.5	60.3	75.5	78.6	71.4
(3) Reemployed	75.8	76.8	71.6	82.1	84.1	79.0

a. Workers ages twenty to sixty-one displaced from full-time nonagricultural wage and salary jobs.

In sum, the personal losses entailed by displacement include not only a decline in weekly earnings but also a high probability of group health insurance loss. To this extent, the earnings loss estimates given in the previous section understate total compensation loss due to displacement.

CONCLUSION

Data from the "Displaced Worker Survey" describe a turbulent labor market in which job displacement due to plant shutdowns, business relocations, or related factors is fairly widespread. The household survey data do not permit us to determine whether this displacement stems from technological change, imports, or other economic factors. The data do show that displacement is disproportionately concentrated in goods-producing industries, where import competition and the introduction of labor-saving technologies are likely greatest.

How great are the short- and long-term costs of adjustment for displaced workers? In answering this question we should bear in mind the story of the man with one foot in a bucket of boiling water and the other in a bucket of ice. When asked how he felt, he replied that on average he was quite comfortable.

So it is with the displaced worker. The median duration of joblessness following displacement is rather long for blue-collar workers. It is considerably shorter for displaced white-collar and service workers. It is also shorter for men than for women. For all groups, however, the median experience masks a very wide variation, with many workers finding new jobs within a few weeks, while others experience a spell of joblessness of a year or more. The foot-in-a-bucket analogy is even more apt when we turn to reemployment earnings, where median losses are less than 15 percent—and smaller still for workers who return to full-time jobs—but with a very large dispersion.

What are the implications of these findings for labor market policy? They tell us that the problems addressed by federal and state adjustment programs are real and that job displacement is a widespread phenomenon in labor market—indeed, probably more widespread than previously anticipated. Moreover, a sizable fraction of these workers—perhaps one-quarter—seem to suffer very large economic losses as a result of displacement and have an enduring decline in their earnings capacity. In the language of current labor market policy they are "dislocated"; in the language of labor economics, they are "structurally unemployed" or, perhaps more accurately, "structurally subemployed." The wide range of postdisplacement experience also demonstrates the importance of targeting adjustment assistance to those workers facing the greatest difficulties. We have shown that variables such as race, education, and area unemployment play a significant role in explaining postdisplacement experience. Nevertheless, the unexplained variance remains large and more information is required to permit timely identification of workers most likely to be "dislocated" amid a much larger pool of displaced workers.

Unfortunately, the DWS survey data cannot tell us what types of adjustment assistance work and what do not. Only careful follow up analysis of the various current and future Title III programs can do that. In sum, much like a health survey, the DWS surveys can tell us a great deal about the incidence of an illness and its symptoms and severity, but very little about its cure. The latter requires practical experience, trial and error, experimentation—and most important—commitment to finding a treatment.

NOTES

1. A special survey of workers receiving Trade Adjustment Assistance (TAA) was undertaken in the mid-1970s. The TAA program provided adjustment assistance to workers who lost jobs due to international trade (this determination was made after an investigation by the U.S. Department of Labor). The resulting microdata file was not useful in studying the labor market experience of displaced workers because most of the workers in the survey eventually returned to their former employers (Carson and Nicholson 1981).

2. See Flaim and Sehgal (1985), Devens (1986), and U.S. Department of Labor (1986). The omitted categories are completion of a seasonal job, self-employed business failure, and "other."

3. The major differences between our selection criteria and that employed by the BLS are that (a) the BLS selects workers with three or more years of tenure on the old job, we do not; (b) the BLS selects workers displaced over the previous five years but we drop workers

displaced in the previous year; (c) the BLS includes agricultural, private-household, part-time, and workers over age sixty-two and we do not. Points b and c are discussed in the text. Regarding a, we decided not to adopt a tenure cutoff because (a) the BLS choice of three years is arbitrary; (b) many workers in our sample may have low tenure precisely because they face high risk of displacement; (c) our multivariate statistical work has shown that tenure on the former job is not a major explanatory variable concerning hardship following displacement; and (d) as far as we are aware, tenure on the former job does not currently play a major role in allocating Title III or other adjustment assistance to displaced workers. None of the major conclusions of this report would change, however, if the BLS tenure criteria were adopted. The major effect would be to reduce the displacement totals in Table 1–1.

4. An alternative, but not mutually exclusive, explanation focuses on recall bias. The overall level of 1981–82 displacements is lower in the 1986 as compared to the 1984 DWS survey, which may be because respondents forget or otherwise under report layoffs in the more distant past or underestimate the subsequent passage of time (such as report a 1982 layoff as 1983). Such recall biases may be greater for nonshutdown layoffs, as compared to more visible and memorable plant shutdowns.

5. The displacement totals in Table 1–2 and all subsequent tables are slightly higher than the sum of annual displacements in Table 1–1. This is because the age criteria in Table 1–1 is age in year of displacement, while all subsequent tables use age at the time of the survey. Using the latter criteria in Table 1–1 would have biased year-to-year comparisons of displacement flows.

6. Because the labor force tabulations from the "Displaced Worker Survey" are not seasonally adjusted, the relevant comparisons are the unadjusted January 1984 and January 1986 unemployment rates.

7. In 1981, for example, 23.38 million workers experienced one or more spells of unemployment at some time during the year. In the same year 1.65 million workers were displaced from one or more jobs and experienced at least one or more weeks of joblessness following displacement. If the average spell of unemployment for these displaced workers was three times the average for all unemployed workers, the displaced workers would have accounted for 21 percent of total weeks of unemployment in that year.

8. Multivariate semilog earnings equations with January 1984 earnings as the dependent variable were estimated for the four sex-occupation groups in Table 1–10. The independent variables were natural log of former earnings, age, age $\geq$ 50 spline, education, race, tenure on old

job, industry unionization of former job, craft occupation (blue-collar only), expected layoff, shutdown, received UI, exhausted UI, area unemployment, reemployed in same three-digit industry, reemployed in same three-digit occupation, four years of displacement dummy variables. For details, see Podgursky and Swaim (1987c).

REFERENCES

Bendick, Marc Jr., and Judith Radlinski Devine. 1981. "Workers Dislocated by Economic Change: Do They Need Employment and Training Assistance?" In National Commission for Employment Policy, *Seventh Annual Report: The Federal Interest in Employment and Training.* Washington, D.C.: U.S. G.P.O., 175–226.

Bluestone, Barry, and Bennett Harrison. 1982. *The Deindustrialization of America* New York: Basic Books.

Bowers, Norman. 1980. "Probing the Issues of Unemployment Duration." *Monthly Labor Review* 103 (July): 23–32.

Burtless, Gary. 1983. "Why Is Insured Unemployment So Low?" *Brookings Papers on Economic Activity* 1: 225–49.

Corson, Walter, and Walter Nicholson. 1981. "Trade Adjustment Assistance for Workers: Results of a Survey of Recipients under the Trade Act of 1974." In Ronald Ehrenberg, ed., *Research in Labor Economics* 4. Greenwich, Conn.: JAI Press, 417–67.

Craypo, Charles, and William Davisson. 1983. "Plant Shutdown, Collective Bargaining, and Job and Employment Experiences of Displaced Brewery Workers." *Labor Studies Journal* 7: 195–215.

Crosslin, Robert L., James S. Hanna, and David W. Stevens. 1984. "Identification of Dislocated Workers Utilizing Unemployment Insurance Administrative Data: Results of a Five State Analysis." Washington, D.C.: National Commission for Employment Policy, RR-84-03.

Devens, Richard Jr. 1986. "Displaced Workers: One Year Later." *Monthly Labor Review* 109 (July): 40–43.

Ehrenberg, Ronald G., and Ronald L. Oaxaca. 1986. "Unemployment Insurance, Duration of Unemployment, and Subsequent Wage Gain." *American Economic Review* 66 (December): 754–66.

Feldstein, Martin. 1983. "The Economics of the New Unemployment." *Public Interest* 53 (Fall): 3–42.

Flaim, Paul O., and Ellen Sehgal. 1985. "Displaced Workers of 1979–1983: How Have They Fared?" *Monthly Labor Review* 108 (June): 3–16.

Folbre, Nancy R., Julia L. Leighton, and Melissa Roderick. 1984. "Plant Closings and Their Regulation in Maine, 1971–1982." *Industrial and Labor Relations Review* 37 (January): 185–96.

Gainer, William J. 1986. "GAO's Preliminary Analysis of U.S. Business Closures and Permanent Layoffs during 1983 and 1984." OTA/GAO Workshop on Plant Closings, April 30–May 1. Washington, D.C.: U.S. General Accounting Office.

Gilpatrick, Elinor. 1966. *Structural Unemployment and Aggregate Demand.* Baltimore: Johns-Hopkins Press.

Gordon, Avery F., Paul Schervish, and Barry Bluestone. 1985. "The Unemployment and Reemployment Experience of Michigan Auto Workers." Boston College: Social Welfare Research Institute.

Grodus, Jeanne Prail, Paul Jarley, and Louis A. Ferman. 1981. *Plant Closings and Economic Dislocation.* Kalamazoo, Mich.: W.E. Upjohn Institute.

Hamermesh, Daniel S. 1987. "The Costs of Worker Displacement." *Quarterly Journal of Economics* 102 (February): 51–75.

Harris, Candee. 1984. "The Magnitude of Job Loss from Plant Closings and the Generation of Replacement Jobs: Some Recent Evidence." *Annals of the American Academy* 475 (September): 15–27.

Horvath, Francis. 1982. "Forgotten Unemployment: Recall Bias in Retrospective Data." *Monthly Labor Review* 105 (March): 40–43.

Jacobson, Louis S. 1978. "Earnings Losses of Workers Displaced from Manufacturing Industries." In U.S. Department of Labor, Bureau of International Labor Affairs, *The Impact of International Trade and Investment on Employment.* Washington, D.C.: U.S. G.P.O., 87–98.

———. 1984. "A Tale of Employment Decline in Two Cities: How Bad Was the Worst of Times?" *Industrial and Labor Relations Review* 37 (July): 557–69.

Leonard, Jonathan. 1985. "On the Size Distribution of Employment and Establishments." Mimeo. Berkeley: Institute of Industrial Relations. University of California.

———. 1987. "In the Wrong Place at the Wrong Time: The Extent of Frictional and Structural Unemployment." In Kevin Lang and Jonathan Leonard, eds., *Unemployment and the Structure of Labor Markets.* New York: Basil Blackwell, 141–63.

Levy, Robert A., James Jondrow, and Louis Jacobson. 1985. "The Causes and Consequences of Displacement in Steel and High Technology Industries." Alexandria, Va.: Center for Naval Analyses.

Lillien, David M. 1982. "Sectoral Shifts and Cyclical Unemployment." *Journal of Political Economy* (August): 777–93.

Marston, Steven T. 1976. "Employment Instability and High Unemployment Rates." *Brookings Papers on Economic Activity* 1 (1976): 169–210.

Neuman, George R. 1978. "The Labor Market Adjustment of Trade-Displaced Workers: Evidence from the Trade Adjustment Assistance

Program." In Ronald Ehrenberg, ed., *Research in Labor Economics* 2. Greenwich, Conn.: JAI Press, 353–81.

Podgursky, Michael, and Paul Swaim. 1987a. "The Duration of Joblessness Following Plant Shutdowns and Job Displacement." *Industrial Relations* 52 (Fall): 213–226.

———. 1987b. "Health Insurance Loss: The Case of the Displaced Worker." *Monthly Labor Review* 110 (April): 30–33.

———. 1987c. "Job Displacement and Earnings Loss: Evidence from the Displaced Worker Survey." *Industrial and Labor Relations Review* 41 (October): 17–29.

———. 1986a. "Job Displacement, Reemployment, and Earnings Loss: Evidence from the Displaced Worker Survey." Washington, D.C.: National Commission for Employment Policy, RR-86-18.

———. 1986b. "Labor Market Adjustment and Job Displacement: Evidence from the January, 1984 Displaced Worker Survey." Washington, D.C.: U.S. Department of Labor, Bureau of International Labor Affairs.

———. 1986c. "Plant Shutdowns and Job Displacement: How Do New England Workers Fare?" *New England Economic Indicators*. Boston: Federal Reserve Bank of Boston, 3–5.

Ruhm, Christopher. 1986. "The Economic Consequences of Labor Mobility." Mimeo. Boston University Department of Economics.

Sabelhaus, John, and Robert Bednarzik. 1985. "Earnings Losses of Dislocated Workers," Economic Discussion Paper 16. Washington, D.C.: U.S. Department of Labor, Bureau of International Labor Affairs.

Shapiro, David, and Steven H. Sandell. 1985. "Age Discrimination in Wages and Displaced Older Men." *Southern Economic Journal* 52 (July): 90–92.

Sider, Hal. 1985. "Unemployment Duration and Incidence: 1968–82." *American Economic Review* 75 (June): 461–72.

United States Congressional Budget Office. 1982. *Dislocated Workers: Issues and Federal Options*. Washington, D.C.: G.P.O.

United States Department of Labor, Bureau of Labor Statistics. 1987. *Analysis of Mass Layoff Data*. Mimeo. Washington, D.C.: G.P.O.

———. 1986. "Reemployment Increases among Displaced Workers." *News* 86–414 (October 14).

United States Office of Technology Assessment. 1986. *Technology and Structural Unemployment: Reemploying Displaced Adults*. OTA-ITE-250. Washington, D.C.: G.P.O.

Young, Kan, Ann Lawson, and Jennifer Duncan. 1986. "Trade Ripples across U.S. Industries: Effects of International Trade on Industry Output and Employment." Washington, D.C.: U.S. Department of Commerce Office of Business Analysis.

2 TECHNOLOGICAL CHANGE AND THE EXTENT OF FRICTIONAL AND STRUCTURAL UNEMPLOYMENT

Jonathan S. Leonard

Structural and frictional unemployment are usually considered among the unpleasant and exogenous facts of economic life about which little can be done. As technology advances and the composition of demand changes, employment must also shift. In the process of adjusting to a new equilibrium, some people will endure spells of unemployment. Usually, this is considered part of a healthy reequilibration process, and the resulting unemployment is seen as part of the underlying "natural" rate of unemployment. Recent oil and trade shocks that have reduced manufacturing employment focus attention on how the economy adjusts to structural changes. This chapter analyzes the nature of this adjustment process and shows the magnitude of gross flows of employment across industries and establishments. It dissects the flow of job creation and destruction and develops an empirical view of the dynamics of establishment size in relation to employment and unemployment. Only when the ongoing rate of job turnover in the economy is established can we begin to judge how flexible the economy is and how great technological change would have to be to strain the economy's ability to adapt.

At least since the Luddites forcefully expressed their opinions, many people have believed that technological change contributes to unemployment. With the advent of the computer age, fears have increased either that there will be little productive work left for people to do or that as the pace of technological change quickens, the volatility of employment will increase as industries go in and out of technological style. The latter hypothesis depends on change generating unemployment. In a flexible economy, this need not necessarily be the outcome. With good information and low adjustment costs, workers may shift across industries without experiencing frequent or long unemployment. This chapter shows the normal level of turnover of jobs in the U.S. economy, and demonstrates that in a normal year a substantial fraction of all jobs are destroyed and created. Most of this flux is within, not across, industry lines.

If technological change were a driving force in employment growth and decline, and if the degree of technological change or the impact of technological change on employment varied substantially across industries, then we would expect to see sharp differences in employment patterns among establishments in different industries. In fact, there is greater employment variation within industries than across industries. Part of the employment variation within industries may be due to temporary cost advantages reaped from technological advances, but this is not the sort of technological change that has a pervasive effect across an entire industry. The cross-industry shifts, which have attracted the most attention, and where one might expect technological effects to dominate, ignore the source of most job flux.

In a sense, the technological unemployment cup is both half empty and half full. If all the employment variation observed here is fundamentally caused by technological change, then technological change, as measured here, may account for roughly 2.2 percentage points of unemployment in an average year between 1977 and 1982. At the same time, an economy that loses one in nine jobs and creates one in eight jobs in an average year already has experience with great job volatility, which suggests considerable flexibility to respond to additional technological change.

The population of establishments analyzed here is described in the next section. An overview of the economy of the state studied here, and of the growth and decline of employment, is then provided, as well as new evidence on the instability of jobs and the transient nature of demand shocks at the establishment level. The penultimate section tests for the existence of industry, area, or year effects on establishment growth rates, and the final section presents the chapter's conclusions.

POPULATION CHARACTERISTICS

The sample studied here is drawn from a complete survey of establishments in the state of Wisconsin. Although the industrial composition of the Wisconsin economy is not exactly representative of the U.S. economy, it is not a bad approximation. For example, manufacturing employment accounts for 34.5 percent of private nonagricultural wage and salary employment in Wisconsin in 1980, compared to 27.4 percent of employment in the nation. There is not much reason to expect Wisconsin employers to be any more insulated from or exposed to technological change. The underlying data are collected as part of the administration of the unemployment insurance (U.I.) program. This is not a sample. It is (in theory and by law) the population of establishments in the state. The Wisconsin Department of Industry, Labor, and Human Relations prepares annual files from the March *Unemployment Compensation Contribution Reports*. These reports must by law be filed by all establishments paying at least *one* employee $1,500 in any quarter of the year.[1] The data are the primary source of federal employment statistics. In the majority of cases, these establishments are the sole operating asset of a firm, so there is not much distinction between establishments and firms. Where possible, companywide reports for multiestablishment companies and for companies that acquired other companies between 1977 and 1982 have been eliminated from the sample studied here. Transfers of ownership are treated not as a continuation of a single business but rather as a death and a birth. This obviously is not appropriate for some applications. In particular, job gain and loss rates and the variance of growth rates may be overstated.

Births and deaths have relatively little impact on job creation and destruction because they are concentrated in the smallest establishments. It is, however, possible that plant closings have a disproportionately large effect on unemployment. In a 1978–79 version of this data set that counts transfers as a continuation of business rather than as a birth and death, 11 percent of all job losses and 18 percent of all job gains were accounted for by deaths and births, respectively (Wisconsin 1984: 145, Table 2–2). The greater part of the gross job flows, 89 percent of losses and 82 percent of gains, occurred through the contraction or expansion of going concerns.[2]

The population studied here has 124,711 establishments with 1,198,638 employees in 1978. That averages 9.6 employees per establishment, not a great surprise to those familiar with *County Business Patterns* data. The distribution is, of course, highly skewed. More than 80 percent of employment is in establishments with ten or more employees, but at the same time, more than 80 percent of the establishments employ fewer than ten employees. Most of the institutional analysis of business deals with big business. Fewer than 2 percent of the establishments studied here have more than 100 employees. The large establishment is not the typical establishment, but surprisingly little is known about the small establishments that predominate.[3]

GROWTH AND DECLINE: RATIO OF CELL MEANS

What happens when the unemployment rate in a state doubles in three years? Perhaps one pictures a cataclysmic event—war, natural disaster, the invention of the steam engine, or at least an oil shock—some major disturbance causing the rapid extinction of a large proportion of all jobs. In Wisconsin, the state unemployment rate doubled in three years from 5.0 percent in 1979 to 10.0 percent in 1982, which indeed is the period following the second oil shock (see Table 2–1). The number of unemployed people also nearly doubled to 235,630 over these years. What does it take to double the unemployment rate and put an extra 120,000 people out on the street?

It takes only an average annual decline in employment of less than 1.2 percent between 1979 and 1982 (line 7). This is a loss of 79,000 jobs. The remaining third of the additional unemployed in these years is accounted for by the 40,000 person increase in the labor force. During the earlier period, 1977 and 1980, total employment grew. Despite this, the unemployment rate also rose during these years because the growth rate of employment fell more than

Table 2–1. Overview of the Wisconsin Economy, 1977–82.

	1977	1978	1979	1980	1981	1982
1. Unemployment rate	6.3 %	5.9 %	5.0 %	7.1 %	9.5 %	10.3 %
2. Number unemployed (000)	136.76	132.28	115.83	169.14	223.97	235.63
3. Growth rate of number unemployed	—	−3.3 %	−12.4 %	46.0 %	32.4 %	5.2 %
4. Labor force (000)	2,170.8	2,242.0	2,316.6	2,382.2	2,357.0	2,356.3
5. Growth rate of labor force	—	3.3 %	3.3 %	2.8 %	−1.1 %	−0.0
6. Employment (000)	2,033.7	2,109.5	2,199.7	2,214.2	2,134.0	2,120.9
7. Growth rate of employment	—	3.7 %	4.3 %	0.7 %	−3.6 %	−0.6 %
8. Inflation rate of CPI index, Milwaukee	—	5.5 %	13.3 %	16.9 %	11.2 %	7.2 %
9. U.S. unemployment rate	6.9 %	6.0 %	5.8 %	7.0 %	7.5 %	9.5 %

Source: Lines 1–8: Wisconsin State Department of Industry, Labor and Human Relations, Employment and Economic Indicators, 1977–82 (May, June, July publications). Line 9: Economic Report of the President (1985: Table B-33, p. 271).

the growth rate of the labor force. Under such conditions, it does not take great declines in the employment growth rate to produce an increase in the unemployment rate.

Between 1977 and 1980, sample employment increased by 15.6 percent. In the next year it fell by 3.5 percent. Table 2–2 shows that total employment grew by 10 percent during 1977–78, 4 percent during 1978–79, 2 percent during 1979–80, and 2 percent during 1981–82; employment declined by 3 percent during 1980–81. The rate of employment growth thus dropped by thirteen percentage points between the 1977–78 and 1980–81 periods.

The net employment growth rate is usually all that can be observed. Here it averages 2.8 percent annually among all establishments. But this turns out to be the sum of two large numbers.[4] Growing establishments average 30 percent growth in each year of growth. Shrinking establishments average 21 percent shrinkage. The employment weighted average of these two (and of the stable) yields the observed 2.8 percent net growth.

Distributing establishments by growth rates shows that mean growth does not decline because the entire distribution of growth rates shifts down. The employment growth rate declines not because all establishments are growing slower (they are not) but rather because shrinking establishments shrink faster and because at about the middle of the distribution, establishments that were

Table 2–2.   Growth Rates of Employment, 1977–82.

	Industry		
	All	Nonmanufacturing	Manufacturing
Mean employment per establishment, 1977–82	9.70	6.54	42.76
Ratio of employment:			
1978/1977	1.10	1.11	1.08
1979/1978	1.04	1.03	1.04
1980/1979	1.02	1.02	1.01
1981/1980	0.97	0.98	0.94
1982/1981	1.02	1.01	1.04
Mean	1.03	1.03	1.03
1982/1977	1.14	1.14	1.14

growing start to shrink. It is primarily this shift of only 5 percent of the establishments that lowers aggregate employment growth. The observed aggregate fluctuations occur not because of a widely shared response by establishments to changing incentives, but rather because of a more concentrated change by a small proportion of establishments.

The large changes in the share of all employment in growing or shrinking establishments are apparent[5] in Table 2–3. As the growth rate of total employment declined from 9.1 percent (1978) to −3.7 percent (1981), the share of employment in growing plants declined from .59 to .31. Meanwhile, the share of employment in shrinking plants nearly doubled from .34 (1970) to .61 (1981). These shifts can account for most of the decline in the growth rate of employment between 1977 and 1981.

MEAN OF RATIOS

Carrying the process of disaggregation one step further, take as the growth rate the mean of establishments' growth rates rather than the growth rate of mean employment. Table 2–4 weights each establishment equally, whereas Table 2–2 weighted each establishment's growth by its initial employment.

Comparing Table 2–4 to Table 2–2, we observe that the average establishment grows faster than does total (or average) employ-

Table 2–3. Proportion of Employment in Growing, Shrinking, and Stable Establishments.

	Proportion of Employment in Establishments That Are:		
	Growing	Shrinking	Stable
1977–78	.585	.338	.077
1978–79	.557	.369	.074
1979–80	.462	.462	.076
1980–81	.314	.606	.080
1981–82	.333	.579	.088
Average	.450	.471	.079

Table 2–4. Growth Rates of Employment, 1977–82—Mean of Establishment Ratios.

	Industry		
	All	Nonmanufacturing	Manufacturing
Mean employment per establishment, 1977–82	9.70	6.54	42.76
Mean Ratio of Establishment Employment:			
1978/1977	1.11	1.11	1.13
	(.70)	(.66)	(.99)
1979/1978	1.07	1.06	1.11
	(.75)	(.63)	(1.51)
1980/1979	1.05	1.05	1.10
	(.75)	(.58)	(1.68)
1981/1980	1.03	1.03	1.02
	(.69)	(.58)	(1.35)
1982/1981	1.03	1.03	1.07
	(.85)	(.66)	(1.89)
Mean	1.06	1.06	1.09
1982/1977	1.15	1.14	1.26
	(1.80)	(1.04)	(5.10)

Note: Cross-section standard deviation in parentheses.

ment. This occurs because the small grow faster. Note the large standard variation (in parentheses) of growth rates across establishments. This is particularly true in manufacturing, where coefficients of variation greater than one are common. Although the average growth rate changes over the years by less than 10 percentage points, the standard deviation of growth in the cross-section can exceed 180 percentage points. This reveals considerable heterogeneity in growth rates across establishments.

Comparing the six-year average annual growth rates with the growth rate over six years in Table 2–4, we see again evidence of regression to the mean; growth is concentrated among the small. In an average year, the average growth rate is 6 percent. But this does not take place in the same establishments year after year. It does not compound. Each year a new set of small establishments accounts for much of this growth, for after six years the average establishment has grown by only 15 percent, which is less than one would expect from the compounding of the average 6 percent annual growth rate. Growth and decline tend to be transient rather than chronic conditions—a point we shall later develop further. Over these same six years, the average growing establishment has doubled in size, the average shrinking establishment has been reduced to a third its original size, and only one-third of all establishments have maintained their original employment level.

JOB TURNOVER

Short durations of employment and high frequencies of disemployment are typically thought of in terms of the characteristics of people. The statistics in Table 2–5 (lines 7 and 10) reveal tremendous turnover of jobs themselves. New jobs equal to *13.8* percent of the previous year's base are created each year, while *11.0* percent are destroyed. The difference between these two flows, 2.8 percent net employment growth, is all that is usually observed. Of course, the gross flows analyzed here are themselves only the tip of the iceberg. They include only job destruction and creation that changes the net size of an establishment between one March and the next and ignore all other.[6] But even the tip of the iceberg looks surprisingly large. About one in every nine jobs disappears each year. More than one in every eight jobs are created every year. This does not occur during a great depression or a great boom; this magnitude of gross job flow is experienced in the average year between 1977 and 1982.

We can now reexamine the state economy in light of gross rather than net employment flows. Between 1977 and 1978, two- and one-third jobs were created for every one destroyed (Table 2–5, line 12). Three years later, between 1980 and 1981, only seven-tenths of a job was created for every one destroyed. Both the decline in jobs created and the increase in jobs destroyed contribute to the increase in the unemployment rate observed over these years.

It would be of great interest to know whether similarly large gross flows existed in earlier years and how they affected the "natural" rate of unemployment. Apparently, in the past either gross flows were smaller or they were accommodated with less unemployment and less inflation. The short time period observed here

Table 2-5. The Wisconsin Economy Revisited: Gross Flows, 1978–1982.

	1978	1979	1980	1981	1982
1. Unemployment rate	5.9%	5.0%	7.1%	9.5%	10.3%
2. Number unemployed	132,280	115,830	169,140	223,970	235,630
3. Growth rate of number unemployed	−.033	−.124	.460	.324	.052
4. Employment (sample)	1,198,638	1,242,423	1,260,652	1,216,805	1,245,694
5. Growth rate of employment	.099	.036	.015	−.035	.024
6. Jobs created	187,186	150,931	148,269	115,072	221,583
7. Share of jobs created	.172	.126	.119	.091	.182
8. Growth rate of jobs created	—	−.19	−.02	−.22	.93
9. Jobs destroyed	79,439	107,146	130,040	158,919	192,694
10. Share of jobs destroyed	.073	.089	.105	.126	.158
11. Growth rate of job destruction	—	.35	.21	.22	.21
12. Ratio of job birth to death	2.36	1.41	1.14	.72	1.15
13. Gross turnover rate	.245	.215	.224	.217	.340

cannot answer such questions. The gross turnover rate is the sum of the job creation and the destruction rates and is used as a measure of labor market turbulence. This rate ranges from .22 to .34 (Table 2–5, line 13) but shows no obvious pattern. On the basis of these statistics, one could not say that greater churning in the labor market was associated either with greater or less employment growth.

These statistics from establishments can, under certain assumptions, be used to make inferences about the distribution of job durations—the lifetime of the job itself. These may then be compared to data reported by workers on job tenure—the lifetime of a worker-job match. Assuming stationarity and stable distributions, the average duration of a job is the inverse of the death rate. Under these assumptions, the average job in this sample lasts 9.1 years (completed spell). Hall (1982; 720) reports that the expected median tenure of a worker in 1978 was 7.7 years (completed spell).[7] A job that dies must cause either a quit or a fire and so truncate job tenure. It seems likely that short job durations contribute to short job tenure and so add to unemployment, although nothing more precise than this can be said on the basis of the mean durations and tenures at hand.

Stronger evidence of the relationship between job turnover and unemployment comes from a more direct comparison of the job destruction rate reported here with the transitions from employment to nonemployment reported by individuals in the CPS. Poterba and Summers (1985) correct this series for reporting errors and find that between 1977 and 1982 the average monthly probability of moving from employment to nonemployment is .019 (Poterba and Summers 1985: Table V, Total Adjusted and Raked). I find here that .11 of all jobs disappear in an average year over the same period. This is a monthly rate of about .009. If few of the incumbants in disappearing jobs manage to find new employment without an intervening spell of nonemployment, then this comparison suggests that, depending on the magnitude of measurement error, roughly half of the transitions from employment to nonemployment reported by individuals could be accounted for by the disappearance of their jobs.

This may have important implications for the "natural" rate of unemployment. To illustrate, suppose that the year-to-year employment changes measured here capture only half of all job turnover during a year and that only half of this turnover is associated with any unemployment. (Both of these assumptions are guesses.) Then in an average year, we expect about 11 percent of all jobs to be destroyed and result in unemployment. Dynarski and Sheffrin (1986) report that an average completed spell of unemployment lasts 10.3 weeks, or one-fifth of a year. Using this duration in a rough calculation, job loss could account for about 2.2 percentage points, or more than a quarter of Wisconsin's 7.6 percent average unemployment during 1978–82. The companion paper by Podgursky analyzes further the subsequent experience of displaced workers. Neither standard analyses (in terms of personal characteristics) nor standard policies are likely to be of much use in understanding or preventing the problem of workers who are caught in the wrong place at the wrong time. Neither manpower nor aggregate demand policies address this fundamental instability of jobs.

Nonmanufacturing jobs are sometimes thought of as more stable than those in manufacturing. Two dimensions of stability should be distinguished: stability in a steady-state and stability over the cycle. The first four columns of Table 2–6 show the proportions of jobs created and destroyed each year in the nonmanufacturing and manufacturing sectors of the Wisconsin sample. In nearly all years, both the job creation and the job destruc-

Table 2–6.   Job Turnover in Wisconsin by Sector, 1978–82.

| | Proportion of Jobs | | | | | |
| | Nonmanufacturing | | Manufacturing | | Manufacturing, BLS[a] | |
Sector	Gained	Lost	Gained	Lost	New Hires	Layoffs
1978	.19	.084	.14	.054	.26	.11
1979	.14	.11	.11	.057	.31	.11
1980	.13	.12	.10	.082	.26	.14
1981	.11	.14	.055	.11	.16	.26
1982	.17	.16	.20	.15	.16	.20

a. These are twelve times the average of the April through March monthly rates published in the BLS, *Employment and Earnings,* vols. 24–29 (1977–82: Table D-4), for the Wisconsin manufacturing sector. Because the federal government discontinued the series, the 1982 figures are for the eight months through November 1981.

tion rates are higher outside of manufacturing. By this measure, manufacturing jobs are more stable. They are also more cyclically sensitive. The rates of job gain and loss change more over the cycle in manufacturing than outside.

The last two columns of Table 2–6 present new hire and layoff rates in Wisconsin manufacturing derived from *Employment and Earnings*. These are the sum of the reported monthly rates. The new hire and layoff rates were selected from among other components of accessions and separations because they were presumed to be more closely tied to job gain and loss. The rates of job creation and destruction calculated here range between one-third and three-quarters of the new hire and layoff rates. This suggests that a substantial portion of new hires and layoffs are accounted for by job creation and destruction.[8]

Technological change is typically thought of as affecting different industries to different degrees. For example, the rate of technological change in the furniture industry is not generally thought to be as great as that in the electronic machinery industry. If technological changes play a substantial role in explaining the rates of job loss and gain observed here, then we would expect to see different rates in different industries. Tables 2–7 and 2–8 present annual job loss and gain rates by industry. Employment in industries typically characterized as undergoing great technological change (such as electrical machinery or chemicals) is not obviously more volatile than in industries usually thought of as embodying more mature, unchanging technologies (such as furniture, lumber, or stone, clay, and glass).

THE DYNAMICS OF ESTABLISHMENT SIZE

This section examines the nature of the time path of changes in establishment size. The correlation of the logarithm of establishment size for establishment i in year t (S_{it}) and of the first difference of this, $D_{it} = S_{it} - S_{it-1}$, are analyzed here.

Establishment size can be modelled as the sum of transient and cumulative innovations.

$$S_{it} = w_{it} + \mu_{it} \qquad (2.1)$$

Table 2–7. Proportion of Jobs Lost, 1978–1982.

Industry	Year				
	1978	1979	1980	1981	1982
Mining and construction	.14	.18	.22	.25	.30
Food and kindred products	.09	.08	.09	.06	.17
Textiles	.03	.11	.13	.17	.13
Apparel	.07	.15	.13	.12	.18
Lumber	.08	.15	.15	.18	.25
Furniture	.07	.08	.11	.15	.13
Paper	.03	.04	.03	.06	.12
Printing and publishing	.04	.04	.06	.07	.09
Chemicals	.05	.06	.04	.06	.11
Petroleum and coal	.03	.08	.07	.13	.04
Rubber and plastics	.05	.04	.18	.11	.17
Leather	.04	.08	.14	.08	.11
Stone, clay, and glass	.05	.07	.08	.17	.21
Primary metals	.06	.02	.08	.15	.14
Fabricated metal	.05	.04	.06	.13	.15
Machinery, except electrical	.06	.02	.07	.14	.13
Electrical equipment	.04	.09	.10	.11	.15
Transportation equipment	.04	.02	.11	.09	.27
Instruments	.03	.13	.06	.09	.28
Miscellaneous manufacturing	.06	.09	.11	.13	.10
Transportation and utilities	.05	.07	.08	.10	.13
Wholesale and retail trade	.09	.11	.12	.14	.16
Finance, insurance, and banking	.06	.07	.06	.08	.10
Services	.09	.11	.11	.13	.15

and

$$w_{it} = w_{i,t-1} + \epsilon_{it} \tag{2.2}$$

where

S_{it} = logarithm of establishment i size in year t
μ_{it} = white noise, $E(\mu_{it} \cdot \epsilon_{it}) = 0$
w_{it} = random walk component.

Table 2–8.   Proportion of Jobs Gained, 1978–1982.

Industry	1978	1979	1980	1981	1982
Mining and construction	.29	.20	.15	.13	.14
Food and kindred products	.11	.10	.14	.11	.32
Textiles	.19	.06	.08	.01	.05
Apparel	.26	.10	.08	.07	.18
Lumber	.32	.10	.11	.06	.09
Furniture	.13	.08	.08	.05	.15
Paper	.10	.12	.03	.03	.10
Printing and publishing	.12	.09	.08	.06	.12
Chemicals	.12	.05	.08	.06	.10
Petroleum and coal	.05	.14	.09	.05	.10
Rubber and plastics	.18	.13	.12	.07	.16
Leather	.09	.04	.02	.08	.12
Stone, clay, and glass	.13	.10	.12	.06	.12
Primary metals	.16	.13	.06	.04	.08
Fabricated metal	.12	.10	.10	.05	.25
Machinery, except electrical	.15	.14	.10	.03	.34
Electrical equipment	.13	.11	.22	.06	.11
Transportation equipment	.10	.06	.04	.07	.18
Instruments	.13	.10	.24	.04	.08
Miscellaneous manufacturing	.13	.11	.10	.06	.10
Transportation and utilities	.13	.11	.10	.08	.12
Wholesale and retail trade	.19	.13	.13	.12	.19
Finance, insurance, and banking	.16	.10	.10	.10	.11
Services	.21	.15	.14	.13	.19

The first difference $(S_{it} - S_{it-1})$ of the logarithm of size may now be expressed as

$$D_{it} = \epsilon_{it} + \mu_{it} - \mu_{t-1} \tag{2.3}$$

where ϵ_{it} is the innovation in the random walk component of size, and $(\mu_t - \mu_{t-1})$ is a moving average component. Positive autocorrelation of the ϵ_1 indicates the persistence of shocks or lags in adjustments. If the ϵ_i are serially uncorrelated, then this model predicts that growth rates ($D_{it} = \Delta S_{it}$) more than two years apart are uncorrelated and follow a random walk. It also predicts that growth rates in adjoining years will be negatively correlated:

$$COR(D_{it}, D_{1,t-1}) = \frac{-\sigma_u^2}{\sigma_\epsilon^2 + 2\sigma_\mu^2} \qquad (2.4)$$

$$= \frac{-1}{\dfrac{\sigma_\epsilon^2}{\sigma_\mu^2} + 2} \qquad (2.5)$$

In this model the ratio of lasting to transient errors is identified from the correlation of the logarithm of growth rates two years apart. A test of the fit of this model is provided by its prediction of negatively correlated growth rates in neighboring years and uncorrelated growth rates in years further apart.

Table 2–9 presents a correlation matrix for the logarithm of size and its first difference, the logarithm of growth rate. Unlike the rest of the analysis in this chapter, these correlation matrices are calculated only for the subsample of establishments with positive employment in all years. The growth rates are smaller than in the full sample. Note also that the cross-section standard-deviation of size hardly changes over time and that the lowest growth rate (in 1982) is associated with the highest cross-section standard-deviation of growth rates.

Table 2–9 shows a number of pieces of evidence pointing to a regression to the mean in size. The elements of the upper right corner of the table are all negative. In every case larger size is associated with slower growth in each subsequent year. By the same token, larger size is associated with faster growth in each previous year. Large establishments have recently grown and will soon shrink, on average. Small establishments have recently shrunk and will soon grow, on average. The latter statistical artifact is the foundation for the belief that small establishments are the fountainheads of employment growth. (See Leonard 1985 for further development.)

The lower right quadrant of Table 2–8 shows the correlations of growth rates with themselves over time. All but one of the correlations are negative, and all of the significant correlations are

Table 2–9. Correlation Matrices for the Logarithm of Firm Size (S_t) and for the First Difference ($D_t = S_t - S_{t-1}$) of the Logarithm of Firm Size, 1977–82.

	Mean	σ	S_{78}	S_{79}	S_{80}	S_{81}	S_{82}	D_{78}	D_{79}	D_{80}	D_{81}	D_{82}
S_{77}	1.93	1.34	.966	.949	.932	.918	.898	−.118	−.046	−.051	−.069	−.076
S_{78}	1.99	1.35		.967	.950	.935	.914	.142	−.104	−.055	−.073	−.078
S_{79}	2.02	1.35			.966	.951	.930	.084	.150	−.115	−.076	−.079
S_{80}	2.02	1.36				.967	.947	.079	.089	.143	−.142	−.077
S_{81}	2.01	1.35					.963	.075	.085	.077	.114	−.138
S_{82}	1.97	1.35						.073	.085	.079	.049	.132
D_{78}	.058	.349							−.225	−.017	−.018	−.008
D_{79}	.033	.344								−.237	−.015	−.003
D_{80}	.001	.350									−.258	.006
D_{81}	−.017	.347										−.239
D_{82}	−.035	.366										

Note: $N = 49,508$ firms with positive employment in all years. All of these correlations are significant well beyond conventional levels, with the following exceptions: (D_{82}, D_{78}) at .06, (D_{82}, D_{79}) at .53, and (D_{82}, D_{80}) at .16.

negative. The strongest pattern is for growth rates one year apart. These average a correlation of $-.24$. Above-average growth in one year is likely to be followed (and proceeded) by significantly below-average growth in the next year. If the establishment grows, it likely shrank in the recent past and will shrink in the near future. There is certainly not complete persistence of shocks to growth rates. But neither is there complete adjustment from a shock after one year. What adjustment occurs is primarily in the first year. An employment growth rate 100 percent above average one year is likely to be followed by one 25 percent below average next year, which is then followed by a random walk. This also explains why the average changes we previously observed between 1977 and 1982 were much less than the compounding of the annual average changes.

This correlation of first differences in size can now be interpreted in terms of equation (2.3). As predicted by this process, growth rates one year apart are negatively correlated; those more than one year apart are close to uncorrelated. That $E(\epsilon_{it}, \epsilon_{i,t+k}) \cong 0$ for $k \geq 2$, suggests that establishments quickly adjust and that shocks are not persistent. The one-year-apart correlation is roughly .25, which corresponds to $\sigma_{\epsilon}^2 = 2\sigma_{\mu}^2$. Half of the variance in growth rates then represents real shocks, and at most half represents a moving average $(MA(1))$ process of transient errors. Because a pure measurement error process is $MA(1))$ in growth rates and implies $COR(D_{it}, D_{1,t-1}) = -0.5$, this provides a bound on the role of measurement error in the results reported here.

There are other possible explanations besides transient real shocks for the half of growth-rate variance that follows an $MA(1)$. This component of variance could all be measurement error. An alternative explanation is that target employment follows a random walk. Actual employment may differ from the target by an error that persists less than one year. Both explanations are consistent with an $MA(1)$ process.

THE NONEXISTENCE OF INDUSTRY SHOCKS

Among the most basic economic models of establishment growth is one that posits that the growth rates of establishments should depend on which industry or region they are in. Structural change implies nontransient shifts of employment across industry and/or regional lines. Technological change is usually assumed to have pervasive effects within any one industry, but different effects in different industries. It has become commonplace to speak of the industry or region shocks suffered by the economy since at least 1973 and to attribute to them problems of both the level and the variation of unemployment. Certain industries or regions are widely recognized as being in growth or decline, and it is usually assumed that such trends are widely shared by establishments within the particular industry or region. This last assumption is challenged by the evidence to be presented here.

There certainly are industries or regions that have experienced a trend of growth or decline, but it is a mistake to infer from this aggregate experience that such growth or decline is widely shared by establishments within these groups. For two-digit SIC industries and for counties in Wisconsin, industry or region trends are largely irrelevant for the average establishment in an industry or region.

The purely idiosyncratic components of variation in establishment growth rates can be reduced by grouping and taking averages of growth rates within industry, by county, by year cells. Table 2–10 shows two pooled time-series cross-section regressions for the mean and variance of growth rates within cells on a set of twenty-five industry dummies, seventy-one county dummies, and four year dummies. The dependent variable in the first regression is the average growth rate of employment for establishments in an industry-county-year cell. In the second regression it is the within-cell variance of the establishment growth rates. Cyclical effects common to all industries will be captured by the year dummies, but otherwise the growth-rate regression is not meant to indicate differing cyclical sensitivities across industries. Rather, its purpose is to indicate whether establishments in different industries have, on average, different mean growth rates between 1977 and 1982. This is taken here as a measure of structural change.

Judging from the R^2 (.02) the complete set of industry and county variables capture little of the variance of establishment growth rates. Although the F-statistic of the first equation is marginally significant of the 5 percent level and that of the second equation is significant at the 1 percent level, individually, most of the coefficients are not significantly different from zero. The ex-

Table 2–10. Regressions of Within-Cell Mean and Variance of Growth Rates.

	Mean of Cell Growth Rate	Within Cell Variance of Growth Rate
Intercept	1.099	−1.733
	(.11)	(11.21)
Year 1979	−.021	1.138
	(.03)	(3.14)
Year 1980	−.029	3.529
	(.03)	(3.14)
Year 1981	−.110	−.102
	(.03)	(3.14)
Year 1982	−.033	4.345
	(.03)	(3.14)
SIC20 Food	.045	3.541
	(.06)	(6.13)
SIC21 Tobacco	−.001	1.143
	(.28)	(26.97)
SIC22 Textiles	−.025	−1.488
	(.09)	(9.23)
SIC23 Apparel	.279	15.489
	(.07)	(6.90)
SIC24 Lumber	.028	−0.056
	(.06)	(6.10)
SIC25 Furniture	.029	0.100
	(.07)	(6.86)
SIC26 Paper	.035	.544
	(.07)	(7.10)
SIC27 Printing and publishing	.037	−.309
	(.06)	(6.10)
SIC28 Chemicals	.027	−.082
	(.07)	(7.21)
SIC29 Petroleum	.061	1.789
	(.11)	(10.91)
SIC30 Rubber and plastic	.206	19.485
	(.07)	(6.65)
SIC31 Leather	−.009	1.364
	(.08)	(7.54)
SIC32 Stone, clay, and glass	−.007	−.418
	(.06)	(6.28)

Table 2–10. (Continued)

	Mean of Cell Growth Rate	Within Cell Variance of Growth Rate
SIC33 Primary metal	.152	10.543
	(.07)	(6.95)
SIC34 Fabricated metal	.050	−.125
	(.06)	(6.31)
SIC35 Machinery	.073	−.247
	(.06)	(6.15)
SIC36 Electrical equipment	.170	2.704
	(.07)	(7.00)
SIC37 Transportation equipment	.005	1.11
	(.07)	(6.85)
SIC38 Instruments	.070	.945
	(.07)	(7.28)
SIC39 Miscellaneous manufacturing	.023	.367
	(.07)	(6.47)
SIC4- Transportation and public utilities	.033	−.176
	(.06)	(6.10)
SIC5- Wholesale and retail trade	.018	−.077
	(.06)	(6.08)
SIC6- Finance, insurance, and real estate	.020	−.198
	(.06)	(6.08)
SIC7- Personal, business, repair, and entertainment services	.029	−.069
	(.06)	(6.10)
SIC8- Health, education, and legal services	.056	.104
	(.06)	(6.08)
R^2	.02	.01
F-statistic	1.33	1.01
Mean of dependent variable	1.08	2.16
S.E.E.	.85	82.58

Standard error in parentheses.
Correlation of residuals from two equations: 0.9140.
Note: $N = 6,920$. Based on 124,737 underlying plant observations. Omitted industry is construction and mining (SIC = 1). Both equations include dichotomous variables for seventy-one counties, of which only two were systematically different from zero in each regression.

ceptions run contrary to expectations. The four industries with significantly different growth rates are apparel (.28), rubber and plastic (.21), primary metal (.15) and electrical equipment (.17). All of these industries show higher than average growth rates, yet with the exception of the last, total employment fell in all these industries in Wisconsin between 1977 and 1982 (U.S. Department of Labor 1977 to 1982).

The variance of growth rates within nearly all industries and counties is greater than the variance across industries and counties. Knowing the industry or county an establishment is in does not contribute significantly to knowledge about its growth rate.[9] For the average establishment (not the average worker), there is neither an industry nor a county effect. The risk (that is, layoff risk) that a worker faces is first establishment specific and second (that is, reemployment probability) industry or region specific. In most applications, information on the average worker (or the employment weighted average establishment—aggregate employment) will be more appropriate than information on the average establishment. The first method of reconciling the nonexistence of industry or county effects observed here, with their existence taken for granted everywhere else, is to note the difference between weighted and unweighted averages. This in turn suggests that what are typically labeled as industry effects really tend to affect only the largest establishments within an industry. For many purposes, this suffices. Moreover, cross-industry shifts are likely to cause more unemployment than the cross-establishment shifts within an industry that dominate here. A related explanation is that there is, for unknown reasons, large variation in growth rates across establishments. What show up as changes in aggregate industry growth rates come about because a relatively small proportion of establishments shift from growing to shrinking, or vice versa.

Competition provides a second explanation for the nonexistence of industry effects. Suppose product demand is fixed, markets are competitive, and establishments gain small randomly arriving cost savings through technological progress. This yields an expected negative correlation of growth rates within an industry because one establishment's gain must be another's loss. If technological change is driving these growth rates, it must be affecting different establishments within the same industry very differently.

There is only one significant calendar-year effect in Table 2–10. The average establishment may not be much influenced by its industry or region, but it is influenced by the year. However, given the degrees of freedom here, this is not a very powerful result. The business cycle surely exists, but it does not greatly and similarly affect most establishments. In particular, the declines in total employment growth rate from 9.9 (1978) to 3.6 (1979) and to 1.5 (1980) are not accompanied by significant reductions in the mean growth rate of establishments. The exception is 1981, when mean growth rates fall significantly by 11 percentage points. Otherwise, one would not have significant evidence that a recession or boom had occurred by observing the unweighted average establishment in Table 2–10.

Table 2–10 pools across years and so averages out changes over the cycle, but the main result can also be observed in unpooled regressions on single years (not shown). Out of twenty-five industry dummies, from one to four are significantly different from zero in a single year between 1977 and 1982. Similar results are found for counties. Although the different cyclical sensitivities of total employment in different industries is well known, this does not generally carry over to the average growth rates of establishments.

The second equation in Table 2–8 is a regression of the variance of establishment growth rates within industry, area, year cells on a set of dichotomous variables, indicating industry, area and year. Again, with few exceptions, there is no general evidence of significant industry, area, or year effects on the variance of establishment growth rates. The exceptions may well be caused by reporting errors in the raw data. It is interesting to note that years of high unemployment rates or of employment decline are not associated with significantly greater variance of growth rates across establishments within cells.

David Lilien (1982) has advanced the argument that cyclical increases in the unemployment rate are caused by structural change, measured by the employment-share weighted variance of the logarithm of industry growth rate across one- or two-digit SIC industries. For example, he reports this variance of log growth rates at .00081 in 1981. The logical argument made by Lilien to tie this variation causally to unemployment carries through with at least as great force to further disaggregated measures. What happens when we expand his measure to include frictional unemployment by calculating the variance of the logarithm of employment growth across individual establishments?

This measure takes on the following values: .118 (1977–78), .113 (1978–79), .115 (1979–80), .114 (1980–81), .127 (1981–82). These are unweighted. Evidently, the cross-industry measure includes only a small part of the variation in growth rates across establishments. Here we observe a total variance 140 times the cross-industry variance measured by Lilien. Obviously, the within industry variance accounts for all but a negligible part of this. By this measure, then, frictional sources are of far greater importance than structural sources of unemployment. The total variance shows an upward trend between 1978 and 1981. More often than not, it moves in the same direction as the unemployment rate, although the unemployment rate increases most in a year (1980–81) that this variance actually declines. With only five time-series observations, the concordance of these data with Lilien's hypothesis cannot be precisely judged.

A distinct hypothesis is that, because of different cyclical sensitivities, faster mean employment growth is associated with greater variance in establishment growth rates. This would imply that the predicted values and residuals from the variance regression are positively correlated with those from the mean growth regression in Table 2–7. The observed values are actually strongly positively correlated ($r \cong .9$). Cells with high (or higher than expected) mean employment growth rates also have a high variance of growth rates across establishments within the cell. As the mean of the distribution of growth rates shifts up, the variance tends to increase.

CONCLUSION

This chapter has attempted to provide some new empirical evidence on the nature and magnitude of structural and frictional shifts in employment across industries and establishments. The findings from this analysis of the private employers of Wisconsin over one business cycle provide some perspective from which to judge the impact of technological change.

About one-ninth of all jobs are destroyed and more than one-eighth created each year on average between 1977 and 1982. Huge gross flows are hidden beneath the usual net flow data. Gross employment flows range from three to seventeen times greater than net employment flows. Jobs themselves are more unstable than previous aggregate statistics have revealed. As much as half of the transitions of workers from employment to nonemployment may be accounted for by the destruction of jobs. Such job loss may account for roughly 2.2 percentage points, or more than a quarter of Wisconsin's average unemployment during 1978–82.

There are few strong industry effects on employment growth at the establishment level. Rather there is substantial diversity among establishments within an industry. The across-sector variation in the logarithm of employment growth rates, used by Lilien (1982) to measure structural change, is just the tip of the iceberg. One hundred forty times greater is the total variation across establishments, nearly all of which is within industry—not across. By this measure, employment shifts across establishments within an industry are of far greater magnitude than shifts across industry lines. Increases in this growth rate variance are at best weakly associated with increases in the unemployment rate.

Establishments appear to adjust their employment quickly. Whatever adjustment occurs is largely completed within the first year. This is followed by a movement in the other direction that suggests both measurement error and overshooting the employment target. Employment growth rates one year apart are negatively correlated and thereafter nearly follow a random walk.

This chapter has shown surprisingly large gross employment flows based on the population of establishments in one state. Between 1977 and 1982, 11.0 percent of the previous year's employment is destroyed and 13.8 percent is created each year. Gross job turnover ranging from one in three to one in five jobs occurs in these years. The level of employment at establishments is characterized by substantial volatility that shows some positive cyclical variation but little industry effect.[10] Roughly one-quarter of the "natural" rate of unemployment may be accounted for by these largely idiosyncratic fluctuations in labor demand within establishments. An economy that loses one in nine jobs and creates one in eight jobs yearly would appear to be one with considerable flexibility to absorb technological change.

NOTES

I thank Bill Dickens, Kevin Lang, David Lilien, Richard Freeman, and participants at the University of California at Irvine Conference on "Unemployment and the Structure of Labor Markets" for their comments. I also thank the Wisconsin State Department of Development for their cooperation. This work was partially supported by an Olin Fellowship at the National Bureau of Economic Research. Nothing here represents the official policies of the State of Wisconsin, the National Bureau of Economic Research, or the National Academy of Science.

1. Establishments using only self-employment or unpaid family labor are not required to file reports and are exempt from unemployment insurance taxes. Therefore, one-worker establishments are likely to be underreprsented here. However, one-person establishments with an office address and a telephone number are likely to be included. Through 1977, agricultural establishments, railroads, and nonprofit organizations were exempt from U.I. coverage. Beginning in 1978, only railroads, nonprofit establishments with one to three employees, and agricultural establishments with less than ten employees were excluded. Of these changes, only the nonprofits are of substance. To maintain a consistent series, nonprofit and government employment were excluded from the data used here in all years. These exclusions include 25 percent of state employment. Foreign (out-of-state) employment is also excluded.

2. Where possible, large establishments reporting the greatest percentage change in employment were checked against published *County Business Patterns* data. If the published data ruled out such large changes, the observations were dropped from the sample. This occurred in fewer than seventy cases, but other reporting errors cannot be precluded. In particular, establishments that may have incorrectly reported stable employment were not checked.

3. If the results to be analyzed here are thought of as coming from a population, there is no need or scope for statistical inference. The results presented here are in this case the true population parameters calculated without sampling. In a broader sense, the establishments analyzed here may be thought of as a sample from a larger population across states or time, or each establishment's employment may be thought of as including a deviation from target. In both these latter cases, the usual rules of statistical inference apply.

4. Because vacancies average only 1.7 to 3.7 percent of the workforce (Abraham 1983) and are typically filled within a few months, such

turnover is assumed to have no effect on the establishment side measures of job gain or loss. In other words, I assume that workers who quit or are fired are all quickly replaced and so do not affect the measures of job gain or loss calculated here.

5. Because of a regression to the mean phenomena, the shrinking establishments tend to start larger than the growing establishments. Table 2–3 shows the share of the previous year's employment accounted for by establishments that grew since the previous year. Although growing establishments account for 23 percent of all establishments, they account for an average of 4 percent of all employment in the year prior to their growth. Similarly, shrinking establishments account for 21 percent of all establishments but 47 percent of all employment in the year prior to their decline. In part because of an integer constraint in the way employment is counted here, the stable establishments are primarily one- and two-person establishments. Stable establishments then account for about two-thirds of the establishments but only 8 percent of the jobs each year.

6. Overcounts of job loss and gain when ownership of an establishment changes hands appear to be a relatively minor problem with the data used here. A version of this data that made great efforts to correct for this still shows an average 10 percent yearly job gain and 11 percent yearly job loss between 1978 and 1981 (Wisconsin Department of Development 1984: 133).

7. This lends itself to a competing risks formulation. If a worker quits or is fired before the job is done, we know only that job duration (life of the job, not the job-employee match) exceeded job tenure (life of the employee-job match).

8. It is reasonable to expect greater variations in the level of employment where wages are more rigid. Leonard (1986) shows that annual variation in employment is not greater in unionized plants than in their nonunion counterparts. If wage rigidity is to contribute to the explanation of establishment level employment volatility, then it is probably a pervasive institution not isolated to the union sector.

9. This heterogeneity across establishments within an industry and region may also help explain the difficulties encountered by compensating differentials studies that utilize industry level data to measure—for example, a worker's risk of becoming unemployed (see Murphy and Topel 1987). Moreover, this substantial idiosyncratic part of unemployment risk should be diversifiable. In this sample, the correlation of growth rates across establishments is too low to be a barrier to insurance against layoff.

10. In other work (Leonard 1985), technology does have a strong impact on employment growth. Within industries, the faster growing es-

tablishments are those with higher proportions of nonclerical white-collar employees in their work forces. Skill-intensive employers appear better able to adapt and thrive.

REFERENCES

Abraham, Katharine. 1983. "Structural/Frictional vs. Deficient Demand Unemployment." *American Economic Review* 73 (4) (September):708–24.

Abraham, Katharine, and Katz, Lawrence. 1984. "Cyclical Unemployment: Sectoral Shifts or Aggregate Disturbances?" Cambridge, Mass.: National Bureau of Economic Research, Working Paper No. 1410. July.

Clark, Kim, and Lawrence Summers. 1979. "Labor Market Dynamics and Unemployment," *Brookings Papers on Economic Activity* 1:13–60.

Dickens, William T. and Kevin Lang. 1985. "A Test of Dual Labor Market Theory," *American Economic Review.*

Dynarski, Mark and Steven Sheffrin. 1987. "New Evidence on the Cyclical Behavior of Unemployment Durations." In K. Lang and J. Leonard, eds., *Unemployment and the Structure of Labor Markets.* Oxford: Basil Blackwell.

Gibrat, R. 1930. *Les Inegalités Economiques.* Paris: Ricueil Sirey.

Hall, Robert E. 1982. "The Importance of Lifetime Jobs in the U.S. Economy." *American Economic Review* 72 (4) (September):716–24.

Leonard, Jonathan S. 1985. "On the Size Distribution of Employment and Establishments." Cambridge, Mass.: National Bureau of Economic Research, Working Paper No. 1951.

———. 1986. "Employment Variability and Wage Rigidity: A Comparison of Union and Non-Union Plants." Unpublished paper, University of California, Berkeley. February.

Lilien, David M. 1980. "The Cyclical Pattern of Temporary Layoffs in United States Manufacturing." *Review of Economics and Statistics* (February):24–31.

———. 1982. "Sectoral Shifts and Cyclical Unemployment." *Journal of Political Economy* 90 (4) (August):777–93.

Murphy, Kevin, and Robert Topel. 1987. "Unemployment Risk and Earnings." In K. Lang and J. Leonard, eds., *Unemployment and the Structure of Labor Markets* Oxford: Basic Blackwell.

Poterba, James M., and Lawrence H. Summers. 1985. "Reporting Errors and Labor Market Dynamics." Cambridge, Mass.: National Bureau of Economic Research, unpublished paper. September.

Steindl, Josef. 1965. *Random Processes and the Growth of Firms: A Study of the Pareto Law.* New York: Hafner.

U.S. Department of Commerce, Bureau of the Census. 1977–82. *County Business Patterns, Wisconsin.* Washington, D.C.: U.S. Government Printing Office.

U.S. Department of Labor, Bureau of Labor Statistics. 1977–82. *Employment and Earnings,* vols. 24–29. Washington, D.C.: U.S. Government Printing Office.

Wisconsin Department of Development. 1984. *The Job Generation Process in Wisconsin: 1969–1981.* Madison: State of Wisconsin.

3 THE EFFECT OF TECHNICAL PROGRESS ON PRODUCTIVITY, WAGES, AND THE DISTRIBUTION OF EMPLOYMENT
Theory and Postwar Experience in the United States

Robert M. Costrell

WHAT ARE THE ISSUES?

This chapter provides a theoretical and empirical analysis of three related issues: the effect of technical progress on the growth of productivity, the growth of real wages, and the distribution of employment among various sectors of the economy. It is well known that aggregate productivity growth has slowed markedly over the last fifteen or twenty years. This event, which has been widely studied, has raised the following questions: Which sectors have contributed most to the productivity slowdown? Has the slowdown in those sectors been associated with an abatement of technical progress or of capital-deepening, or has it reflected a measurement problem? Has the shift of resources among sectors played a significant role in the productivity slowdown?

In a related development, aggregate real hourly wage growth has also slowed, and similar questions have been raised. Which sectors have been involved? What has been the role of technical progress, its labor-saving bias, and capital-deepening? Has the shift of employment from high-wage to low-wage sectors been a large factor in the aggregate real wage slowdown?

Finally, there has been concern over the employment shifts themselves, not only for those who actually shift sectors but also

73

for those who are unable to enter or move up to the high-wage jobs held by the previous generation. What have been the causes and consequences of these shifts? Specifically, what has been the role of uneven rates of technical progress and of its labor-saving bias? How have these compared with other sources, such as trade imbalances? And what has been the typical wage penalty associated with these shifts?

Our examination of these issues is presented in two parts. The first part provides a theoretical analysis of technical progress in a multisectoral economy with exogenous wage differentials between sectors. The second part examines the postwar data on hours of employment, wages, and productivity at the one-digit level of aggregation (twelve nonfarm business sectors), supplemented by data on capital at the broad level of manufacturing versus nonfarm nonmanufacturing.

THEORY

The Theory of One Sector

The following sections review the theory of the effect of technical progress on productivity and wages in a simple one-sector model.[1] First the basic equations for the growth rates of productivity and factor prices are set out, taking the growth rate of capital as given. Then growth theory is used to tie long-run capital accumulation to the rate and bias of technical progress. Notation is summarized in Table 3–1.

Productivity Growth. Let output be governed by a constant returns production function

$$Q = F(K, L; t)$$

where the quantity of output is Q, capital and labor[2] inputs are K and L, and t represents time and the state of technology. It should be noted that we are assuming each firm operates on the production frontier, in a technically efficient fashion, with no "fat," or at least with a constant degree of inefficiency.[3]

Table 3–1 Notation.

In multisectoral model, subscript i denotes sector i, and unsubscripted variables denote aggregates

· superscript denotes time derivative, such as $\dot{x} = dx/dt$

∧ superscript denotes proportional growth rate, e.g., $\hat{x} = \dot{x}/x = (dx/dt)/x$

$K \equiv$ capital input

$L \equiv$ labor input

$t \equiv$ time and the state of technology

$Q = F(K, L; t) \equiv$ output (of sector, or Divisia index of aggregate)

$F_K, F_L \equiv \partial Q/\partial K, \partial Q/\partial L \equiv$ marginal products of capital and labor *(MPK, MPL)*

$\theta_K, \theta_L \equiv KF_K/F, LF_L/F \equiv$ output elasticities of capital and labor

$T \equiv (\partial Q/\partial t)/Q \equiv$ rate of technical progress, total factor productivity growth

$B \equiv$ the labor-saving bias of technical progress (if positive)

$q \equiv Q/L \equiv$ output per unit of labor, average product of labor *(APL)*

$k \equiv K/L \equiv$ capital/labor ratio

$\sigma \equiv$ elasticity of substitution between capital and labor

$s \equiv$ average propensity to save

$r \equiv$ real return to capital

$W \equiv$ average nominal wage rate

$w \equiv$ average real wage rate

$w_n \equiv$ average real wage of nondisplaced workers

$w_d \equiv$ average real wage of sectors with declining employment shares, weighted by rate of decline

$w_g \equiv$ average real wage of sectors with growing employment shares, weighted by rate of growth

$w_{gap} \equiv (w_g - w_d)/w \equiv$ wage gap between growing and declining sectors, relative to average real wage

$p \equiv$ price (of sector, or Divisia index of aggregate)

$\alpha_i \equiv p_iQ_i/\Sigma p_iQ_i \equiv$ sector i's share of nominal output

$\lambda_i \equiv L_i/L \equiv$ sector i's share of employment

$\mu_i \equiv K_i/K \equiv$ sector i's share of capital inputs

$\delta_i \equiv \lambda_iW_i/W = \lambda_iw_i/w \equiv$ sector i's share of wage payments

$v \equiv$ value added per unit of labor

$D \equiv \Sigma|\dot{\lambda_i}|/2 \equiv$ relative displacement per year

---

We can then differentiate with respect to time, and denote proportional rates of change by $\hat{\ }$, giving us

$$\hat{Q} = \theta_K \hat{K} + \theta_L \hat{L} + T$$

where

$$\theta_K \equiv KF_K/F \qquad \text{and } \theta_L \equiv LF_L/F$$
$$F_K \equiv \partial F(K, L; t)/\partial K \text{ and } F_L \equiv \partial F(K, L; t)/\partial L$$

and

$$T \equiv [\partial F(K, L; t)/\partial t]/F$$

Here, F_K and F_L are the marginal productivities of capital and labor, θ_K and θ_L are the output elasticities with respect to capital and labor (the distributive shares under marginal productivity payments[4]), and T is the rate of technical progress, also known as the growth rate of total factor productivity. It immediately follows that output per unit of labor (the average product of labor), $q \equiv Q/L$, rises at the rate

$$\hat{q} = \hat{Q} - \hat{L} = \theta_K \hat{k} + T \tag{3.1}$$

where $k \equiv K/L$ is the capital/labor ratio, since $\theta_K + \theta_L = 1$. This is the standard method of decomposing labor productivity growth into the effects of capital-deepening, $\hat{k}$, and technical progress, T, due to Solow (1957).[5]

Factor Price Growth.   The demand for capital and labor are governed by the cost-minimizing decisions of firms, predicated on the prices of labor and capital. Formally, firms choose the capital-labor ratio to satisfy

$$F_L(k; t)/F_K(k; t) = w/r$$

where w and r signify the real wage and real return to capital. The marginal productivities depend only on the capital-labor ratio and the state of technology. Differentiating over time it can be shown that

$$\hat{k} \equiv \hat{K} - \hat{L} = \sigma(\hat{w} - \hat{r}) + B$$

where

$$\sigma \equiv F_K F_L / F F_{KL} \quad \text{and } B \equiv \sigma[(\partial F_K / \partial t)/F_K - (\partial F_L / \partial t)/F_L]$$

Here, σ represents the elasticity of substitution between labor and capital, a measure of the ease with which capital and labor can be substituted for one another, in response to changes in factor prices. This number is positive but is usually believed to be less than unity, representing relatively limited substitution possibilities.

B is known as the bias of technical progress. It represents the effect of technical progress on the firms' desired capital-labor ratio, apart from any change in factor prices. If $B > 0$, firms will choose to increase the capital-labor ratio over time, and technical progress is said to have a labor-saving bias. Conversely, if $B < 0$, the bias is capital-saving, which is usually considered less likely.

Now, letting the real wage equal the marginal productivity of labor, it can readily be shown that its growth rate is

$$\hat{w} = T + (\theta_K / \sigma)(\hat{k} - B) \tag{3.2}$$

Similarly, the growth rate of the real return to capital is given by

$$\hat{r} = T - (\theta_L / \sigma)(\hat{k} - B)$$

and the difference between the two is

$$\hat{w} - \hat{r} = (\hat{k} - B)/\sigma$$

Technical progress per se ($T > 0$) tends to raise the real wage and return to capital, but if the bias is labor-saving ($B > 0$), it favors capital over labor. On the other hand, capital-deepening tends to raise the wage and reduce the return to capital.

Labor's Share: The Link between Wages and Productivity. Wages and productivity will grow at the same rate unless labor's share changes. That is, the difference between the growth rate of wages and productivity is the growth rate of labor's share

$$\hat{w} - \hat{q} = \hat{\theta}_L = [(1 - \sigma)\hat{k} - B]\theta_K / \sigma \tag{3.3}$$

Capital-deepening raises wages more than productivity and therefore raises labor's share, provided $\sigma < 1$. On the other hand, if the bias of technical progress is labor-saving ($B > 0$), it reduces the share of wages. To keep labor's share from falling in the face of labor-saving technical progress, we must have a certain rate of capital-deepening. Fortunately, long-run considerations lead us to believe that technical progress, and especially biased technical progress, stimulate such capital-deepening.

The Long-Run Effect of Technical Progress on Capital-Deepening. The long-run effects of technical progress on capital-deepening are the subject of growth theory.[6] Consider Figure 3–1, due to Solow (1956). For a given average propensity to save, s, savings per head are given by

$$sQ/L = s \cdot q(k)$$

an increasing function of k. The ray out of the origin, nk, represents the savings per head required to maintain the capital-labor ratio in the face of labor force growth at the constant rate n. Long-run equilibrium (steady-state) initially obtains at the intersection, point A, with capital-labor ratio k_0. At any lower $k < k_0$, savings per head exceed nk, so k grows, while the converse holds at any $k > k_0$.

Now, if a round of technical progress occurs, raising productivity at the initial capital-labor ratio, k_0, savings per head will increase to point B. Since this exceeds the investment required to maintain k_0, k rises until a new equilibrium is achieved at point

Figure 3–1.   The Solow Growth Model

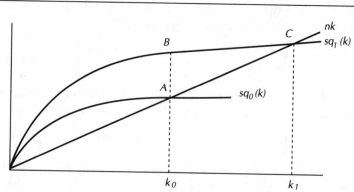

C, with k_1. Hence, in the long run, technical progress stimulates capital-deepening.

This means that the long-run effect of technical progress on productivity is even more favorable than the short-run effect, through the additional effect on capital-deepening. It also means that the long-run effect on wages is more favorable than the short-run. Moreover, if $\sigma < 1$, the induced capital-deepening raises wages more than it raises productivity—that is, it tends to raise the share of labor and reduce the profit rate. It is only the labor-saving bias of technical progress that prevents this.

Now, suppose the average propensity to save is not a constant but increases with the share of capital (a widely held view). At any given capital-labor ratio, the labor-saving bias of technical progress tends to raise capital's share, so that raises the average propensity to save. This provides a further stimulus to savings per head, shifting the curve higher than B, so there is additional capital-deepening before a new steady-state equilibrium is attained. Hence, biased technical progress is especially favorable to capital-deepening, which mitigates the effect of bias on labor's share.

A Multisectoral Model with Factor Price Differentials

The following sections extend the theory to a multisectoral model, where factor market imperfections prevent the equalization of wages and returns to physical capital among the various sectors. Again, we derive expressions for the growth rates of aggregate productivity, real wages, and labor's share. Sectoral shifts of employment and capital find their places in these equations. A method is presented for estimating the wage loss due to relative displacement, both in the aggregate and for the average relatively displaced worker. The multisectoral equations developed here will guide the empirical work in the second half of this chapter.

Factor Price Differentials. Before turning to a formal analysis, it may be worth reviewing recent research on the sectoral dispersion of wages, which is the basis for the model to be developed. Dickens and Katz (1987), for example, find standard deviations of average log wages across industries, for given occupations, ranging from

13 percent (clerical) to 37 percent (sales). After attempting to correct for individual differences, using standard human capital, demographic, and geographic variables, these standard deviations are reduced but still range from 8 to 24 percent. Similarly, Reiff (1986) finds that human capital variables explain approximately 15 percent of the variance of individual earnings, while industry and occupation account for about 25 percent. Finally, Bell and Freeman (1985) find that the dispersion of wages among industries increased dramatically in the 1970s.

Some of the residual variance across sectors may reflect unmeasured skill differentials[7] or compensating differentials for onerous work, in which case sectoral shift does not entail true wage loss for the individuals or society. We should, however, consider the wage and productivity implications of employment shifts among sectors with wage differentials that reflect labor market imperfections—such as labor market discrimination, union wage premiums, or wage premiums offered to avoid the threat of unionization. Similarly, capital shifts among sectors with different returns to physical capital,[8] due to entry barriers, have productivity implications as well. Finally, it should be noted that recent research (such as Salinger 1984, Freeman 1983, Dickens and Katz 1987) finds evidence that labor market and capital market imperfections are themselves correlated—that is, that unions or the threat of unions have been most effective at raising wages where concentration and barriers to entry have been important.

At the present state of knowledge, however, factor price differentials across sectors remain something of a "mystery" (Dickens and Katz 1987). We shall not shed any further light on their causes but confine ourselves to their theoretical and empirical implications.

Aggregate Productivity Growth and Sectoral Shift. The theory of a multisectoral economy suggests that growth in aggregate output and real income should be measured using weights that reflect consumer preferences as indicated by market prices. Specifically, the preferred measure is the Divisia index of output growth:

$$\hat{Q} \equiv \frac{\Sigma p_i dQ_i/dt}{\Sigma p_i Q_i} = \Sigma \alpha_i \hat{Q}_i$$

where the weights $\alpha_i \equiv p_i Q_i / \Sigma p_i Q_i$ represent each sector's share of nominal output and income.[9]

The growth rate of real income and output per unit of labor—that is, the growth rate of aggregate labor productivity—is

$$\hat{q} \equiv \hat{Q} - \hat{L}$$
$$= \Sigma \alpha_i (\hat{Q}_i - \hat{L})$$
$$= \Sigma \alpha_i (\hat{q}_i + \hat{L}_i - \hat{L}), \text{ since } q_i \equiv Q_i / L_i$$
$$\hat{q} = \Sigma \alpha_i \hat{q}_i + \Sigma \alpha_i \hat{\lambda}_i$$

where $L \equiv \Sigma L_i$, the total labor input, and $\lambda_i \equiv L_i / L$, sector i's share of total employment. This expression shows that aggregate productivity growth can be expressed in two components: a weighted average of sectoral productivity growth and a composition effect based on changes in the distribution of employment.

More specifically, the composition effect can be reexpressed using the definitions of value added per unit of labor in sector i and in the economy as a whole:

$$v_i \equiv p_i Q_i / L_i, \qquad v \equiv \Sigma p_i Q_i / L = \Sigma \lambda_i v_i$$

Therefore, the composition effect can be written

$$\Sigma \alpha_i \hat{\lambda}_i = \Sigma \lambda_i (v_i / v) \hat{\lambda}_i = \text{Covariance } (v_i / v, \hat{\lambda}_i)$$

by analogy with the probabilistic concept of covariance, using λ_i as the probability density function.[10] The composition effect reflects whether sectoral employment growth is positively or negatively correlated with value added per unit of labor. Aggregate productivity growth is therefore

$$\hat{q} = \Sigma \alpha_i \hat{q}_i + Cov(v_i / v, \hat{\lambda}_i) \tag{3.4}$$

Equation (3.4) lies behind the popular argument that any employment shifts away from high value-added sectors (such as manufacturing) to low value-added sectors (such as services) reduce aggregate productivity.[11] Although this argument has obvious algebraic appeal, further consideration must be given to the ultimate reasons for the employment shifts and for the differences in value-added per unit of labor.

Now, value added per unit of labor in sector i, relative to the economy as a whole, can be expressed as

$$v_i/v = \theta_L(w_i/w) + \theta_K(r_i/r)(k_i/k)$$

where θ_K and θ_L are the aggregate shares of capital and labor. This can therefore vary across sectors due to unequal capital-labor ratios, k_i, as well as unequal factor prices, w_i and r_i. We now intend to show that it is only the inequality of factor prices that matters.

Consider a shift of employment away from sectors that are high in value added due to their capital-intensity. In terms of (3.4), this gives us a negative composition effect on productivity. However, the reallocation of resources away from particularly capital-intensive sectors frees up a certain amount of capital and allows the economy to increase the capital-intensity of all (or most) sectors. This, in turn, raises the productivity of most sectors. Therefore, the negative composition effect is offset by a rise in sectoral productivity advance, the first term in (3.4).

More formally, (3.4) can be informatively rewritten, with the aid of sector i's analogue of (3.1) and a certain amount of manipulation:

$$\hat{q} = T + \theta_K \hat{k} + \theta_K Cov(r_i/r, \hat{\mu}_i) + \theta_L Cov(w_i/w, \hat{\lambda}_i) \quad (3.5)$$

where

$$T \equiv \Sigma\alpha_i T_i, \ \mu_i \equiv K_i/K, \ w \equiv \Sigma\lambda_i w_i, \ r \equiv \Sigma\mu_i r_i$$

and the covariances use capital and employment weights, μ_i and λ_i, respectively.[12] The first two terms are straightforward generalizations of equation (3.1): Productivity growth is governed by the aggregate rates of technical progress and capital-deepening. The last two terms represent shifts of resources among sectors with different factor prices, but there is no term for shifts among sectors with different capital-intensities: They have no significance for aggregate productivity growth.

Now, if input markets were competitive, such that $r_i = r$ and $w_i = w$, then the last two terms would vanish and sectoral shift would have no significance at all for aggregate productivity growth. However, if capital and/or labor shift toward sectors with low factor prices, then the covariances are negative, representing a drag on productivity growth.[13] Equations (3.4) and (3.5) will inform our empirical discussion of productivity growth in the second half of this chapter.

Aggregate Wage Growth and Sectoral Shift.   To develop an expression for the growth rate of the average real wage,[14] we begin with the growth rate of the average nominal wage, $W \equiv \Sigma \lambda_i W_i$

$$\hat{W} = \Sigma \delta_i \hat{W}_i + Cov(w_i/w, \hat{\lambda}_i)$$

where

$$\delta_i \equiv \lambda_i W_i / W = \lambda_i w_i / w$$

is sector i's share of total wages, and where we note that the relative nominal wage W_i/W is identical to the relative real wage w_i/w regardless of the deflator. The appropriate deflator for the rest of the expression is the Divisia price index, which grows at the rate

$$\hat{p} = \Sigma \alpha_i \hat{p}_i$$

Consequently, aggregate real wage growth

$$\hat{w} = \hat{W} - \hat{p}$$

can be expressed as

$$\hat{w} = \Sigma \delta_i \hat{w}_i + Cov(w_i/w, \hat{\lambda}_i) \tag{3.6}$$

where

$$\hat{w}_i = \hat{W}_i - \hat{p}$$

the real wage growth of a worker who remains in sector i. Thus, aggregate real wage growth is a weighted average of the gains of such workers, plus a covariance term we have already seen, concerning employment shifts, which will be discussed further in the section on "Wage Losses from Relative Displacement," below. Expression (3.6) will be calculated in the second half of the chapter, where it will be found that the covariance term has been quite small, although it has risen recently.

For the remainder of this section, we focus on the first term in

(3.6), the weighted average of the wage growth of nondisplaced workers. Let us rewrite it as

$$\hat{w}_n \equiv \Sigma \delta_i (\hat{W}_i - \hat{p})$$
$$\equiv \Sigma \delta_i (\hat{W}_i - \hat{p}_i) + \Sigma \delta_i (\hat{p}_i - \hat{p})$$
$$\hat{w}_n = \Sigma \delta_i \hat{F}_{Li} + Cov(\theta_{Li}, \hat{p}_i)/\theta_L \qquad (3.7)$$

where the covariance uses α_i as weights. This equation is calculated in the second half of the chapter, and the covariance is rather small. Hence, real wage growth is primarily governed by the first term, a weighted average of the purchasing power of each sector's wage over its own output, which is that sector's marginal productivity of labor, under the assumption we have been making of competitive output markets.

Now, using the analogue of equation (3.2), each sector's marginal productivity of labor can be expressed in terms of the rate and bias of that sector's technical progress and rate of capital-deepening. Thus, we can write

$$\hat{w}_n = \Sigma \delta_i T_i + \Sigma \delta_i \theta_{Ki} (\hat{k}_i - B_i)/\sigma_i + Cov(\theta_{Li}, \hat{p}_i)/\theta_L \qquad (3.8)$$

The three terms of this expression can also be calculated (though the second term cannot be broken down without further information to disentangle B_i and σ_i), provided we have data on capital. The key point here is that the first term is a weighted average of the rates of technical progress.

In fact, if we make the further assumption (which we will not impose on the data in the second half of the chapter) that wages rise at the same rate in all sectors (that is, the wage gaps are proportionally constant), and similarly for the returns to capital, equation (3.8) simplifies to

$$\hat{w}_n = T + \theta_K [(\hat{k}_i - B_i)/\sigma_i] \qquad (3.9)$$

where the bracketed term is identical for any sector i under the given assumptions. This immediately parallels equation (3.2) for the one-sector model. Specifically, the first term shows that technical progress, $T \equiv \Sigma \alpha_i T_i$, directly benefits nondisplaced workers to the same extent as the economy as a whole, as given in (3.5). Similarly, the second term suggests that labor-saving bias tends to reduce wages, while capital-deepening tends to raise them.

Labor's Share and Sectoral Shift. The difference between the growth rate of real wages and productivity is given by the behavior of labor's share:

$$\hat{\theta}_L = \Sigma \delta_i \hat{\theta}_{Li} + Cov(\theta_{Li}, \hat{a}_i)/\theta_L \qquad (3.10)$$

a weighted average of the growth rate of each sector's share of labor, plus a covariance term telling us whether value added is shifting toward those sectors where labor's share is high or low. This expression will be measured in the second half of the chapter.

If we again impose the assumption that the wage gaps are proportionally constant, along with those for the return to capital, we have

$$\hat{\theta}_L = \theta_K \{[(\hat{k}_i - B_i)/\sigma_i] - \hat{k} + Cov(w_i/w, \hat{\lambda}_i) - Cov(r_i/r, \hat{\mu}_i)\} \qquad (3.11)$$

where, again, the bracketed term is independent of i. Aside from the covariance terms, the parallel with the one-sector expression (3.3), is evident. The covariance terms tell us that labor's share is raised by resource movements toward sectors with high labor shares due to high wages and low returns to capital. Resource movements to sectors with high labor shares due to low capital-labor ratios also raise the aggregate share of labor, and that will show up in the first term: Such movements allow each sector to enjoy a more rapid rate of capital-deepening.

Wage Losses from Relative Displacement. As mentioned above, the covariance between employment growth rates and wage levels represents the wage losses from relative displacement.[15] More precisely, consider the following fundamental relationship, to be shown:

$$Cov(w_i/w, \hat{\lambda}_i) = D \cdot w_{gap} \qquad (3.12)$$

Here D is a measure of the amount of relative displacement per year and w_{gap} is the loss associated with displacement. Hence, the covariance is the total effect of relative displacement on wages.

Specifically,

$$D \equiv \Sigma|\dot{\lambda}_i|/2 \qquad (3.13)$$

measures the proportion of employment shares that have shifted out of one sector and into another (dividing by two avoids double counting), a common and natural measure of net relative displacement per year. This statistic, of course, is sensitive to the level of disaggregation. Our one-digit data in the second half of this chapter shows that this has recently risen to 0.8 percent of the total hours worked per year, mostly coming out of manufacturing and going into services.

The wage gap is calculated as follows. First, consider the sectors with declining employment shares. We measure the average predisplacement wage as a weighted average from these sectors, call it w_d. The weights used are the rates at which employment share is eroding—that is, the amount of relative displacement per year from that sector:

$$w_d \equiv \Sigma w_j(-\dot{\lambda}_j)/\Sigma(-\dot{\lambda}_j)$$

where $\dot{\lambda}_j = d\lambda_j/dt = \lambda_j\hat{\lambda}_j$, and j is such that $-\dot{\lambda}_j > 0$—that is, the summations cover only the declining sectors. Hence, if we assume that a given declining sector's average wage is representative for those who are relatively displaced from that sector, then w_d measures the average predisplacement wage for those relatively displaced from all declining sectors. Similarly, our measure of the postdisplacement wage is a weighted average from the growing sectors:

$$w_g \equiv \Sigma w_k\dot{\lambda}_k/\Sigma\dot{\lambda}_k$$

where k is such that $\dot{\lambda}_k > 0$—that is, the summations cover only those sectors for which employment share is growing.

The wage gap between growing and declining sectors, as a fraction of the overall average wage, is

$$w_{gap} = (w_g - w_d)/w \qquad (3.14)$$

This gap is estimated in the second half of the chapter, using one-digit data. Over the most recent business cycle, it has reached -15 percent, since service wages fall short of manufacturing.

Now our fundamental relationship is readily established:

$$Cov(w_i/w, \hat{\lambda}_i) \equiv \Sigma(w_j/w)\dot{\lambda}_j + \Sigma(w_k/w)\dot{\lambda}_k$$
$$= D \cdot w_{gap}$$

since

$$D \equiv \Sigma\dot{\lambda}_k \equiv \Sigma(-\dot{\lambda}_j) \equiv \Sigma|\dot{\lambda}_i|/2$$

That is, the wage loss to those who are displaced, w_{gap}, times the amount of displacement, D, gives us the total wage effect of displacement, as measured by the covariance.

Using our results from the second half of the chapter, the recent wage gap of -15 percent, which affects 0.8 percent of the hours worked per year, gives total wage losses from displacement of only -0.12 percent of total wages per year, a rather small drag on average real hourly wage growth, in (3.6). The drag on productivity growth, in (3.5), will be even less, -0.08 percent per year, given $\theta_L = 0.65$. The point, however, is that these losses are highly concentrated on that small part of the workforce each year, which either loses a high-wage job or loses the hope of entering or moving up to such a job, as did the previous generation.

In this section, we have shown how to estimate the losses from displacement for those involved and for the economy as a whole. There is, of course, direct household survey evidence on the costs of displacement in recent years,[16] which is more informative than our inferences from highly aggregated establishment data. However, our method will allow us to compare the recent losses with earlier periods, to see if anything unprecedented is occurring.

The Determinants of Relative Displacement. Having established the significance of the covariance term for the affected parties, if not for aggregate productivity and wage growth, we now discuss its causes. As an algebraic matter

$$\hat{\lambda}_i \equiv \hat{L}_i - \hat{L} = (\hat{Q}_i - \hat{Q}) + (\hat{q} - \hat{q}_i) \qquad (3.15)$$

an expression that will guide our empirical work in the second half of the chapter. It tells us that the proximate determinants of relative displacement are shifts in output and intersectoral differences in productivity growth. These output and productivity effects are not in general independent, but they help us to categorize the channels through which the more fundamental determinants of relative displacement operate.

To get at the more fundamental or exogenous determinants of

displacement, a two-sector model was analyzed.[17] We consider the exogenous variables to be the rate and bias of technical progress in each sector, the rate of aggregate capital-deepening,[18] and exogenous demand shifts. These exogenous determinants affect the rate of displacement through three types of output effects and two types of productivity effects.

Call the sector of declining employment share manufacturing and the growing sector services. Then the three types of output shifts are as follows:

1. *Exogenous demand shifts.* Obviously, exogenous shifts in demand away from domestic manufacturing toward services will displace labor toward services. In this category, we would include exogenous changes in consumer tastes, which Fuchs (1980) has argued are important, such as demographic changes that have increased the demand for health services and retail trade (particularly eating and drinking establishments), while the introduction of new consumer electronics might work in the opposite direction. Furthermore, we should also consider exogenous shifts in nonconsumer demands: investment goods, government demand, and net exports. Taking these in turn, displacement would be slowed by exogenous increases in investment demand for equipment (versus structures)—for example, due to new technologies or tax incentives; and displacement would also be slowed by large government programs of military procurement. Finally, displacement would be accelerated by shifts in world demand away from U.S. manufactured goods, that is, the emergence of huge trade deficits in manufactures, which have been widely associated with U.S. budget deficits.

2. *Income effects.* Another type of output shift works through a rise in per capita income and the difference between the income elasticities of demand for services and manufactured goods. It has long been conjectured in some quarters that the demand for services, such as health services, grows faster than for manufactured goods as per capita incomes grow. If so, then anything that raises per capita incomes accelerates the displacement toward services. Specifically, the exogenous determinants of per capita income are aggregate technical progress, $T \equiv \Sigma a_i T_i$, and aggregate capital-deepening.

3. *Price effects.* Output also shifts between sectors as demand responds to changes in relative prices, which are themselves governed by exogenous developments. Specifically, if technical progress is particularly rapid in manufacturing, this reduces the relative price of manufactured goods, expanding demand and slowing the employment shift toward services. Relative prices may also respond to changes in the relative costs of labor and capital, which in turn depend on whether the labor-saving bias of technical progress outweighs the rate of aggregate capital-deepening. If so, this will tend to reduce the demand for labor, reducing the wage, relative to the cost of capital. Then, if the service sector is more labor-intensive, this will reduce the price of services relative to manufactured goods, thereby increasing demand and attracting employment. In practice, this effect on relative prices is likely to be less important than the effect of uneven technical progress, explained above.

We now consider the two types of productivity effects:

1. *Uneven rates of technical progress.* Particularly rapid technical progress in the manufacturing sector directly displaces labor towards the service sector. Historically, this has been the leading hypothesis in explaining the shift, due to the work of Baumol (1967) and Fuchs (1968). An obvious question is whether this productivity effect outweighs the demand effect of the change in relative prices, which works in the opposite direction, as explained above.

2. *Uneven capital-deepening effects.* Which sector's productivity benefits most from aggregate capital-deepening? If the manufacturing sector finds it easiest to substitute capital for labor (higher σ), then it will tend to engage in more capital-deepening. Furthermore, this will have a greater effect on its productivity if its output elasticity with respect to capital (θ_K) is greater. If so, then the effect of rapid capital accumulation and/or slow population growth will tend to raise productivity more rapidly in manufacturing, displacing labor toward services.

   Also, if the manufacturing sector has a pronounced labor-saving bias, then it will be more inclined to engage in capital-deepening, thereby raising productivity and displacing labor toward services.

Uneven capital-deepening effects can also result from demand shifts. For example, if demand shifts exogenously away from manufacturing, and if capital is immobile, then the decline in manufacturing employment will raise the sector's capital/labor ratio. Consequently, in this case an exogenous demand shift will not only be registered as an output shift (the first term of equation [3.15]), but will also have an additional effect through uneven productivity growth (the second term). That is, if capital is immobile (as it certainly is over short periods), then exogenous demand shifts will lead to disproportionately large employment shifts, since that is the only factor that can adjust. To take this a step further, consider the fact that capital is heterogeneous, for example, with respect to vintage (a fact not reflected in the models of this chapter). Then a demand shift away from manufacturing will lead to scrapping of older vintages, which will increase average productivity of the workers remaining in the sector. Again, the demand shift will have a disproportionate effect on employment (not fully captured in the first term of [3.15]), since workers in the marginal plants will be the first to go. Logically, this effect should be registered as the result of capital-deepening on newer vintages (new capital/total employment rises as total employment falls) and "shallowing" on older vintages (as they are scrapped). However, the homogeneous treatment of capital means this effect would likely be measured as total factor productivity growth, even though it does not reflect technical progress. Some observers believe this has occurred in the steel industry, for example.[19]

POSTWAR EXPERIENCE

We now examine postwar data on the distribution of wages and employment across the twelve sectors of the nonfarm economy. In particular, we identify the wage gap between sectors with declining and growing shares of employment and show that it has widened to about −15 percent over the most recent business cycle. The popular characterization of the recent net shift from durables manufacturing into the service sector receives considerable support from these data. The causes of the employment shifts are examined using comparable data on real value added. There is some evidence of both output shifts and uneven productivity growth, though it is impossible to be clear on the exogenous determinants.

Although the wage gap has been large, and net employment shifts have accelerated, nonetheless, the effects on aggregate wage and productivity growth have remained relatively minor. Rather, the continuing slow growth of average wages and productivity reflects behavior within sectors more than shifts among them. Specifically, real wage growth remains slow in all sectors, and is closely associated with the continuing slow growth in measured productivity outside of durables manufacturing.

Description of the Data

Our main source of data is the BLS Office of Productivity and Technology, which has provided data on output, prices, wages, and hours of employment across the twelve major sectors of the private nonfarm economy, at approximately the one-digit level of aggregation. These sectors are listed in Table 3–2. The output data represent real value added, derived from the BEA National Income and Product Accounts. Certain adjustments are made by the BLS, which exclude about one-quarter of GNP: agriculture ($1\frac{1}{2}$ percent), general government ($10\frac{1}{2}$ percent), imputed rents to owner-occupied housing ($6\frac{1}{2}$ percent), private household employees and nonprofit institutions ($3\frac{1}{2}$ percent, primarily in the service sector), and net income from investment and employment abroad (the "rest-of-the-world" sector, which distinguishes GNP from GDP, and which has been decreasing in recent years to about 1 percent of GNP in 1985). Aside from agriculture, these exclusions are made because of the difficulty in either obtaining reliable measures of the associated labor inputs ("rest-of-the-world" and owner-occupied housing) or of measuring output independently of labor inputs (general government, household employment, and nonprofit institutions).[20]

The labor input measure for this subset of the economy is drawn primarily from the payroll data of the BLS establishment survey, but it is augmented by household data *(Current Population Survey)* on the self-employed and unpaid family workers. This adjustment is significant in retail trade and services. The labor input measure is hours paid.

The measure of "wages" is actually the somewhat broader concept of "compensation." It not only covers direct labor payments (wages and salaries, commissions, tips, and so forth), but also employer contributions to public and private insurance, pensions, and health and welfare plans.

At the one-digit level of disaggregation, the BLS is not currently confident of the available data on capital outside of manufacturing and does not publish them. However, for the broad aggregate sectors of manufacturing and nonfarm nonmanufacturing in the private business economy,[21] the BLS does provide the relevant data, including capital. We have analyzed these data and report on these results, where relevant, to supplement our main findings.

Employment Shares

Shifts in the distribution of employment have captured the attention of social observers for some time now. Sir William Petty commented in the seventeenth century on the movement from agriculture to manufacturing to services, and Colin Clark (1951) went so far as to propose that this empirical regularity by codified as Petty's Law. All of the OECD countries have indeed experienced large shifts in accord with Petty's Law over the last century. In the United States, manufacturing appears to have reached its peak proportion of nonfarm employment by 1919, while the service-sector employment share has grown ever since. We will now show, however, that the shift appears to have accelerated in recent years.

Figure 3–2 depicts the postwar behavior of employment shares (the λ_i's of previous notation) of the four largest one-digit sectors. The unit of analysis is hours of all persons. Throughout the postwar period, the employment share of services has steadily risen, while nondurables manufacturing has declined. The behavior of retail trade and durables has been more erratic and sensitive to the business cycle, but durables have distinctly declined in recent years.

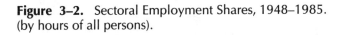

Figure 3–2. Sectoral Employment Shares, 1948–1985. (by hours of all persons).

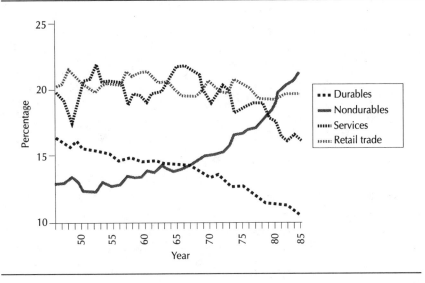

To provide more detail on the service sector and retail trade, we have examined data on production or nonsupervisory workers.[22] Within the service sector, the largest component has been health services, which have accounted for 28 percent of the hours employed since 1973. Business services have been the high-growth area, rising from 14 percent of nonsupervisory service sector hours to 20 percent, since 1973.

Within retail trade, the share of nonsupervisory hours accounted for by eating and drinking establishments has risen from 21 to 30 percent over the last two decades. It should also be noted that retail trade has experienced particularly rapid growth in part-time employment.[23] This means that the relative stability of employment in retail trade, using hours as our unit of analysis, significantly understates the growing portion of the working population employed in that sector. Although part-time employment offers an attractive degree of flexibility to many, household survey data show that a rising portion has been involuntary.[24]

In any case, the ensuing analysis will be based on one-digit data for hours of employment of all persons (including the self-employed). Table 3–2 shows the distribution of employment among

Table 3–2.   Distribution of Employment, by Hours of All Persons, 1948–1985.

	$\lambda_i \equiv$ Employment Shares (percentage)						
	1948	1957	1966	1973	1979	1985
Durables	18.9%	20.2%	21.4%	19.9%	18.6%	15.7%
Nondurables	16.1	14.7	14.3	13.2	11.7	10.3
Services	13.0	13.1	13.8	15.2	17.3	20.8
Retail trade	20.3	20.4	19.7	19.6	19.0	19.2
Wholesale trade	6.6	6.8	7.0	7.2	7.6	7.7
Finance/insurance/real estate	4.4	5.1	5.5	6.4	7.0	7.8
Construction	6.1	6.9	6.8	7.3	7.4	7.4
Transportation	7.6	6.0	5.2	4.8	4.6	4.3
Government enterprises	1.8	2.0	2.3	2.3	2.3	2.3
Communications	1.6	1.8	1.7	1.9	1.8	1.7
Mining	2.2	1.7	1.2	1.1	1.5	1.4
Utilities	1.2	1.3	1.2	1.2	1.2	1.3

	$\dot{\lambda}_i \equiv$ Percentage Points Growth Per Year					
	1948–85	1948–57	1957–66	1966–73	1973–79	1979–85
Durables	−.09	.15	.13	−.22	−.22	−.47
Nondurables	−.16	−.16	−.04	−.15	−.25	−.22
Services	.21	.01	.08	.19	.35	.58
Retail trade	−.03	.00	−.08	.00	−.10	.03
Wholesale trade	.03	.02	.03	.02	.07	.03
Finance/insurance/real estate	.09	.08	.04	.12	.11	.13
Construction	.03	.08	−.01	.07	.02	.00
Transportation	−.09	−.17	−.09	−.06	−.02	−.06
Government enterprises	.01	.02	.03	.01	−.01	.01
Communications	.00	.01	−.01	.03	.00	−.02
Mining	−.02	−.06	−.06	−.02	.06	−.01
Utilities	.00	.00	−.01	.01	.00	.01
D ≡ Relative displacement	.38	.39	.30	.44	.61	.79

Data for Tables 3–2 to 3–5 and 3–7 to 3–9 from printouts of BLS Office of Productivity and Technology. See Table 3–1 for definitions.

all sectors for selected years, representing comparable (nonrecession) points in the business cycle. The top panel gives the employment shares, as in Figure 3–2. The bottom panel gives the change in shares, in percentage points per year. The first column of the bottom panel tells us that over the postwar period as a whole, manufacturing lost about a quarter of a point per year (largely in nondurables), while the service sector gained almost as much.[25]

The pattern, however, has not been constant throughout the period. The decline in nondurables has accelerated since the 1950s. The share of employment in durables has had a particularly marked turn-around, growing through 1966 and declining since then, culminating in the loss of half a percentage point per year over the last business cycle. Indeed, absolute employment has failed to reach its prior cyclical peak for the first time. The behavior of the service sector shows the mirror image: rapid acceleration since 1966, culminating in growth in excess of half a percentage point per year over the last business cycle.

These patterns of relative displacement are summarized by the statistic D, given in equation (3.13), above. The proverbial bottom line of Table 3–2 shows that at the one-digit level of disaggregation, shares of employment shifted between sectors at the rate of .3 to .4 percentage points per year up to 1973 and have risen to about .8 percentage points per year over the most recent business cycle. These figures are based on nonrecession endpoints, such as 1979 and 1985, to avoid cyclical patterns. Alternatively, regression analysis based on the full postwar time series, controlling for unemployment, finds the same rise in D. Indeed, this analysis points to an even more pronounced rise following 1982, to a cyclically controlled rate of 1.2 percent per year. This represents over one million persons per year, although, of course, this is *relative* displacement, rather than *absolute* displacement.[26] On the other hand, these figures are *net* displacement, which necessarily understates the *gross* flows between sectors. Also, these figures necessarily understate the results we would find with more disaggregated data.

Wage Differentials

Figure 3–3 depicts the relative wages of the four largest sectors, using the most comprehensive measure, hourly labor compensa-

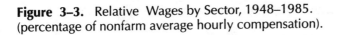

Figure 3–3. Relative Wages by Sector, 1948–1985.
(percentage of nonfarm average hourly compensation).

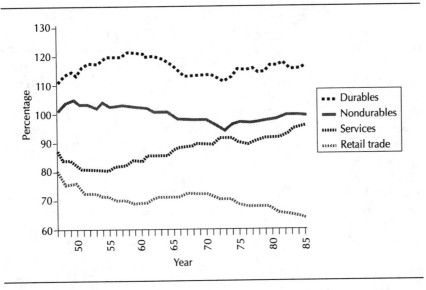

tion of all persons, as defined above. Throughout the postwar period, the ranking of these four sectors has remained unchanged: Durables pay the highest, followed by nondurables, services, and retail trade pays the lowest. The gap between nondurables and services, however, has narrowed, while it has widened between services and retail trade. Table 3–3 provides data on the smaller sectors as well.

As discussed above, to get more detail on the service and retail trade sectors, we examined data on the wages (versus total compensation) of production and nonsupervisory workers. Within the service sector, both health service workers (the largest component) and business service workers (the fastest-growing component) are paid about the same as the service sector average. This may be surprising because business services are sometimes believed to be a high-wage area of computer and data processing services, as well as management and public relations services. However, numerically more important areas within business services are temporary help supply services, and services to buildings, while detective and protective services rank only slightly below.[27] Consequently, business services are not paid all that much better than the rest of the service sector.

Table 3–3.   Relative Wages by Sector, 1948–1985.[a]

	(Percentage of Nonfarm Average)					
	1948	1957	1966	1973	1979	1985
Durables	111%	119%	114%	111%	115%	116%
Nondurables	103	103	98	95	98	100
Services	83	81	88	92	91	96
Retail trade	76	71	72	71	69	65
Wholesale trade	122	114	112	114	109	110
Finance/insurance/ real estate	103	104	106	105	103	113
Construction	125	121	128	125	120	109
Transportation	108	116	113	122	124	115
Government enterprises	96	89	97	106	102	108
Communications	105	108	117	130	133	140
Mining	124	124	113	117	128	130
Utilities	111	118	126	125	129	142

a. $W_i/W = w_i/w \equiv$ Relative hourly labor compensation of all persons

Within retail trade, eating and drinking establishments (the rapid-growth component) provide the lowest wages, about a third less than the rest of retail trade, which is otherwise the lowest. Furthermore, the relative wage here has deteriorated in recent years, as has the rest of retail trade.[28]

As stated above, the ensuing analysis will not proceed at this level of detail because the data we have for hourly compensation of all persons is at the one-digit level. We will not take any account of the skill differentials among sectors, nor of the different degrees of safety and comfort (such as mining). These obviously impart an upward bias to the wage differentials we are interested in. On the other hand, a downward bias is introduced by the assumption that all part-time work is voluntary, which particularly affects the retail trade sector, as discussed above.

Wage Effects of Relative Displacement

We have shown how to calculate an estimate of the average wage prior to displacement from the declining sectors, w_d, and after entering the growing sectors, w_g, as well as the wage gap between them, given in (3.14). Table 3–4 reports these calculations. Over the intervals considered, the pre- and postdisplacement wages were quite close up until 1979. Since 1979, however, a wage gap of -15 percent has materialized. This would appear to be good evidence of the smoke behind the fire of public concern with sectoral shift.[29] Furthermore, as discussed above, the amount of relative displacement has also risen to 0.79 percent per year. The product of these two figures gives the total wage effect of relative displacement of -12.0 hundredths of a percent of total wages, per year. This figure, given on the fifth line of Table 3–4, is the covariance (scaled up by 100) between relative wages and employment growth, discussed at length above. Its recent acceleration reflects the rise in both the rate of displacement and the wage gap.[30]

It is not difficult to ascertain the sectors involved. The total wage loss from relative displacement can be allocated among the declining sectors j by the calculation

$$-\dot{\lambda}_j(w_g - w_j)/w$$

That is, $-\dot{\lambda}_j$ of the workforce is relatively displaced from sector j per year, which experiences a wage change of $w_g - w_j$[31] (assuming their postdisplacement wage is the same as those displaced from other sectors). This represents the contribution to the total wage effect of those leaving declining sector j. Similarly, from the postdisplacement viewpoint, the wage loss experienced by those entering growing sector k is

$$\dot{\lambda}_k(w_k - w_d)/w$$

The calculations in Table 3–4 show that three-quarters of the recent wage loss from relative displacement has been experienced by those leaving durables, and about seven-eighths of it has accrued to those entering services.

These figures provide striking confirmation of popular perceptions. The only aspect of the popular story that is not exhibited by these data is the McDonald's syndrome because retail trade plays only a small role in the deterioration from 1979–85 and actually makes a positive contribution prior to 1979 by declining in employment. As mentioned above, however, these data are inadequate to investigate the alleged McDonald's syndrome for a few reasons. First, the one-digit level of aggregation masks within re-

Table 3-4. Wage Effects of Relative Displacement, 1948–1985.

	1948–57	1957–66	1966–73	1973–79	1979–85
	Wages of the Relatively Displaced (as Percentage of Average Wage)				
Postdisplacement $\equiv w_g/w$	111.9 %	104.4 %	103.8 %	100.2 %	96.4 %
Predisplacement $\equiv w_d/w$	110.1	102.7	107.4	99.1	111.6
Wage gap $\equiv (w_g - w_d)/w$	1.8	1.7	−3.6	1.1	−15.2
Relative displacement $\equiv D$	.39	.30	.44	.61	.79
Total wage effect $\equiv D \times$ Wage gap	.7	.5	−1.6	.7	−12.0
	Wage Effects by Declining Sector				
Durables	1.4		−1.9	−2.7	−9.0
Nondurables		.2	1.1	1.0	−.5
Services					
Retail Trade		2.6	.1	3.1	
Wholesale Trade					
Finance/insurance/real estate					
Construction	−.1			−.1	
Transportation	−.9	−.8	−.5	−1.3	
Government enterprises					
Communications		−.1		−.1	
Mining	−.7	−.8	−.2	−.1	−.7
Utilities		−.2		−.1	−.5

Wage Effects by Growing Sector

Sector					
Durables	.8	1.8			-10.4
Nondurables					-1.2
Services	-.3	-1.4	-3.3	-2.7	-.1
Retail trade					-.5
Wholesale trade	.2	.3	.1	.8	
Finance/insurance/real estate	-.5	.1	-.2	.5	
Construction	1.1		1.3	.5	
Transportation					
Government enterprises	-.4	-.3	-.1		-.1
Communications	-.1		.4		
Mining				1.5	
Utilities			.1		.2

For definitions, see Table 3–1 and text.
Relative displacement is also found in Table 3–2.
The total wage effect is $100 \times Cov(w_i/w, \hat{\lambda}_i)$, found also in Table 3–7.
Wage effects by declining sector j are $-\lambda_j(w_g - w_j)/w$.
Wage effects by growing sector k are $\lambda_k(w_k - w_d)/w$.

tail trade some of the shift toward eating and drinking establishments. Second, these data on all persons include the decline of self-employment in retail trade. If we confine ourselves to payroll employees, about 30 percent of the total hourly wage loss has accrued to those entering retail trade, and about 40 percent, after 1982.

Output and Productivity Effects on Employment Shares

What lies behind the recent shift out of durables into services? Equation (3.15) decomposes employment shifts into their proximate determinants: output shifts and intersectoral differences in productivity growth. Table 3–5 partitions the employment growth figures from the bottom of Table 3–2, $\dot{\lambda}_i \equiv \lambda_i \hat{\lambda}_i$, into these two effects. The output effect on sector i's employment growth is given by

$$\lambda_i(\hat{Q}_i - \hat{Q})$$

where $\hat{Q}$ is a measure of aggregate output growth.[32] Similarly, the component of employment growth that is due to slow productivity growth is

$$\lambda_i(\hat{q} - \hat{q}_i)$$

where $\hat{q}$ measures aggregate productivity growth.

As a first pass, the employment shift out of manufacturing, both for the postwar period as a whole, and particularly for the recent shift out of durables, appears to be associated with relatively rapid productivity growth, rather than net shifts in demand. Indeed, these data are consistent with the widely noted constancy of manufacturing's share in real GNP. However, there are several points to be made before drawing any conclusions based on this constancy.

First, if we examine the service sector, we find that most of the recent employment growth has been due to rapid output growth, and somewhat less to slow productivity growth. This may seem inconsistent with the manufacturing data, but they are reconciled by considering other sectors, particularly construction. Taken literally, Table 3–5 tells us that rapid productivity growth in manufacturing displaced workers into construction (where measured productivity growth has not only been slow, but strongly *negative*), while slow output growth in construction displaced workers into services. This account is not terribly appealing, because construction plays such a critical role in it, yet construction's employment share was constant.[33] Clearly what we are interested in is the behavior of manufacturing's productivity growth relative to services, not an aggregate that includes construction. Similarly, manufacturing's constant share of aggregate output is less telling than its output relative to services. Making these comparisons, we find that *both* output shifts and the productivity gap contributed to the employment shift from manufacturing to services.

Second, we should consider the possibility of measurement error. It is widely recognized that the output data for the service sector are somewhat less reliable than for manufacturing.[34] Specifically, if service sector output growth has been underestimated, as some believe, then by definition, service productivity growth has also been underestimated. Furthermore, measures of manufacturing output and productivity growth based on real value added are somewhat greater in recent years than those based on real shipments.[35] To the extent that these two possible measurement errors are significant, they would reduce the productivity gap and increase the role of output shifts in explaining the employment shift.

Moving on from the issue of the relative importance of output shifts and productivity gaps, let us consider these effects in more detail, in an attempt to get at the exogenous determinants of relative displacement. First, what lies behind the output shifts? Some relevant data are presented in Table 3–6 and Figure 3–4. These data give the shares of real final demand (real GNP less inventory accumulation) accounted for by durable goods, according to the sector of demand. Although durables' share of real final demand does not correspond precisely to the share of real value added originating in durables manufacturing (the subject of Table 3–5), it, too, has remained constant over the most recent business cycle.

Durables' share of real final demand is broken down into consumption, investment, exports, imports (entered negatively, since they reduce demand for U.S. production of durables), and government demand for durables (the defense component of this is bro-

Table 3–5. Output and Productivity Effects on Employment Shares, 1948–1985 (percentage points per year).

	1948–85	1948–57	1957–66	1966–73	1973–79	1979–85
Durables						
Employment growth	−.09%	.15%	.13%	−.22%	−.22%	−.47%
Output growth	−.01	.13	.06	−.15	−.13	−.02
Slow productivity	−.08	.02	.07	−.07	−.09	−.45
Nondurables						
Employment growth	−.16	−.16	−.04	−.15	−.25	−.22
Output growth	−.01	−.04	.04	.07	−.09	−.06
Slow productivity	−.15	−.12	−.09	−.22	−.16	−.16
Services						
Employment growth	.21	.01	.08	.19	.35	.58
Output growth	.14	−.12	.09	.22	.32	.35
Slow productivity	.07	.13	−.01	−.03	.03	.23
Retail Trade						
Employment growth	−.03	.00	−.08	.00	−.10	.03
Output growth	−.02	.00	−.12	.03	−.05	.07
Slow productivity	−.01	.00	.05	−.03	−.05	−.05
Wholesale Trade						
Employment growth	.03	.02	.03	.02	.07	.03
Output growth	.07	.05	.06	.12	.03	.10
Slow productivity	−.04	−.03	−.03	−.10	.04	−.07
Finance/Insurance/Real Estate						
Employment growth	.09	.08	.04	.12	.11	.13
Output growth	.02	.05	−.02	.06	.08	−.06
Slow productivity	.07	.03	.06	.06	.03	.19

Construction						
Employment growth	.03	.08	-.01	.07	.02	.00
Output growth	-.12	.11	-.05	-.37	-.17	-.25
Slow productivity	.16	-.03	.04	.44	.19	.25
Transportation						
Employment growth	-.09	-.17	-.09	-.06	-.02	-.06
Output growth	-.11	-.24	-.08	-.03	.00	-.16
Slow productivity	.02	.07	-.02	-.02	-.02	.10
Government Enterprises						
Employment growth	.01	.02	.03	.01	-.01	.01
Output growth	-.02	-.05	.00	-.04	.00	-.01
Slow productivity	.04	.07	.03	.05	-.01	.02
Communications						
Employment growth	.00	.01	-.01	.03	.00	-.02
Output growth	.06	.06	.04	.08	.06	.05
Slow productivity	-.05	-.04	-.05	-.05	-.07	-.07
Mining						
Employment growth	-.02	-.06	-.06	-.02	.06	-.01
Output growth	-.02	.00	-.03	-.02	-.04	-.03
Slow productivity	.00	-.05	-.02	.00	.10	.02
Utilities						
Employment growth	.00	.00	-.01	.01	.00	.01
Output growth	.03	.06	.02	.04	-.01	.02
Slow productivity	-.03	-.06	-.03	-.03	.01	-.01

Employment growth is λ_i, from Table 3-2.
It is broken down into $\lambda_i(\hat{Q}_i - \hat{Q})$ and $\lambda_i(\hat{q} - \hat{q}_i)$, the effects of output growth and slow productivity, respectively. See text.

Table 3–6. Durable Goods as Percentage of Real Final Demand, 1948–1986.

	1948	1957	1966	1973	1979	1986
Total	15.2%	16.2%	16.3%	17.6%	19.0%	19.1%
Consumption	5.6	6.0	6.7	8.1	8.4	10.0
Investment	7.1	5.6	6.6	7.5	8.3	9.0
Exports	2.2	2.4	2.5	3.7	4.3	4.0
Imports	−0.9	−1.1	−2.0	−3.3	−3.6	−6.7
Government	1.1	3.4	2.6	1.5	1.7	2.8
Defense				1.1	1.2	2.1

	1948–86	1948–57	1957–66	1966–73	1973–79	1979–86
Change in Total	3.9%	1.0%	.1%	1.2%	1.5%	.1%
Change in consumption	4.4	.3	.7	1.5	.2	1.7
Change in investment	1.9	−1.5	1.0	.9	.8	.7
Change in exports	1.8	.2	.1	1.2	.6	−.3
Change in imports	−5.8	−.2	−.9	−1.3	−.4	−3.1
Change in government	1.7	2.3	−.8	−1.1	.2	1.1
Change in defense						.9

Source: Bureau of Economic Analysis, U.S. Department of Commerce, National Income and Product Accounts, Tables 1.2, 3.8A, 3.8B, 4.2, and 5.3.
Note: Calculations based on 1982 prices.

Figure 3-4. Real Final Demand For Durables, 1948-1986.

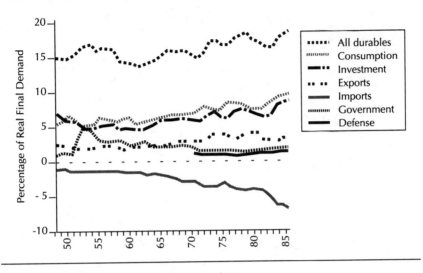

Source: NIPA Tables 1.2, 3.8A, 3.8B, 4.2, and 5.3.

ken out since 1972). Most recently, we can see the dramatic rise in imports of durables, along with the modest (although unprecedented) decline in exports of durables. Half of this has been offset by the rise in consumption of durables, followed by government demand (primarily defense) and also by a rise in investment demand for durables.

The rise in defense spending would be considered an *exogenous demand shift*. Exogenous policy may also account for some of the rise in investment demand for durables, since tax revisions during the period have been favorable toward equipment, at the expense of structures (which declined as a share of real final demand). Possibly of greater importance was the exogenous introduction of new commodities, especially since so much of the equipment demand has been for computers.

The rise in imports of durables and the deterioration of exports are widely viewed as rooted in U.S. fiscal policy, which stimulated U.S. demand and also raised the value of the dollar until 1985. On the other hand, the rise in durables consumption certainly reflects the fall in the relative price of durables, both domestic (depicted in Figure 3-5) and foreign. In this respect, the rise in consumption

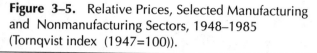

Figure 3–5. Relative Prices, Selected Manufacturing and Nonmanufacturing Sectors, 1948–1985 (Tornqvist index (1947=100)).

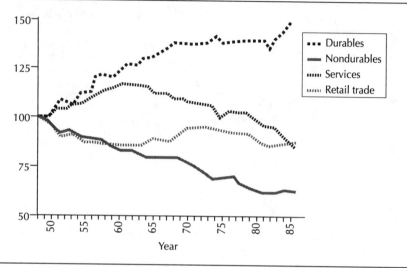

is not entirely independent of the rise in imports, especially since low import prices tend to discipline domestic prices for traded durables.

Turning to the output of services, which has risen, we again see offsetting changes. The price effects, depicted in Figure 3–5, should have reduced demand for services, unlike what we experienced.[36] However, some of the rise in prices may have been blunted in the health service area by the rise in third-party payments.[37] In any case, it would appear that adverse price effects have been more than offset by income effects (such as Summers 1985 finds for health services), and by exogenous demand shifts (such as those cited by Fuchs 1980, particularly demographic changes which also favor health services). It has also been argued that business demand for services may have been raised by changes in the organization of production, as functions that were formerly in-house for manufacturing firms have been subcontracted out to business services, such as services to buildings.[38]

We now turn to the issue of uneven productivity growth. To disentangle the sources of the productivity gap, we first need reliable data on capital. Lacking this at the one-digit level, we examined the BLS data on the broader aggregates, manufacturing versus nonfarm nonmanufacturing.

1. *Differential capital-deepening effects.* There is evidence that manufacturing enjoyed a significantly higher rate of capital-deepening, which our formal models would attribute to either a greater ease of substitution (high σ) or to greater labor-saving bias (high B). It may, however, also reflect a demand-induced decline of manufacturing employment leaving relatively immobile capital behind in the short run. Whatever the reason, the Solow technique of equation (3.1) ascribes relatively little of the productivity gap to uneven capital-deepening, since manufacturing has a relatively low distributive share of capital (θ_K). If, however, the distributive share of capital understates the output elasticity, as would be the case if manufacturing enjoyed increasing returns to scale,[39] then we would attribute some of the productivity gap to uneven capital-deepening.

2. *Uneven rates of total factor productivity growth.* Aside from uneven capital-deepening, there is evidently a large residual in the productivity gap, known as the gap in total factor productivity growth. There are at least three possible explanations for this. First, it may be uneven rates of technical progress. Second, it may represent a tightening of slack in manufacturing, in response to foreign competition. Finally, it may be a statistical artifact of scrapping older vintages of capital in some manufacturing industries, as resources are "rationalized," also in response to foreign competition.

   In any case, rapid growth in manufacturing's total factor productivity relative to nonmanufacturing (for whatever reasons) tends to reduce the relative price of manufactured goods, as seen in Figure 3–5. The effect on demand, therefore, offsets to some degree this productivity effect.

To summarize, then, the available evidence suggests that the main determinants of the recent acceleration in the employment shift out of manufacturing into services have been the gap in total factor productivity growth between manufacturing and services, and the adverse effect of recent trade imbalances on the demand for our manufactured goods (and these may be related). The employment shift would have been even more pronounced had there not been a defense buildup and a durables consumption boom (possibly due to an import-induced drop in the relative price of durables).

Aggregate Real Wage Growth and Its Slowdown

This section examines the behavior of the average real wage in the nonfarm business economy over the postwar period. We begin by comparing different measures of it, based on different deflators. We then consider the role played by the employment shifts we have just discussed, to sectors with low wage rates. The slow growth in real wages, which continues across sectors, is shown to be primarily attributable to the continued slow growth in labor's imputed marginal productivity within selected sectors.

Perhaps the most widely followed measure of real wage growth is the BLS figure for real compensation per hour of all persons (including the self-employed) in the nonfarm business sector. The BLS deflates nominal compensation per hour by its Consumer Price Index (CPI) to obtain real compensation per hour. As the first line of Table 3–7 shows, this measure of real compensation growth has steadily slowed over the postwar period from about $3\frac{1}{2}$ percent per year over 1948–57 to no growth over the most recent business cycle.[40] This latest development has occasioned considerable concern. However, although the real wage slowdown shows up in a variety of other measures, the magnitude of the slowdown is somewhat sensitive to the choice of deflator.

Consider the second line of Table 3–7, which shows real wage growth, using the implicit deflator for the consumption component of GNP. Two of the main differences between the CPI and the implicit GNP consumption deflator arise from the necessities of the BLS program of monthly surveys, as opposed to the BEA's quarterly GNP figures. First, the bundle of consumption goods covered by the GNP deflator is somewhat broader than by the CPI; and second, the bundle covered by the GNP deflator changes quarterly, since it is a Paasche index, while the CPI is a Laspeyres index, based on a bundle of goods purchased by a representative urban consumer in 1972–73. Also, the CPI's treatment of housing costs, which account for 35 percent of the index, were widely considered to overestimate the rate of inflation from the mid-1970s until it was changed in 1983. Most economists prefer the GNP consumption deflator as a measure of the cost of living. This deflator gives roughly the same figure for the postwar period as a whole, but with a significantly less pronounced slowdown.

The third line in Table 3–7 uses a third deflator, for the output of the nonfarm sector.[41] This may be a somewhat less accurate guide to the cost of living than the consumption deflator because it includes nonconsumption goods and exports, while it excludes imports as well as farm output. However, since 1966 these two indices have been quite close. We will use this latter deflator in our discussion, since it is the same deflator used for productivity analysis, in the next section. This means the link between the two is labor's share of value added, discussed in the following section.

As equation (3.6) shows, aggregate real wage growth can be decomposed into the covariance term representing the wage effects of relative displacement, and a weighted average of the real wage growth within the various sectors (that is, the real wage growth of the nondisplaced). Lines 4 and 5 of Table 3–7 present these data.[42] As discussed earlier, the wage effect of relative displacement to low-wage sectors plays virtually no role in aggregate wage growth until the most recent period, and even now it is still relatively small.[43] The behavior of aggregate real wages is primarily governed by what happens to the real wages of those who remain within their sector,[44] which is given next in Table 3–7. These data show that the real wage slowdown has hit workers in all sectors since 1973.

Equation (3.7) shows that the wage growth of the nondisplaced can be decomposed into a covariance term concerning relative prices and a weighted average of the purchasing power of each sector's wage over its own output—that is, the marginal productivity of labor (MPL), under the assumption of competitive output markets. Table 3–7 shows that the relative price effect has been small, although it has changed sign over the postwar period. The dominant factor in aggregate real wage growth has been the growth of MPL's across sectors. Table 3–7 gives the weighted average of these growth rates, followed by the growth rate within each sector.

The bottom of Table 3–7 examines the real wage slowdown between 1948–66 and 1979–85 and in between. The change in

Table 3-7. Aggregate Wage Growth and Slowdown, 1948–1985.

	1948–85	1948–57	1957–66	1966–73	1973–79	1979–85
	Aggregate Real Wage Growth, by Deflator (annual average rate)					
By CPI	1.86%	3.43%	2.47%	2.10%	.36%	-.18%
By GNP consumption deflator	1.93	3.06	2.18	2.31	.93	.43
By nonfarm output deflator[a]	1.95	2.64	2.57	2.47	.72	.59
Displacement effect	-.02	.01	.01	-.02	.01	-.12
Nondisplaced	1.97	2.63	2.57	2.48	.72	.71
	Real Wage Growth by Sector (annual average rate)					
Durables	2.07%	3.41%	2.05%	2.09%	1.27%	.84%
Nondurables	1.86	2.60	2.05	2.02	1.20	.97
Services	2.35	2.37	3.55	3.00	.62	1.50
Retail trade	1.53	1.84	2.76	2.31	.29	-.44
Wholesale trade	1.67	1.96	2.37	2.66	.03	.67
Finance/insurance/real estate	2.18	2.72	2.78	2.37	.34	2.09
Construction	1.56	2.25	3.16	2.20	-.02	-1.01
Transportation	2.12	3.36	2.35	3.55	.98	-.61
Government enterprises	2.26	1.85	3.49	3.79	.07	1.43
Communications	2.72	2.90	3.53	3.87	1.19	1.41
Mining	2.07	2.62	1.56	2.94	2.18	.89
Utilities	2.61	3.38	3.24	2.43	1.18	2.17

Nondisplaced[b]	1.97	2.63	2.57	2.48	.72	.71
Relative price effect	.05	.12	.07	.23	−.09	−.13
Weighted MPL growth	1.91	2.51	2.50	2.25	.80	.84

MPL Growth by Sector (annual average rate)

Durables	2.44%	2.05%	2.27%	3.15%	1.54%	3.37%
Nondurables	3.03	3.97	2.76	4.15	2.51	1.25
Services	1.20	.24	2.30	2.37	.79	.04
Retail trade	1.82	3.41	2.39	1.51	.75	−.01
Wholesale trade	2.50	2.97	3.09	3.21	.00	2.58
Finance/insurance/real estate	1.81	2.51	1.78	2.29	.80	1.22
Construction	−.35	2.59	2.17	−4.41	−2.30	−1.83
Transportation	1.53	1.41	2.88	3.32	1.19	−2.06
Government enterprises	.28	−.82	1.14	.59	.83	−.27
Communications	4.02	3.21	3.81	5.85	6.19	1.26
Mining	.71	4.06	4.32	2.48	−9.34	−1.75
Utilities	3.06	4.64	4.13	4.88	.84	−.80

Table 3-7. (Continued)

	Change: 1948-66 to 1979-85	Change: 1948-66 to 1966-73	Change: 1966-73 to 1973-79	Change: 1973-79 to 1979-85
Change in aggregate wage growth	-2.02	-.14	-1.74	-.13
Change in displacement effect	-.13	-.02	.02	-.13
Change in relative price effect	-.22	.14	-.32	-.04
Shift to rapid MPL-growth sectors	-.15	-.07	-.04	-.05
Change in weighted MPL growth	-1.50	-.17	-1.41	.08
Contribution by sector:				
Durables	.25	.23	-.36	.38
Nondurables	-.24	.11	-.21	-.14
Services	-.21	.13	-.22	-.12
Retail trade	-.41	-.20	-.11	-.10
Wholesale trade	-.03	.01	-.26	.21
Finance/insurance/real estate	-.06	.01	-.10	.03
Construction	-.35	-.58	-.10	.04
Transportation	-.23	.08	.19	.04
Government enterprises	-.01	.01	-.12	-.18
Communications	-.07	.05	.01	-.03
Mining	-.07	-.03	.01	-.12
Utilities	-.08	.01	-.17	.13

a. See equation (3.6).
b. See equation (3.7).

aggregate wage growth is broken down into three terms, the first three of which concern different aspects of sectoral shift. The first term tells us that for the period as a whole, very little of the slowdown was due to the recent appearance of wage losses from relative displacement (although that does account for what little slowdown occurs in the most recent interval). The next line shows us that up until 1973, prices grew most rapidly in sectors where labor's share was high, which tended to raise aggregate wages, but the trend has reversed itself since then. The next line shows the effect of employment shifts to sectors where MPL growth has been slow. These three shift effects account for a quarter of the real wage slowdown.

The main source of the wage slowdown was the slowdown in MPL growth, particularly over the interval 1973–79. For the period as a whole, the MPL slowdown hit all sectors except durables. The slowdown in retail trade, construction, nondurables, transportation, and services account for over two-thirds of the total real wage slowdown. As we shall see momentarily, the aggregate productivity slowdown is also largely attributable to the slowdown in these sectors.

If we had adequate data on capital, the slowdown of MPL growth could be broken down into the effects of total factor productivity and capital-deepening net of labor-saving bias, as suggested by equation (3.8). Our examination of the broad manufacturing versus nonmanufacturing data suggests that capital-deepening has consistently exceeded the labor-saving bias in both sectors, contributing significantly to MPL growth, throughout the postwar period. The major source of the real wage slowdown has been the slowdown (to negative growth) of total factor productivity growth in the nonmanufacturing sector.

Aggregate Productivity Growth and Its Slowdown

The first line of Table 3–8 reports aggregate labor productivity growth over the postwar period and the intervals we have been considering.[45] Ideally, we would like to estimate the various components of aggregate productivity growth indicated in equation (3.5): aggregate technical progress, aggregate capital-deepening,

Table 3–8. Aggregate Productivity Growth and Slowdown, 1948–1985 (annual averages).

	1948–85	1948–57	1957–66	1966–73	1973–79	1979–85
Aggregate productivity growth[a]	1.93	2.54	2.91	1.99	.60	.79
Labor displacement effect	−.01	.00	.00	−.01	.00	−.08
T, k-growth, K-displacement	1.94	2.53	2.91	2.00	.59	.87
Aggregate productivity growth[b]	1.93	2.54	2.91	1.99	.60	.79
Value added shifts	.02	.00	−.04	.08	.15	−.06
Weighted APL growth	1.91	2.54	2.95	1.91	.45	.85
APL Growth by Sector						
Durables	2.31	2.24	2.54	2.17	1.04	3.51
Nondurables	2.96	3.15	3.47	3.47	1.89	2.36
Services	1.45	1.37	2.98	2.03	.40	−.35
Retail trade	1.92	2.37	2.65	2.01	.86	1.11
Wholesale trade	2.39	2.77	3.31	3.21	.10	1.76
Finance/insurance/real estate	.75	1.74	1.77	.82	.14	−1.70
Construction	−.34	2.77	2.25	−4.40	−2.00	−2.48
Transportation	1.48	1.32	3.23	2.31	1.05	−1.46
Government enterprises	.04	−1.50	1.43	−.47	.95	−.06
Communications	4.86	4.86	5.67	4.64	4.13	4.62
Mining	1.45	4.92	4.52	1.92	−7.26	−.23
Utilities	4.11	7.14	5.43	4.25	.08	1.45

	Change from 1948–66 to 1979–85	Change from 1948–66 to 1966–73	Change from 1966–73 to 1973–79	Change from 1973–79 to 1979–85
Change in productivity growth	−1.94	−.73	−1.39	.19
Change in value added shifts	−.05	.10	.07	−.21
Shift to rapid APL-growth sectors	−.18	−.04	−.07	−.08
Change in weighted APL growth	−1.70	−.78	−1.39	.48
Contribution by sector:				
Durables	.17	−.05	−.22	.44
Nondurables	−.13	.02	−.20	.05
Services	−.31	−.01	−.19	−.10
Retail trade	−.18	−.06	−.14	.03
Wholesale trade	−.11	.02	−.28	.15
Finance/insurance/real estate	−.40	−.10	−.08	−.22
Construction	−.30	−.43	.16	−.03
Transportation	−.19	.00	−.07	−.13
Government enterprises	.00	−.01	.02	−.02
Communications	−.02	−.02	−.02	.02
Mining	−.10	−.09	−.25	.24
Utilities	−.14	−.06	−.13	.05

a. See equation (3.5).
b. See equation (3.4).

and the effects of labor and capital allocation over sectors with different factor prices. Without reliable capital data at the one-digit level, the only item we can break out of (3.5) is the effect of labor displacement, shown to be negligible on the second line of Table 3–8.[46]

For want of a better alternative, the rest of Table 3–8 is based on equation (3.4), which decomposes productivity growth into a weighted average of each sector's growth in the average productivity of labor (APL), and a composition effect reflecting shifts of employment among sectors with different levels of value added per hour. As discussed above, this composition effect is not entirely meaningful, but line 5 shows that it is negligible, as some (but not all) previous researchers have also found.[47]

The bottom half of Table 3–8 shows that the productivity slowdown began over 1966–73.[48] It was most pronounced over 1973–79 and has only partially recovered since.[49] Insofar as the value added shifts are relevant, they would seem to have gone against the tide, as the emergence of positive value added shifts slightly eased the slowdown up until 1979, while the recent negative value added shifts retarded the productivity recovery over 1979–85. There have, however, been shifts to sectors with slow productivity growth rates, particularly the service sector, in a process analyzed by Baumol (1967). These two shift effects accounted for no more than an eighth of the slowdown from 1948–66 to 1979–85 (.23 out of 1.94). The lion's share is accounted for by productivity slowdowns within all of the sectors except durables, which has posted a significant recovery since 1979. Over half of the productivity slowdown comes from the slowdown in finance/insurance/real estate, services, and construction. Each of these sectors is now experiencing productivity growth that is not only slower than before but is actually *negative*. Indeed, productivity growth has been negative in construction for the last two decades. Slowdowns in retail trade and transportation account for another fifth of the slowdown. These are many of the same sectors that accounted for most of the real wage slowdown, due to their slowdown in marginal productivity growth.

The quality of these data has been the subject of considerable concern. It is widely recognized that the data for the construction, service, and finance/insurance/real estate sectors are less reliable than for manufacturing, and many believe that APL growth is underestimated in these sectors.[50] If so, then, MPL growth is also, since, by definition, growth in the implicit deflators has been overestimated. Hence, any downward bias in the measurement of productivity growth translates immediately into a downward bias for real wage growth. It should, however, be pointed out that the bias may be less pronounced for real wage growth, since some of the critical sectors there are more reliably estimated (such as retail trade versus finance/insurance/real estate).

Still, as argued by Rees (1980), the chairman of the National Research Council Panel to Review Productivity Statistics, this does not mean that the wage and productivity slowdowns are necessarily statistical illusions. For that to be the case, the downward bias in these sectors would had to have grown worse over the period in question, when, in fact, some of the biases were being ameliorated.

To summarize, the productivity slowdown does not appear to be attributable to sectoral shifts to any significant degree. Nor has it been attributable to a failure of capital-deepening, according to the available data. In fact, there is considerable debate and uncertainty concerning the causes of the slowdown. In any case, it is clear that the measured slowdowns of productivity and real wages are closely related. Whatever discrepancy may exist between the two is captured by the behavior of labor's distributive share, to which we now turn.

Labor's Share

Table 3–9 shows that over the postwar period as a whole, labor's share has been constant, so the growth rate of aggregate real hourly wages has been equal to that of aggregate productivity. As equation (3.10) shows, the growth of labor's share in aggregate output depends on its growth within sectors and on shifts of value added among sectors. The first column of the bottom panel shows there has been no significant trend in either. Within sectors, therefore, the effects of capital-deepening have been offset by labor-saving bias. Among sectors, the shift of value added from durables to services has had little effect on labor's share because the shares within these sectors have been quite comparable since the 1960s.

The most significant growth in labor's share occurred over

Table 3–9. Labor's Share of Aggregate Output, 1948–1985

	1948	1957	1966	1973	1979	1985
Nonfarm economy	65.4%	66.0%	64.0%	66.2%	66.7%	65.9%
Durables	74.4	73.1	71.3	76.4	78.7	78.0
Nondurables	61.2	65.9	61.8	64.9	67.3	63.0
Services	89.0	80.4	75.6	77.5	79.3	81.1
Retail trade	69.3	76.1	74.4	71.8	71.4	66.7
Wholesale trade	59.4	60.4	59.3	59.2	58.9	61.9
Finance/insurance/real estate	31.8	34.1	34.2	37.9	39.4	46.9
Construction	90.4	88.8	88.2	88.2	86.6	90.0
Transportation	70.6	71.2	69.0	74.1	74.8	72.1
Government enterprises	83.8	89.1	86.8	93.5	92.8	91.7
Communications	61.7	53.2	45.0	49.0	55.4	45.3
Mining	40.1	37.1	36.4	37.9	33.4	30.5
Utilities	43.5	34.7	30.9	32.3	33.8	29.5

	1948–85	1948–57	1957–66	1966–73	1973–79	1979–85
Growth rate of labor's share[a]	.02	.10	−.34	.48	.12	−.20
Shift to high-share sectors	−.04	.01	.05	−.03	−.14	−.15
Growth within sectors	.06	.09	−.39	.51	.27	−.05

a. See equation (3.10).

1966–73: The productivity slowdown began over that interval, while the real wage slowdown awaited the next cycle. Prior to that, labor's share fell, and it is falling again in recent years. Consequently, the partial recovery in productivity is not yet being fully shared by labor.[51]

CONCLUSION

We may now summarize the available evidence on the postwar behavior of productivity, real wages, and the distribution of employment. Aggregate productivity growth continues to be slow, despite its recovery in durables. Negative measured productivity growth in finance/insurance/real estate, services, and construction have been the major culprits, and there is no evidence that this is due to negative capital-deepening.

The continued slow growth of real hourly wages has been closely associated with the slow productivity growth. The imputed growth of labor's marginal productivity has been low or negative in several of the same sectors outside of durables.

Employment shares have shifted from high-wage durables to low-wage services at an accelerated rate in recent years, with large losses for those involved. This has been a drag on real wage growth, but less important than the slow productivity growth outside of durables. The causes of the employment shift appear to include both the productivity gap between durables and services, and also the large trade deficit in durables.

All of these adverse developments share a common element: slow productivity growth outside of durables. This limited evidence suggests, therefore, that these sectors offer particularly high benefits for technical progress, though we know little about the costs.

NOTES

Earlier versions of this report were presented to and discussed by the Panel on Technology and Employment. I would like to express my appreciation to Dr. Leonard A. Rapping and Dr. David C. Mowery.

1.   The theory of distribution dates to Hicks (1932) and the theory of productivity was put forth by Solow (1957).

2.   We will not address the critical issue of labor's heterogeneity, and whether technical progress is biased against skilled or unskilled labor.

3.   This may be an inappropriate assumption if, as some believe, the recent productivity growth in manufacturing reflects a reduction of "fat" in the face of foreign competition.

4.   See Hall (1986) for a discussion of productivity growth accounting when marginal cost pricing does not obtain.

5.   It is at this point that the assumption of a constant degree of inefficiency is critical. If that assumption fails, then measured growth in total factor productivity, T, will be wrongly ascribed to technical progress.

6.   A more formal analysis of the growth model in this section is found in an expanded version of this chapter, available from the author.

7.   Williamson and Lindert (1980) assume, in their empirical work, that all wage dispersion reflects skill differentials. In the calculations below, the opposite simplifying assumption is made.

8.   The r_i's below denote returns to physical capital, which are not equalized in the same fashion as returns to financial capital. That is, while financial capital is highly mobile, there may well be significant barriers to mobility of physical capital. The difference is reflected in the valuation ratio of the replacement cost of physical capital to the value of a firm's capital on the securities markets, known as Tobin's q.

9.   The difference between the Divisia index and the official (Laspeyres) measure of real output (e.g., real GNP) is that the Divisia weights reflect continuously changing prices and quantities, while the Laspeyres weights use fixed, base-year prices. In practice, these differences are often rather small. Similarly, the Divisia price index, used below, differs both from Laspeyres price indexes, such as the CPI, which use fixed base-year quantities, and also from the Paasche price indexes, such as the implicit GNP deflator. See Diewert (1978).

10. The λ_i-weighted means are $E(\hat{\lambda}_i) = 0$ and $E(v_i/v) = 1$. Equivalently, the composition effect can be expressed as the unweighted (uniform pdf) $Cov(v_i/v, \dot{\lambda}_i)$, since the unweighted mean of $\dot{\lambda}_i$ is zero. However, the unweighted mean of v_i/v is not in general unity, which renders the interpretation less appealing.

11. See, for example, Thurow (1979, 1980a, 1980b). Actually, Thurow's calculations are based on constant dollar value added, rather than current dollar value added, which is the theoretically appropriate

measure. Thurow's calculations will be sensitive to the choice of base year.

12. As in note 10 above, we could equivalently use unweighted covariances with $\dot{\mu}_i$ and $\dot{\lambda}_i$, but, again, the unweighted means of r_i/r and w_i/w will not generally equal unity.

13. Of course, the causes of sectoral shift may include technical progress and capital-deepening, so the covariance terms are not independent of the first two terms, as will be discussed below.

14. Expressions analogous to those in this section can be readily inferred for the behavior of the real returns to physical capital.

15. We do not consider the losses during the period between jobs, which may be quite long—that is, the issue of technological unemployment.

16. See Podgursky (1988).

17. The model dates to Jones (1965), although our specification separates out the rate and bias of technical progress in each sector. More detail is available in the extended version of this chapter.

18. This assumption means that the two-sector model does *not* incorporate the growth-theoretic argument discussed above, which ties the rate of aggregate capital-deepening to the rate and bias of technical progress.

19. The points in this paragraph came out of discussions with James Klumpner, though they do not necessarily reflect his views.

20. The U.S. Bureau of Labor Statistics (1983:36) points out that only about 1 percent of the remaining output is measured by deflating current-dollar output by an index of labor and materials inputs. However, there appear to be other serious weaknesses in the output measures of the construction industry (4 to 5 percent of GNP) and certain service sectors, as mentioned below.

21. "Private" excludes government enterprises.

22. These have been declining in relative importance, but still constitute two-thirds of employment in manufacturing, and nine-tenths in services and retail trade.

23. In 1948, the workweek in retail trade was 6 percent *above* average, and it remained above average through the 1950s. Since then it has declined to 15 percent *below* average.

24. From 1979 to 1985 that portion has risen from 21% to 29% (See U.S. BLS *Employment and Earnings*).

25. The share of employment in retail trade has not risen. As discussed above, however, figures for payroll employees, rather than hours of all persons, give a very different picture, due to the rapid rise of part-time employment in this sector, and also the decline of self-employment. Over the postwar period as a whole, the share of

employees in retail trade grew by .09 percentage points per year, and it grew by .13 per year over 1979–85. Even this masks the rise in the share of employees in eating and drinking establishments, which has grown by .19 percentage points per year since 1966.

26. That is, population growth and generational turnover mean employment shares can shift between sectors even if no individual workers actually shift.

27. Specifically, these five areas account for the following proportions of business service employees (including both supervisory and non-supervisory), as of August 1985: personnel supply services—21 percent (of which over three-quarters is temporary help supply services); services to buildings—15 percent; computer and data processing services—11 percent (of which a third is computer programming and software); management and public relations services—11 percent; and detective and protective services—10 percent. These categories therefore account for two-thirds of business service employment. (See U.S. BLS *Employment and Earnings* Table B–2.)

28. This may be partially due to the fact that the minimum wage has not been raised since 1981. Also, underreporting of tips may be a problem.

29. Further analysis, based on regression coefficients controlling for unemployment, finds that the wage gap was also large over 1957–66. Other data sets we have examined give larger wage gaps. For example, confining ourselves to payroll employees raises the recent hourly wage gap to −22 percent. Analysis of weekly data raises the recent wage gap to −46 percent.

30. Calculations corresponding to Note 29 also show a rising total wage effect, of considerably larger magnitude. For example, the total wage effect for payroll employees rose to −16.9 hundredths of a percent of total wages for hourly data and −41.6 for weekly. These figures rose even higher after 1982, to −26.1 and −54.9, respectively.

31. It can readily be verified that summing over the declining sectors j gives us the covariance we are examining.

32. The appropriate measure here is $\hat{Q} = \Sigma \lambda_i \hat{q}_i$, since that is the only measure that will force the sum of the output effects to zero. The corresponding index of aggregate productivity growth is used below.

33. More generally, the problem of interpretation arises from offsetting output and productivity effects within several sectors.

34. Even greater doubts about the construction data make the considerations of the previous paragraph more compelling.

35. I would like to thank Larry Mishel for pointing this out to me.
36. If service output and productivity growth are underestimated, then the change of relative prices is overestimated. This could make it easier to explain the shift in output, but it also increases the shift in measured output, which needs to be explained.
37. See Dalton (1985).
38. It should be noted, however, that this specialization of function is by no means a new phenomenon, as discussed by Carter (1970: 57–68).
39. See Hall (1986).
40. The data for 1986 lift the period's real wage growth slightly above zero.
41. The index we use is a Törnqvist approximation to the Divisia index, using the endpoints indicated and the twelve one-digit sectors. For all of the intervals considered, its growth rate is within .3 percent of the official Laspeyres index of nonfarm prices.
42. Line 4, of course, is drawn from line 5 of Table 3–4, scaled back down by 100. It should also be noted that this component of aggregate real wage growth is independent of the deflator used.
43. Other data sets we have examined, mentioned in note 29, reveal somewhat larger displacement effects, which significantly retard the recovery from the real wage slowdown of the 1970s.
44. This may overstate the case, since it is possible that wage gains of the nondisplaced are disciplined by the losses of the displaced. However, this distributional effect does not appear to be large, as judged by the behavior of labor's share, discussed below.
45. These figures differ slightly from the official (Laspeyres) statistics, since they are Törnqvist estimates, as already mentioned.
46. Our analysis of the broad capital data on manufacturing versus nonmanufacturing provides a complete breakdown of equation (3.5) and gives a negligible capital-displacement effect as well. For the postwar period as a whole, these data suggest that aggregate capital-deepening has accounted for about two-fifths of aggregate productivity growth, leaving three-fifths for total factor productivity growth.
47. See, for example, Gollop (1985). Wolff (1985) provides a survey of such studies.
48. U.S. BLS (1983) analysis of quarterly data suggests the slowdown did not begin until 1973–79.
49. Further analysis, based on regression coefficients controlling for unemployment, finds more of a recovery in productivity since 1982, at about $1\frac{1}{2}$ percent.
50. See, for example, Mark (1983: 99), U.S. BLS (1982: 97), Kendrick

(1985: 118), and Rees (1980). Denison, however, finds no compelling difference in the accuracy between goods and services (1985: 25).

51. Further research indicates that real wage growth has trailed productivity growth by a full percentage point since 1982. This is a result of both the decline in labor's share and the recent rise in consumer prices relative to other goods.

REFERENCES

Baumol, W.J. 1967. "Macroeconomics of Unbalanced Growth." *American Economic Review* 57: 415–26.

Bell, L.A., and R.B. Freeman. 1985. "Does a Flexible Industry Wage Structure Increase Employment? The U.S. Experience." NBER Working Paper No. 1604. Cambridge, Mass.: National Bureau of Economic Research.

Carter, A.P. 1970. *Structural Change in the American Economy.* Cambridge, Mass.: Harvard University Press.

Clark, C. 1951. *The Conditions of Economic Progress,* 3d edition. New York: Macmillan.

Dalton, D.H., Jr. 1985. "Health Services: Overview of a Rapidly Growing Service Industry." In U.S. Department of Commerce, Bureau of Economic Affairs, *The Service Economy: Opportunity, Threat or Myth?* Washington, D.C.: U.S.G.P.O.

Denison, E.F. 1985. "Commentary." In U.S. Department of Commerce, Bureau of Economic Affairs. *The Service Economy: Opportunity, Threat or Myth?* Washington, D.C.: U.S.G.P.O.

Dickens, W.T., and L. Katz. 1987. "Industry Wage Differences and Theories of Wage Determination." NBER Working Paper No. 2271. Cambridge, Mass.: National Bureau of Economic Research.

Diewert, W.E. 1978. "Superlative Index Numbers and Consistency in Aggregation." *Econometrica* 46: 883–900.

Freeman, R.B. 1983. "Unionism, Price-Cost Margins, and the Return to Capital." NBER Working Paper No. 1164. Cambridge, Mass.: National Bureau of Economic Research.

Fuchs, V.R. 1968. *The Service Economy.* New York: Columbia University Press.

———. 1980. "Economic Growth and the Rise of Service Employment." In H. Giersch, ed., *Towards an Explanation of Economic Growth.* Tübingen: Mohr.

Gollop, F.M. 1985. "Analysis of the Productivity Slowdown: Evidence for a Sector-Biased or Sector-Neutral Industrial Strategy." In W.J.

Baumol and K. McLennan, eds., *Productivity Growth and U.S. Competitiveness.* New York: Oxford University Press.

Hall, R.E. 1986. "Market Structure and Macroeconomic Fluctuations." *Brookings Papers on Economic Activity* 2: 285–322.

Hicks, J.R. 1932. *The Theory of Wages.* London: Macmillan.

Jones, R.W. 1965. "The Structure of Simple General Equilibrium Models." *Journal of Political Economy* 73:557–72.

Kendrick, J.W. 1985. "Measurement of Output and Productivity in the Service Sector." In R.P. Inman, ed., *Managing the Service Economy.* Cambridge, England: Cambridge University Press.

Mark, J.A. 1983. "Concepts and Measures of Productivity." In U.S. Department of Labor, Bureau of Labor Statistics, *A BLS Reader on Productivity.* Bulletin 2171. Washington, D.C.: U.S.G.P.O.

Podgursky, M.J. 1988. "Job Displacement and Labor Market Adjustment: Evidence from the Displaced Worker Survey." In *The Impact of Technological Change on Employment and Economic Growth,* edited by D.C. Mowery. Cambridge, Mass.: Ballinger Publishing Company.

Rees, A. 1980. "Improving Productivity Measurement." *American Economic Review: Papers and Proceedings* 70:340–42.

Reiff, B. 1986. "Industry and Occupation Employment Structure and the Income Distribution." Mimeo. Massachusetts Institute of Technology, Cambridge, Mass.

Salinger, M.A. 1984. "Tobin's *q*, Unionization, and the Concentration-Profits Relationship." *Rand Journal of Economics* 15:159–70.

Solow, R.M. 1956. "A Contribution to the Theory of Economic Growth." *Quarterly Journal of Economics* 70:65–94.

———. 1957. "Technological Change and the Aggregate Production Function." *Review of Economics and Statistics* 23:101–08.

Summers, R. 1985. "Services in the International Economy." In R.P. Inman, ed., *Managing the Service Economy.* Cambridge, England: Cambridge University Press.

Thurow, L.C. 1979. "The U.S. Productivity Problem." *Data Resources U.S. Review* I:14–19.

———. 1980a. "Discussion." In Federal Reserve Bank of Boston, *The Decline in Productivity Growth.*

———. 1980b. *The Zero-Sum Society.* New York: Basic Books.

U.S. Department of Commerce, Bureau of Economic Analysis. 1986. *The National Income and Product Accounts of the United States, 1929–80.* Washington, D.C.: U.S.G.P.O.

———. July 1987. *Survey of Current Business,* Washington, D.C.: U.S.G.P.O.

U.S. Department of Labor, Bureau of Labor Statistics. 1982. *BLS Handbook of Methods,* Volume I, Bulletin 2134–1. Washington, D.C.: U.S.G.-P.O.

———. 1983. *Trends in Multifactor Productivity, 1948–81,* Bulletin 2178. Washington, D.C.: U.S.G.P.O.

———. *Employment and Earnings,* various issues. Washington, D.C.: U.S.G.P.O.

Williamson, J.G., and P.H. Lindert. 1980. *American Inequality: A Macroeconomic History.* New York: Academic Press.

Wolff, E.N. 1985. "The Magnitude and Causes of the Recent Productivity Slowdown in the United States: A Survey of Recent Studies." In W.J. Baumol and K. McLennan, eds., *Productivity Growth and U.S. Competitiveness.* New York: Oxford University Press.

II

THE EFFECTS OF TECHNOLOGICAL CHANGE ON SKILLS AND THE DISTRIBUTION OF EARNINGS AND INCOME

4 TECHNOLOGICAL CHANGE, SKILL REQUIREMENTS, AND EDUCATION
The Case for Uncertainty

Kenneth I. Spenner

Several centuries of controversy surround the relationship between technology and work. In the United States, national panels and commissions have periodically considered the role of technology in the economy and society, with one of first commissions on the topic dating back at least to the Great Depression (National Resources Committee 1937).

Given the larger debate, this review concentrates on three questions. First, what do past studies tell us about how technological change alters the skill requirements of work? Second, how does past knowledge apply to the near-term future? Finally, what are the policy implications for education and training?

My thesis is straightforward: Uncertainty dominates the answer to each of the questions. Past research contains considerable gaps in quality and coverage so that judgments about how technological change affects work contain substantial uncertainty. Much of what we do know suggests an uncertain, complicated, and contradictory relationship between technological change and the skill requirements of work. Technology has substantial effects on the composition and content of work in the economy, but these effects vary for different dimensions of skill, for different jobs, occupations, industries, and firms, and for different technologies. The effects involve complicated mixtures of offsetting compositional and content changes, of skill upgrading and downgrading. Most important, the effects of technological change on the skill requirements of work are set in a larger context of market forces, managerial preroga- tives (in implementing technologies), and organizational cultures, all of which condition the effects of technological change. The forces of managers, markets, and organizational cultures are suf- ficient to reverse the effects of a technology on skill upgrading or downgrading. Major arguments about a different technological future also involve uncertainty, in the validity of the argument and in the intrinsic nature of the proposed relationships. Finally, pub- lic and private policies for education and training must attend to the uncertainties. To be avoided are education and training poli- cies that assume a single, simple, or unitary effect of technological change on the number or quality of jobs.

The next section offers a selective review and critique of past studies. The second section considers arguments about a different future, and the final section considers education and training pol- icy in the face of uncertainty. Throughout I use a broad definition of *technology,* one that includes new materials and machines as well as new ways of organizing production, people, and ideas. Thus, technology includes "hard" products and things along with pro- cess and organization, although the large portion of available re- search on technological change and skill requirements of work studies various forms of mechanization and computerization, a serious limitation of current evidence.

PAST STUDIES: REVIEW AND CRITIQUE

Major Theoretical Positions

Three central positions inform the debate on how technological change alters the number and quality of jobs: the upgrading, downgrading, and mixed-change or conditional positions. The ar- guments and evidence span different disciplines, models, and types of evidence. The industrialization thesis and the central premises of neoclassical economic theory form the basis for the upgrading position (Kerr, Dunlop, Harbison, and Myers 1964; Bell 1973; Standing 1984). In simplified form, the argument says that the division of labor in the economy evolves along the lines of greater differentiation and efficiency. Technological changes increase productivity, lower costs, and expand markets, in the process requiring a broader variety of skills and higher average skills from the labor force. In some versions of the thesis, the postindustrial economy increasingly relies on highly automated and high-technology work environments that require new forms of skill: responsibility for monitoring, making adjustments, visualizing the whole of the production process, and responding to emergency situations (Crossman 1960). Such arguments have been made for continuous process and chemical manufacturing (Blauner 1964), petroleum refining (Gallie 1978), applications of robotics in metal-working industries (Miller 1983), automated banking operations (Adler 1983), and virtually any industry that uses advanced mechanization (Hirschhorn 1984).

Downgrading arguments focus on the deterioration in the quality of work because of changes in the nature of the labor process. For example, according to Braverman's (1974) thesis, technology has been a key instrument for fractionating and deskilling jobs. Management uses devices such as scientific management, numerical control, automation, and the redesign of jobs to separate the planning and conception features of work from the execution features of work. The eventual result is a polarized labor force: a growing mass of unskilled and semiskilled jobs and workers at the bottom and an elite of managers and professionals at the top. Braverman includes operatives, sales workers, clerical workers, and even some professions such as engineering and computer programming in the deskilling process (for further examples, see Kraft 1977; Scott 1982; Shaiken 1984).

Other versions of the downgrading position point to proletarianization, a process in which skill downgrading occurs through the elimination of nonworking-class positions and the creation of working-class locations (Wright and Singelmann 1982), or differential growth of high versus low skill occupations and industries (Ginzberg 1982; Levin and Rumberger 1983; Rumberger 1984; Rumberger and Levin 1985; Singelmann and Tienda 1985). These latter arguments do not rest on actual changes in the nature of work but depend on differential growth for sectors of the economy to produce skill polarization and net downgrading.

Economists have identified a final type of downgrading argument that focuses on larger economic and social processes to which technology contributes and of which deskilling is a consequence (Bluestone and Harrison 1982). Structural unemployment that issues from technological change forms a part of this larger process. When displaced and structurally unemployed workers return to work—if at all—it is at a lower skill level, especially for workers who remain in the same community (Ferman 1983; Office of Technology Assessment 1986). Deskilling is a secondary consequence of deindustrialization.

A final position in the larger debate, the mixed-change or conditional position, is more a characterization of the empirical evidence than a well-developed theory. According to this position, the effects of technological change or changes in the labor process are mixed and offsetting (Jaffe and Froomkin 1968). Alternately, the effects depend on level of automation (Bright 1966)—upgrading in the early stages, downgrading in the later stages—or depend on the organization milieu (Davis and Taylor 1976; Webster 1986), the way management chooses to implement the change (Adler 1983), other features of the work environment (Vallas 1987), or larger demographic and economic forces (McLaughlin 1983). The outcome is little net change in skill requirements of work (Horowitz and Herrnstadt 1966; Spenner 1979) or offsetting trends in the composition of the occupational structure as some sectors and jobs experience upgrading and others downgrading (Spenner 1982, 1983, 1985).

In summary, the arguments are diverse. The above synopses illustrate rather than offer complete review. The critique that follows takes the theoretical positions as given.

Critique

Societal, Sector, and Occupational Variations; Aggregate and Case Studies. The economy of a society contains an overall skill level at any point in time that reflects both the mixture of jobs and the distribution of people to jobs. Aggregate studies average skill changes across a large number of occupations or industries, looking for shifts in the overall level. It is roughly the case that aggregate studies have been more the domain of the upgrading tradition and have provided more of the support for skill upgrading (Mueller et al. 1969; Rumberger 1981).

A major study conducted for the National Commission on Technology, Automation, and Economic Progress illustrates the aggregate approach. Horowitz and Herrnstadt (1966) compared all jobs in five industries on detailed skill measures in the second and third editions of the U.S. Department of Labor's (1949, 1965) Dictionary of Occupational Titles (DOT). The time period ranges from just after World War II to the early 1960s. The industries included three from manufacturing (slaughter and meatpacking, rubber tires and tubes, and machine shop trades) and two from nonmanufacturing (medical services and banking). The skill measures included general educational development (mathematical, language and reasoning development required), specific vocational preparation (total training time for an average performance at the job), eleven aptitudes (including verbal, numerical, spatial, motor coordination, and manual dexterity), and twelve work conditions (including variation, repetitiveness, discretion, direction, precision, and working under stress). For new jobs in the 1960s the study assessed average levels of complex work with data, people, and things. The evidence showed that each of the different industries contained mixtures of upgrading and downgrading in the different skill measures. No dominant pattern of upgrading or downgrading appeared in any industry across indicators or in any skill indicator across industries. Spenner (1979) extended the earlier study to a sample of all jobs in the economy with the third and fourth editions of the DOT (1965–77). Skill indicators for levels of involvement with data, people, and things showed little change; if anything, there was a slight upgrading over the twelve-year period. Other recent aggregate comparisons can be found in Berg, Freedman, and Freeman (1978), Rumberger (1981), Karasek et al. (1982), and Wright and Singlemann (1982).

Case studies offer a considerably more detailed picture of skill transformations for a particular industry, occupation, or firm but at a cost of population coverage. Examples include Adler's (1983) study of the banking industry and computerization changes, and Wallace and Kalleberg's (1982) study of the impact of technological change on several printing industry occupations.

A recent example occurs in Vallas's (1987) study of the technological change from mechanical and electromechanical to electronic switching systems in the communications industry. Against a background of overall upgrading in the communications industry because of compositional shifts (that is, more workers in higher skilled occupations) Vallas found mixtures of upgrading and downgrading that differed by occupation and dimension of skill for unionized workers in eight union locals in the state of New York. The study design relied on a synthetic cohort strategy to assess technological change and workers' reports of skills required by their jobs. It would have been difficult to obtain this detailed a picture of consequences of technological change with an aggregate study. In general, case studies show more change and volatility in skill levels as a function of technological changes and offer a more detailed picture of change.

Composition and Content as Tracks of Change. Transformations in skill occur along two tracks. Skill change in a sector or the economy might occur through *compositional shifts:* the creation or elimination of jobs of given skill level and the distribution of persons to jobs. Alternately, skill change might occur through actual changes in work *content* (the technical nature of work and the role relations surrounding work performance).

The social and economic forces that accomplish upgrading or downgrading may operate on one front but not the other or may operate in contradictory ways on the two fronts. There is no necessary isomorphism. In the short run, technological changes may be more efficacious in generating skill shifts via changes in work content. In the long run, technological change may generate more change via compositional shift because of increases in productivity, lower costs, and economic growth. For example, there is some evidence that at upper levels of mechanization—automation narrowly defined—work content is downgraded in some craft fields (Bright 1958, 1966), yet other changes offset the downgrading with upgrading of content or upgrading via compositional shift (compare Wallace and Kalleberg 1982 with Hull et al. 1982). Alternately, changes along a single track may vary over the short and long terms. In a study of the effects of computerization on demands for clerical and managerial labor between 1972 and 1978, Osterman (1986) found lower demand over the short-term followed by increased demand over the long term (that is, several years). Osterman interprets the offsetting effects in terms of a bureaucratic reorganization hypothesis: over the long term firms reorganize as a function of technological changes and assume new functions, products, and roles, generating new demand for labor.

The collective body of evidence in economics and sociology—aggregate and case studies, studies of composition and content shifts for a range of or a single technology—is far short of a comprehensive sampling of time and space in the U.S. economy. Thus, sampling limitations comprise a major source of uncertainty in our knowledge of technology and skill requirements. In general, case studies afford greater coverage to the temporal dimension but with severe restrictions on the coverage of the occupation-industry structure. Aggregate studies are more limited in coverage of the temporal dimension but are more expansive in coverage of the sample space. A few studies offer quantitative projections of skill changes into the future, but these are typically of composition shifts only and without any direct measures of one or more dimensions of skill, or under restrictive assumptions about the nature of content shifts (for example, see Rumberger 1984; Rumberger and Levin 1985).

Concepts and Measures of Skill Requirements.   Several questions illustrate the issues. Is it workers who are skilled or jobs that require skill? Is skill a unidimensional feature of work with equivalent and equally meaningful application of the construct at different historical points and for different technologies? How can skill(s) best be measured? And do conclusions about the effects of technological change on skill requirements hinge on the specific concepts and measures of skill that are used? Unfortunately, the answers to these questions reinforce the conclusion about the uncertainty of the knowledge base.

The idea that skills reside in persons and are best studied at that level has precedent in human capital and related perspectives in economics (Becker 1964; also see Oakley 1954; Rumberger 1983). Workers acquire a stock of capabilities, knowledge, and experiences that translate into productivity and that yield reward. This approach does not directly speak to the skill requirements of jobs because the possession of human capital cannot be equated with its use (Berg 1970). Indeed, the fit between the skill capacities of workers and the skill demands of jobs is notoriously "loose" and has been the subject of appreciable study under the rubrics of overeducation and underemployment (Rumberger 1981; Clogg 1979; Clogg and Shockey 1984; Smith 1986). Thus, the schooling, training, or wage levels of workers cannot be equated with the skill requirements of work, except under a very restrictive set of assumptions (Braverman 1974; Field 1980; Rubinson and Ralph 1984).

The idea that skill is a feature of jobs better lends itself to the study of the effects of technological change for several reasons. Classical and contemporary economic theories provide for positional differences among jobs—for example, in John Stuart Mill's reasoning on positional components of wage inequalities or in Thurow's (1975) theory of job competition, where marginal products adhere in jobs and not people. Second, the supply of available education, training, and skills in people enters the technological equation only indirectly and over the long term, whereas the effects of technological change on the content and composition of jobs, and hence skill requirements, are more immediate and direct. Third, a growing body of research suggests that the structure of work has a greater, more immediate effect on people (that is, their capacities, intellectual development, self-concept, and so on) than vice versa. The primary mechanism through which people shape their work and careers appears to be occupational selection over the longer term rather than "skilled" persons effecting immediate or substantial changes in the structure of their work (Kohn and Schooler 1983). Finally, theoretical perspectives on upgrading and downgrading address the skill in jobs rather than people. The suggestion that we study the skill requirements of jobs is not to suggest that people are unreliable reporters of the skill demands of their work or to gainsay the often tragic consequences of technological transformations for workers, or the importance of studies of related phenomena such as overeducation, skill transferability among jobs, or deindustrialization.

The major measurement strategies for job skills are nonmeasurement, indirect measurement, and direct measurement. The *nonmeasurement* strategy equates occupation groups such as white collar or blue collar with implicit skill levels (for examples, see the National Commission on Technology, Automation, and Economic Progress 1966; Jaffe and Froomkin 1968; Bluestone and Harrison 1982; compare Jones 1980). This strategy contains substantial validity problems because the referent is not clear in the cross-section at one point in time to say nothing of how an unknown referent may have changed between two or more points in time.

The *indirect measurement strategy* takes the schooling levels or wage rates of an occupational group as indirect indication of the skill level of one or more jobs (National Commission on Technology, Automation, and Economic Progress 1966; Wallace and Kalleberg 1982). Validity remains an issue with this measurement strategy because the isomorphism between the indirect indicator and the true skill level depends on a set of assumptions about other factors that generate variation in the indirect indicator. For example, such a use of wage rates requires a complex set of assumptions about constancies in the supply and demand for labor (see Field 1980: Appendix, for an exposition).

A more reliable and valid approach to studying technology and skill requirements involves the *direct measurement* of the dimensions of skill for jobs or workers. There are a variety of approaches to the direct measurement of skills. All of the approaches contain limitations. For example, job titles bear an unclear relationship to skill demands. The title can change but the skill demands may not or vice versa. Additionally, consensus does not exist on the relevant dimensions of skill. The available approaches range from the *Position Analysis Questionnaire* (McCormick, Mecham, and Jeanneret 1977), which measures nearly 200 job features, to the *Universal Skills System* developed at Michigan State University, which purports to measure over 1,400 transferable job skills, to the *Dictionary of Occupational Titles* (U.S. Department of Labor 1965, 1977), to ad hoc systems that assess job skills from workers in jobs or from judgments of expert analysts.

Approaches based on the DOT are the most frequently used (for review and critique, consult, Miller, Treiman, Cain, and Roos 1980; also see Spenner 1980; and Cain and Treiman 1981). The DOT contains measures of over forty variables for over 13,000 third edition jobs and over 12,000 fourth edition jobs based on job analyses conducted by the Department of Labor (U.S. Department of Labor 1972). The measures include general educational development (mathematical, language, and reasoning development required), levels of involvement with data, people, and things; specific vocational preparation (total training time for an average performance at the job); eleven aptitudes (including verbal, numerical, spatial, motor coordination, and manual dexterity), and twelve work conditions (including variation, repetitiveness, discretion, direction, precision, and working under stress). The main advantages of the DOT include its comprehensiveness and national scope. The disadvantages include questions about sampling coverage, reliability, validity, and aggregation to job categories that ignore firm- and industry-level variations in skill requirements. For example, manufacturing jobs are overrepresented; service, managerial, and clerical jobs are underrepresented. Additionally, the construction procedures for the fourth edition may have built in a stability bias (underestimating true change) compared with the third edition estimates (Cain and Treiman 1981; Spenner 1983). Thus, while the DOT is perhaps the best available system, its use requires extreme caution. These limitations add uncertainty to the knowledge base. Other, newer methodologies for the direct measurement of skill requirements exist but these are in experimental or developmental stages (for example, see Albin, Hormozi, Mourgos, and Weinberg 1984).

Dimensions of Skill. Empirical studies that partition job characteristics (for people or jobs) consistently find that substantive complexity defines the central core of variation in work content.[2] Skill as substantive complexity refers to the level, scope, and integration of mental, manipulative, and interpersonal tasks in a job. The subdimensions of mental, manipulative, and interpersonal capture well-known points of interface between people and jobs (U.S. Department of Labor 1972). These subdimensions are important because some recent arguments suggest past technologies primarily affected manipulative tasks whereas current and future technologies (for example, microelectronics or computer-based) affect mental and interpersonal task complexity as well (for example, see Rumberger 1984). Substantive complexity of work includes subdimensions of cognitive, motor, physical, and related demands in a job. The level, scope, and integration subdimensions also capture important empirical variations among jobs. For example, a job that requires integrated mental, interpersonal, and manipulative activities across a wide scope of situations but at modest levels on each task dimension may be more complex in skill demands than a job that requires a high level of performance on one task dimension but in a narrow range of situations and without demands in other task domains (for example, a midlevel manager with a wide range of mental and data tasks may have a more complex job than an engineer whose task demands are more complex on a single dimension).

Theoretical and empirical studies also consistently show autonomy-control to be a second major dimension by which jobs are organized (Spenner 1983). Skill as autonomy-control refers to the discretion available in a job to initiate and conclude action, to control the content, manner, and speed with which a task is done. Whereas formal authority places a job within a formal network of jobs, autonomy-control designates within-role discretion, bounds, and leeway for action as provided by the job.

Across jobs in the economy, the two dimensions (substantive complexity, autonomy-control) are positively correlated, estimates placing the correlation in the range of $r = .5$ to $.7$ (Spenner 1980; Kalleberg and Leicht 1986; Vallas 1987). In general, case studies have afforded more attention to skill as autonomy-control while aggregate studies have given greater consideration to skill defined as substantive complexity. Further, some evidence suggests some technologies and the sum total of technological changes exert contradictory effects on the different dimensions of skill. If the dimensionality of skill requirements is greater than two organizing dimensions, then the possibilities for contradictory and offsetting effects are even more complicated. Some of the uncertainty in the knowledge base springs from no consideration or uneven treatment of the dimensionality of skill requirements.

Summary of Select Aggregate and Case Study Evidence

A compilation of major aggregate studies offers one way to summarize how technological changes alter the skill requirements of jobs, based on past research. Table 4–1 summarizes select aggregate studies that meet several criteria: (1) two or more points in time; (2) a sample that refers to a sizable population of jobs, occupations, or industries; (3) some direct measurement of skill as substantive complexity or skill as autonomy-control. The criteria effectively exclude a large number of aggregate economic studies of the demand for labor but with no direct measurement of skill, or studies that indirectly infer the quality of jobs through wage rates, schooling levels, productivity indices, and so on. The table classifies each study by the sample or population, the time period, whether content or composition shifts are studied, and the skill measures and comments on possible threats to the quality of the inferences.

Table 4–1. Summary of Aggregate Studies of the Effects of Technology on Skill Requirements that Employ Direct Measures of Skill.

Study	Sample/ Population	Time Period	Content/ Composition Shift
Horowitz and Herrnstadt (1966)	All Department of Transportation (DOT) jobs in five industries (slaughter and meat packing, rubber tires and tubes, machine shop trades, medical services, banking)	1949–1965	Content
Spenner (1979)	5 percent sample of fourth edition DOT titles (N=622) matched to third edition titles	1965–1977	Content
Berg (1970)	1950–1960 decennial census distributions; 4000 DOT jobs rated in 1956 and 1965	1950–1960 (1956–1965)	Composition and content
Berg, Freedman, and Freeman (1978)	1950–1970 decennial census distributions; second edition DOT estimates for 1950; third edition estimates for 1960 and 1970	1950–1970 (1956–1965)	Composition and content
Rumberger (1981)	1960 and 1976 census and Current Population Survey (CPS), respectively; employed population fourteen and older; third and fourth edition DOT	1960–1976 (1965–1977)	Composition and content
Rumberger (1981: Table 4); also see, Eckaus (1964) and Rawlins and Ulman (1974)	See Rumberger above; 1940–1950 decennial census distributions	1940–1976	Composition
Reanalysis of Dubnoff (1978) data; see Spenner (1982)	Decennial census distributions for all gainful workers (1900–1930) or all employed workers (1940–1970)	1900–1970	Composition

Skill Measures	Outcomes	Notes—Design Threats
For jobs; 25 DOT indicators; most indicators reflect skill as substantive complexity; two or three indicators may approximate skill as autonomy-control	Mixture of upgrading and downgrading; little net change	Limited to five industries; depends on independence of DOT editions
For jobs; DOT indicators for data, people and things; skill as substantive complexity	Small upgrading; little net change	Depends on independence of DOT editions
For jobs; DOT General Educational Development (GED) indicator; skill as substantive complexity	Small compositional upgrading; for content, 54 percent of jobs had the same GED, 31 percent were higher and 15 percent were lower; apparent content upgrading	Depends on independence of editions; possible validity problems with GED; change in GED categories between editions may overestimate upgrading
For jobs; DOT GED indicator; skill as substantive complexity	Same as Berg (1970) for 1950–1960; small compositional upgrading for 1960–1970	Same as Berg (1970)
For jobs; DOT GED indicator; skill as substantive complexity	Modest compositional upgrading; small content upgrading but with some evidence of proletariani-zation as the number of very highest skill jobs declined	Depends on independence of DOT editions; possible validity problems with GED
For jobs; DOT GED indicator; skill as substantive complexity	Overall 18 percent compositional upgrading over thirty-six years; greatest increase between 1950 and 1960	Depends on independence of DOT editions; possible validity problems with GED
For jobs; DOT indicators for data, people, things, SVP, and combination of the first three indicators; skill as substantive complexity	Little net change; only one of eighteen skill-year or higher order effects significant in loglinear decomposition; for one interaction, evidence of skill polarization in recent years	Depends on the quality of the map of detailed occupations from one census year to another; comparison assumes constant work content to third edition DOT scores over entire time period

Table 4–1. (*Continued*)

Study	Sample/ Population	Time Period	Content/ Composition Shift
Mueller et al. (1969)	National probability sample of 1967 labor force (N=2662)	1962–1967 (retrospective measure of job and machine change over five years)	Content (composition inasmuch as 1967 sample members changed jobs)
Karasek, Schwartz, and Pieper (1982)	National samples for 1969, 1972, and 1977; adult employed labor force working twenty or more hours per week (N=4531)	1969–1977	Content (composition partially adjusted for with demographic controls; otherwise assumed constant)
Wright and Singelman (1982)	Decennial census distributions for thirty-seven industry sectors; the design decomposes 1960–1970 shifts into industry, class, and interaction component; skill levels implicit in class categories	1960–1970	Composition (industry and class shifts)
Sobel (1982)	National samples for 1970, 1973, 1976, and 1977; adult employed labor force working twenty or more hours per week	1970–1977	Composition and content

Skill Measures	Outcomes	Notes—Design Threats
For people; detailed reports of level and type of machinery use over five years; self-reports of "skill required" and "own influence in organizing the work"; mixture of skill as substantive complexity and skill as antonomy-control	For job changers over five years, modest upgrading in mechanization level and skill measures; for those who stayed in the same job but experienced machine change: More Same Less "Skill" 53% 36% 7% "Influence" 34% 54% 7% Across all respondents, small upgrading	For job changers, conflation of compositional upgrading with seniority-career effects; short time interval; validity-reliability of self-report and retrospective report data; compositional shift via demographic replacement ignored
For people, aggregated to 240 occupation categories; four replicated questions combined into single scale (learn new things, "skill," creativity, and repetition); skill as substantive complexity	No change in skill discretion scale scores	Validity-reliability of self-reports; slightly different response categories in 1977 compared with 1969 and 1972
For people; measured through class categories (self-employed, have employees, have subordinates, and level of freedom and decisionmaking in jobs); self-reports taken from 1969 National Survey of Working conditions; skill as autonomy-control	Overall small changes; mixed evidence for upgrading and downgrading in class and industry shifts; for upgrading more managers, for downgrading more workers; industry and class composition shifts tend to operate in opposite directions; some evidence for proletarianization in the class composition shift into the working class	Depends on the validity of class measurement; validity-reliability of self-report; possible skill heterogeneity in class and industry categories; assumes constant work content over the time interval; skill is measured indirectly in class categories
For people; related questions in successive surveys taken to measure supervisory status; skill as autonomy-control	Decline in skill; percent classified as supervisors: 1970 36.1% 1973 34.1% 1976 31.4% 1977 31.1%	Different sampling designs at the time points; nonidentical questions to measure supervisory status at the time points; validity-reliability of self-reports; indirect measure of skill

The most important conclusion centers on the uncertainty and serious limitations in the knowledge base. With a single exception, all studies refer to the post–World War II period. Many studies rely on the DOT as a source of skill measures and are subject to the serious limitations of the DOT. Most studies investigate skill as substantive complexity and ignore variations in skill as autonomy-control. By definition, these studies average firm-specific and technology-specific variations in skill requirements. It is also important to note that popular judgments, conventional wisdom, and the knowledge base available to engineers and managers who design and implement technical innovations, and policymakers who legislate and administer about technical innovations—all share these limitations, uncertainties, and small reservoir of solid data.

In an earlier study, I offered several tentative conclusions and hypotheses that I repeat here (Spenner 1985).

1. The quantity of skill requirement change observed in particular studies depends on the indicator of skill. The GED indicator from the DOT offers the most optimistic upgrading estimates and is probably an anomaly.

2. In the postwar era there is no consistent evidence of dramatic change through content shifts in substantive complexity. All studies for time periods up to the mid-1970s suggest little net change or a small upgrading.

3. Studies of compositional shifts in skill as substantive complexity suggest the possibility of a small upgrading since World War II but approximate stability since the turn of the century. The longer-term conclusion requires a strong set of assumptions about constancies in content shifts (Spenner 1982). There is some limited evidence of possible polarization effects: differential growth of the highest and lowest complexity jobs, where subgroups of men and women differentially gain or lose in terms of new job growth and job elimination (Spenner 1982; Hartmann, Kraut, and Tilly 1986).

4. Evidence on aggregate content shifts in skill as autonomy-control is mixed, with one study suggesting modest upgrading (Mueller ret al. 1969) and the other suggesting slight downgrading (Sobel 1982). Since the studies use different indicators of autonomy-control, a firm conclusion is not possible.

5. Only one study addresses compositional shifts in skill as auton-

omy-control (Wright and Singelmann 1982). The evidence suggests a small net downgrading. The methodology of this study indirectly measures autonomy-control, but the larger conclusion is quite possible given the increased location of jobs in bureaucracies over the course of this century and given other evidence that shows jobs in bureaucratic settings are subject to greater constraints on autonomy-control—even though such jobs may involve higher substantive complexity—compared with jobs in more entreprenuerial settings (Kohn 1971; Spenner 1987).

Compared with the putative wisdom of upgrading and downgrading traditions, the collective evidence from aggregate studies shows no dominant trend in the twentieth century and suggests evolutionary not revolutionary rates of change. In summary, the dominant feature of aggregate study evidence is uncertainty; to the extent a conclusion is warranted it would suggest approximate net aggregate stability of skill requirements or a small upgrading.

The aggregate study by Mueller and colleagues (1969) warrants more detailed summary for several reasons. First, the study provides point estimates of the number of workers and jobs affected by mechanization changes (as a subset of technological changes) in a given time period. The sampled population included the U.S. labor force in 1967 ($N = 2,662$). Respondents provided detailed information on their current jobs in 1967, their jobs five years earlier in 1962, and select intervening work experiences. Detailed information on level and type of mechanization in the jobs and changes in such were coded into standardized categories by engineering students. The design can distinguish workers who experienced upgrading or downgrading at the same job versus workers who experienced upgrading or downgrading as a function of changing jobs. The period in question, 1962–67, involved an expanding economy and rapid economic growth, providing a resource-rich and demand-driven environment conducive to technological changes (although only mechanization changes are measured here). On balance, the workers experienced more upgrading than downgrading, a finding generally consistent with DOT-based results for this period.

Based on this sample, about 10 percent of the labor force experienced one or more mechanization changes in their jobs over the five-year period. Mechanization changes thus directly affected 2 to 3 percent of all jobs per year or about 1.5 to 2 million workers per year. Further, the Mueller study also found that mechanization changes directly generated little unemployment and existing employees were typically restrained. Consistent with more contemporary evidence on recent technological change (Office of Technology Assessment 1986), most displacent and adverse effects were for those already working at a low skill level; most machine-change advantage went to those advantaged in other respects (higher education, higher-status jobs). The final section returns to several of these issues.

A related compilation of select case studies can be found in Table 4–2. The number of case studies across several disciplines is substantial, and I have made no attempt to select a probability sample. Further the criteria are less stringent than those employed for aggregate studies (that is, direct measures).

The case studies show much more volatility in skill requirements as a function of changes in technology and the larger labor process. Few of the case studies give attention to issues of measurement validity or reliability or to other sources of invalidity in design inferences. Comparisons across case studies are difficult because of different samples and methods, and different concepts of skill. Thus, designations of "unclear" or "apparent" in the table mean that I was unable to decipher the entry from the source material. The original study may have had a concrete position on the issue.

If rigorous methodological criteria are applied, then few generalizable and replicable conclusions are available. If the criteria are relaxed, then several general impressions (best viewed as hypotheses) characterize the case studies.

1. Case studies provide the strongest evidence for downgrading. Given the diversity of designs, sample, and method, the instances of downgrading are too varied to have all been artifacts. However, the volatility in skill requirements seen in case studies may occur in part because of sample selection effects, or overstudying changing occupations and work areas and understudying stable occupations and firms.

2. Case studies give more attention to content changes in work, reporting more instances of downgrading compared with ag-

gregate studies. Recall that aggregate studies provided more evidence of upgrading through mixtures of content and compositional shifts.

3. Case studies strongly suggest regional, state, and other geographic variations in skill transformations.

4. Case studies suggest the impacts of technology on skill requirements of jobs are not simple, are not necessarily direct, are not constant across settings and firms, and cannot be considered in isolation of larger classes of variables that I have summarized as managers, markets, and organizational cultures. Several recent case studies well illustrate the more complicated variations (Kelley 1986; Webster 1986; Vallas 1987).

Two of the studies show that a specific technology—numerically controlled machinery in the studies reviewed by Kelley (1986) and dedicated word processing systems in the eight firms studied by Webster (1986)—can have opposite implications for the skill requirements of the same occupations, conditional on managerial discretion and organizational variables. Kelley reviewed eleven studies on the introduction of numerically controlled machines in U.S., western European, and Japanese firms, concentrating on whether the machine operators did more advanced programming of the new machinery as an indicator of upgrading or downgrading. In some firms, operators jobs were upgraded, particularly in the West German firm, involving new programming tasks as part of the technological change; in other firms, there was a clear downgrading, with less complex work, less autonomy-control in the role, and no programming activity. Smaller organizations seemed more likely to augment the computerization with programming tasks. Managerial approach played a major role, in some firms following classic scientific management principles, in other firms following narrow technical criteria, and in yet other firms—the typical upgrading situation—there was advance effort to implement computerization changes around worker-centered participation and control. Webster's study of eight British firms, with a different computerized technology, confirms the important role of managers. This is not to suggest that upgrading or downgrading depends on spur-of-the-moment managerial decisions. More likely, a larger set of longer-term economic and sociological factors generate managerial implementa-

Table 4–2. Illustrative Case Studies of Skill Change.

Study	Sample Population	Time Period	Content/ Composition Shift
Braverman (1974)	All work; concentration on operative, clerical, craft, and service occupations; some participant-observation in England	1900–1974 (also late nineteenth century)	Primarily content
Bright (1958, 1966)	Highly automated manufacturing firms, principally auto engine assembly parts, machine shops, and metal working	1950s to mid-1960s	Primarily content
Faunce (1958)	Random sample of workers from machinery departments of Detroit automobile engine plant (N = 125)	Mid-1950s	Content (Comparison of pre- and post–assembly line experiences)
Stone (1974)	Steel industry; skilled craft and heavy laborers	1890–1920 (secondarily through 1960s)	Primarily content

Skill Measures/ Dimensions	General Outcomes	Notes
Including: repetitiveness, responsibility, scope and variety of tasks, integration of mental, manipulative, and interpersonal task components; authority-supervision relations; skill as substantive complexity and autonomy-control; no direct skill measures	Overall deskilling; separation of conception and execution in work; polarization of jobs vis-a-vis skill requirements; growing mass of working class occupations	Sketchy coverage of composition shifts; unclear whether deskilling conclusions apply equally to all occupations or which fractions thereof
Twelve contributions of workers to tasks, including: physical and mental effort, manipulative and general skills, responsibility, and decisionmaking	Across seventeen defined levels of mechanization, have mixed effects, generally increasing skill requirements up to the middle levels of automation and decreasing thereafter	Sketchy coverage of composition shifts; applications to other areas and technology changes unclear; has been criticized for limited range of mechanization in studied plants and limited skill definition (Adler 1983)
Closeness of supervision, responsibility, control over work pace, attention, relationship with supervisor, interactions with coworkers	Deskilling in less control over work pace, more closely supervised, job requires more alertness and attention (could be interpreted as upgrading); upgrading in that the worker was responsible for a larger share of the production process; increased isolation from coworkers; altered relationship with supervisors	Validity-reliability of self-report; no consideration of composition shifts; short time span confounded with with newness of technology change
Unclear; apparently includes training, experience, dexterity, judgment, and general knowledge of production process; skill as substantive complexity and autonomy-control; no direct skill measures	Substantial deskilling of focal jobs; technology not primary cause but an instrument in larger control process (i.e., employer control over wages and labor unrest)	Quality of time one (nineteenth century) skill levels in the steel industry unknown; no consideration of composition shifts

Table 4–2. (*Continued*)

Study	Sample Population	Time Period	Content/ Composition Shift
Kraft (1977)	Computer programmers (about 100 programmers interviewed, participant-observer study)	1940s–1970s	Primarily content
Glenn and Feldberg (1979)	Clerical work	1870–1880; principally twentieth century	Content and composition
Burawoy (1979)	Engine division of Chicago-based multinational corporation, machine shop occupations; participant-observer study	1944 and 1974	Content
Wallace and Kalleberg (1982)	Printing industry occupations; principally compositors, machine operators, and linotypists	1931–1978	Content and composition as reflected in wage rates

Skill Measures/ Dimensions	General Outcomes	Notes
No explicit definition or direct measures; apparently includes span and cognitive complexity of task, discretion and control over work; skill as substantive complexity and autonomy-control	Deskilling; management strategy to simplify, routinize, and standardize programming task; accomplished through canned programs, structured programming and chief programmer teams; some small fraction of systems analyst and engineer positions are upgraded	Sketchy coverage of composition shifts; applies largely to programmers in large business firms (versus smaller firms, academic positions, and so on)
No explicit definitions or direct measures; apparently includes task complexity and scope, control over work; skill as substantive complexity and autonomy-control	Progressive fragmentation, specialization, and routinization of clerical work roles; coupled with massive growth, substantial deskilling; upgrading for small number of systems analysts and supervisors	Quality of time one (nineteenth century) skill levels of clerical work unknown; study focuses most on secretary, typist, and stenographer, to the exclusion of other clerical roles
No explicit definitions or direct measures; apparently includes task scope and complexity, discretion, and autonomy in work role; skill as substantive complexity and autonomy-control but more as located in workers than job requirements	Larger changes in piece-rate and rate-fixing systems, bargaining relations, and redistribution of hierarchical conflict led to mixtures of upgrading and downgrading (i.e., more autonomy for a number of occupations); more important larger process involves the operations through which the factory social system contains struggles and manufactures consent	At times "skill" equated with experience and training; no consideration of compositional shifts; unclear how interactional dynamics in the labor process of this particular shop (given an important theoretical role) characterize other work settings
Indirect measure: wage rates	Steady, substantial decline in printing industry-skilled–occupation wages *relative* to several comparison occupation groups; regression analyses indicate capital-intensity as the major proximate causal factor	Complex assumptions associated with indirect measure; change in wage rates may be due to factors other than skill change

Table 4–2. (*Continued*)

Study	Sample Population	Time Period	Content/ Composition Shift
Hull, Friedman, and Rodgers (1982)	Printers for three largest New York newspapers in the sample at both time points (N = 408 for 1950; N = 245 for 1976)	1950 and 1976	Content
Adler (1983)	Clerical occupations in four largest French banks; observational study	1930s–early 1980s	Content
Kelley (1986)	Eleven studies that investigated introduction of numerical control technology; U.S., U.K., West German, and Japanese plants; twenty-two establishments, forty-one different blue-collar jobs	Primarily 1970s	Primarily content

Skill Measures/ Dimensions	General Outcomes	Notes
Printers self-reports of the *physical* and *intellectual* demands of new methods of printing; skills as substantive complexity	Percent of printers defining new methods of printing as More Same Less Physical demands 18% 24% 58% Intellectual demands 53% 21% 27% Overall, modest to strong upgrading in lowered physical and increased intellectual demands	Validity-reliability of self-reports; printers most subject to downgrading may not be in the sample in 1976
Same worker contributions as Bright (1958); some adjustment of dimensions for qualitative changes or new skills; skill as substantive complexity and autonomy-control; apparently no direct measures	As banks moved from lower to higher forms of automation, mixture of upgrading and downgrading effects; at highest level of automation, qualitative transformation of work so as to require new categories of skill: greater worker responsibility for production, more abstract tasks and greater interdependence of jobs; impact of technology substantially mediated by market factors, managerial strategies, and social definitions of skill requirements	No consideration of composition shifts; quality of time one skill levels in the banking industry unknown.
Whether workers in affected blue-collar occupations (NC operations) perform any programming tasks; some mixture of skill as substantive complexity and autonomy-control	Mixture of upgrading, downgrading, and skill polarization that was largely establishment-specific; no evidence of singular managerial motives to deskill or upgrade; strong evidence of the role of managerial discretion and organizational variables (i.e., size); three managerial approaches: scientific management, technocratic, and worker-centered	Sketchy or no coverage of composition shifts; uneven skill measures; highly variable design quality in the eleven studies; involves cross-national comparisons; unclear whether there were direct time one and time two skill measures

Table 4–2. (*Continued*)

Study	Sample Population	Time Period	Content/ Composition Shift
Webster (1986)	Eight British firms in Bradford, West Yorkshire that introduced dedicated word processing systems; clericals working in these service and manufacturing firms	1980s	Primarily content
Vallas (1987)	Eight New York and New Jersey locals (N = 802; response rate = 51 percent), representing operators, switching and maintenance craft workers, clerical workers, and customer service representatives in regional telecommunications industry; technology changes include industry shifts from mechanical to electromechanical to electronic switching systems	1984 for content shifts; 1950–1980 for compositional shifts	Composition and content

tion strategies that are more conducive to upgrading or downgrading.

The study by Vallas (1987), although subject to strong assumptions, suggests that the introduction of new microelectronic technology in the telecommunications industry (1) independently affects both dimensions of skill, substantive complexity, and autonomy-control and (2) downgrades some occupations and upgrades others. The occupations that were downgraded lost on both skill dimensions; only outdoor craftworkers responsible for troubleshooting and repair experienced upgrading. In this industry-technology situation, the technical features of work appear to play

Skill Measures/ Dimensions	General Outcomes	Notes
Including: variety, discretion, task scope, repetitiveness, "complexity"; both skill as substantive complexity and autonomy-control	Mixture of upgrading and downgrading; the word processing technology expands the range of options for organizing clerical work; accordingly, some firms fragmented and specialized tasks (deskilling) while other firms expanded clerical jobs (upgrading); central role of management and organizational variables	No consideration of compositional shifts: apparently a short time span between, before, and after observations; unclear whether there were direct time one and time two skill measures
Direct measures of skill as substantive complexity and as autonomy-control taken from workers' self-reports; three levels of coded automation in the cross-section (synthetic cohort design to assess change)	Apparent upgrading between 1950 and 1980 via compositional shift (assumes constant work content); individual and aggregate (local) level effects of automation on work content showed deskilling of substantive complexity and autonomy-control across all sampled jobs; select jobs (outside craft workers) showed content upgrading on both skill dimensions	Strong assumptions of synthetic cohort design to measure change (workers at different technology levels, in 1984 assumed to reflect continuum of temporal change in work content in the industry); limitations of self-report measures; low response rate

a larger role in determining where upgrading versus downgrading will occur.

When juxtaposed what do the aggregate and case studies suggest? The safest conclusion suggests offsetting trends, with a slow evolution in aggregate skill levels but substantial skill requirement change in particular sectors, occupations, and industries. To the extent there is trend in the aggregate and case studies, it might be summarized by a contradictory skill shift hypothesis (see Figure 4–1). The hypothesis raises the possibility that the substantive complexity of work environments has gradually increased across the economy in recent history while the skill levels as autonomy-

Figure 4–1. Summary of Hypothesized Changes in the Skill Level of Work, by Skill Dimension and Nature of Change in Work.

| | Skill Dimension | |
	Substantive Complexity	Autonomy-Control
Content shift	Approximate stability or small upward shift Examples: Mueller et al. (1968); Rumberger (1981)	Possible downward shifts, particularly since World War II Examples: Kraft (1982); Sobel (1982); Stone (1984)
Change track Compositional shift	Approximate stability or small upward shift Examples: Dubnoff (1978); Spenner (1982)	Approximate stability or small downward shifts Examples: Levin and Rumberger (1983); Wright and Singelmann (1982)

Note: Each cell cites illustrative studies. For comprehensive citations and argument, consult Spenner (1983, 1985).

control have gradually decreased. This hypothesis awaits a comprehensive test.

The case of engineering illustrates the contradictory skill shift hypothesis and associated forms of change. Several case studies suggest the engineers of 100 years ago were independent professionals and business people compared with today's engineers, who typically are employees of large firms with a narrowed (but more complex) skill range and less autonomy-control (Braverman 1974: 242–46; Stark 1980). Consider the tremendous increase in the complexity of engineering over the years (largely driven by technology) on the one hand and the substantial fractionation and specialization of engineering on the other hand (aeronautical, astronautical, biological, biomedical, chemical, biochemical, civil, and so on). Further, the number of engineers in the United States grew dramatically from about 7,000 in 1880 to over 136,000 in 1920 to over 1.2 million today (or more depending on the classification). The work content shift in autonomy-control may well have been a skill downgrading for the original 7,000 compared with those that followed, even in the face of a substantive complexity upgrading in content because of technological change. The compositional shift in substantive complexity was massive in the upgrading direction because the substantive complexity of engineering far exceeds the average substantive complexity level for the U.S. labor force, whether assessed in the 1880s or 1980s. The compositional shift in autonomy-control for engineering is less clear; perhaps the dramatic increase of engineers in large firms and bureaucracies reflects a skill downgrading compared with the larger relative share of self-employed engineers before the turn of the century, but there is less direct evidence of this type of change.

Under the contradictory skill shift hypothesis, one reading of the collective evidence is that it approximately matches the trends suggested by engineering. If we consider the concept of an "average" occupation and work environment now compared with the turn of the century, then workers of today face skill demands that are slightly to modestly more substantively complex and, perhaps, jobs and skill levels that afford less autonomy-control compared with the past. These are characterizations of average aggregate shifts. Particular jobs, sectors, and industries may have experienced more dramatic forms of upgrading and downgrading.

In summary, the dominant impression from the existing knowledge base on the relationship between technological change and skill requirements is one of uncertainty. The collective research has produced no simple satisfactory answer to the larger questions. The available data and methodologies are limited. The samplings of skill changes and technologies in time and space are extremely limited. The concepts and measures of skill are poor.

Further, the evidence suggests a complicated relationship between technological change and skill transformations. There are short- and long-term effects, direct and indirect effects, regional, occupational, and other sector variations; and there are many apparent conditioning variables, including managerial discretion, markets, and organizational cultures. The next section considers some of the conditioning relationships and reviews arguments that the future effects of technology on skill requirements might differ from past relationships.

A DIFFERENT FUTURE?

There are several reasons to suggest the future relationship of technology to the skill requirements of work will differ from the past. The most general argument, the postindustrial thesis, suggests that the economy and social system are in a qualitatively different mode compared with the past (for examples, consult Fuchs 1968; Bell 1973; Piore and Sabel 1984; Hirschhorn 1984). The variations in the argument are many: The white-collar service revolution is upon us; the exodus from agriculture has been exhausted; national economies face a new level of integration and dependence on the world economy; we are in the midst of a new industrial revolution, built around computers and microelectronics; industrial/finance/corporate capitalism face a new level of crisis; today's firms handle product, labor, and market uncertainties in fundamentally different ways; and so on. For each of these arguments and others, there are counterarguments (for example, compare Chirot 1986). The debates are ongoing and without a current resolution. If accurate, the arguments about qualitative shifts strongly limit the use of past studies of skill and technology to make judgments about the future.

One specific argument that suggests a different future and the inapplicability of past knowledge comes from the hypothesis that current "high" technology (that is, computers and microelectronics) differs from prior technological innovations. Among the proponents of the thesis, Rumberger and Levin have proffered the most detailed argument and predictions (Rumberger 1983, 1984; Levin and Rumberger 1983; Rumberger and Levin 1985, 1987; also see Adler 1986).

Briefly, their argument suggests past technological innovations principally altered the manual, manipulative, and physical features of work. Many of these changes were concentrated in agriculture, manufacturing, and construction. Current (and future) technologies alter the mental demands of work in addition to physical demands. Rumberger and Levin suggest the effects of high technology on the composition of the labor force are several: (1) displacement of skilled mental labor; (2) job creation but in a polarized fashion, with a small relative and absolute number of "high skill" jobs (that is, measured with high wages and high education in their studies), and a large relative and absolute number of low skill jobs. Further, productivity increases may limit the ability of the economy to produce enough new jobs to offset the displacement. As evidence, they review the recent data and projections of occupation-specific and industry-specific growth and decline, in particular for "high-tech" occupations and industries (under a variety of definitions of high technology). High technology occupations and industries are growing at a rapid rate but begin with a small absolute base. Further, the occupations that will produce the largest number of new jobs are by a substantial margin medium and low skill by any definition (custodians, cashiers, waiters and waitresses, truck drivers, general office and sales clerks). Thus, a major part of the future of high technology is the creation of jobs that are of lower than average skill.

Although the argument sounds convincing, it has not been directly tested for several reasons. First, most of the evidence refers to recent and projected growth levels (relative and absolute) for occupations and industries across the economy. The specific effects of high technology have not been filtered out. The growth levels and projections reflect all sources, high technology, low technology, and otherwise (that is, as generated by demographic changes, productivity increases or declines, foreign competition, larger supply and demand variations in the economy, the movement of production to other countries, and so on). Their assertions may be correct, but overall growth levels of occupations and industries reflect all forces. Changes that are not high technology could alter the projections. Thus, we are left with an interesting hypothesis but little direct data on high versus other technology forms and their compositional shift implications.

Second, the evidence considered by Rumberger and Levin makes no direct adjustments for content shifts in skill requirements caused by high technology, which could offset or augment the compositional shifts they hypothesize. In the absence of direct measures of multiple skill dimensions and some systematic sampling of time and space, their argument is more of an hypothesis awaiting test than a well-established relationship. Finally, let us suppose their thesis is correct: The net effect of high technology is compositional downgrading, even after adjusting for content shift. Uncertainty still enters the picture under the rubric of managerial discretion, markets, and organizational cultures, as these quantities might change or alter the relationship.

Managerial Discretion

Managerial discretion refers to the role that managers play in deciding whether to implement a technological change, what change to implement, when it is implemented, and how it is implemented. Managerial discretion includes the design of new technology, hence the important role of engineers, and firm or establishment policy and procedures with respect to displaced workers, job and task fractionation and redesign. The evidence from several disciplines is clear and unambiguous: Managerial discretion plays a central role in determining the consequences of technological change for the skill requirements of work. Indeed, one explanation for the substantial diversity of outcomes in the empirical literature suggests the highly variable but potent role of management generates a wide range of outcomes. Further, it is possible, but not proven, that managerial discretion plays a larger role than intrinsic features of a technological change in defining skill requirements. Technology defines a range of possibilities; other social and economic processes take over from there.

Two previously mentioned studies illustrate the central role of managerial discretion (see Table 4–2). Kelley's (1986) study reviewed several other studies and generated new data from several firms (eleven studies and firms in all) that involved the introduction of computerized numerical control technology in manufacturing firms. The various plants were located in the United States, the United Kingdom, West Germany, and Japan. One way in which management exercised choice in implementing the technical change involved which jobs and people would be responsible for the various levels of programming, monitoring, and trouble-shooting the new system. The options ranged from specialized roles where the earlier operators would only tend the machines and others would program and trouble-shoot versus a task arrangement in which the more complicated and discretion-based tasks were distributed across roles, including the earlier machine operator, even if this arrangement involved retraining. Kelley found a wide range of skill transformation outcomes, which included the original operative jobs being upgraded in some firms, downgraded in others, and experiencing little net skill change in others. The study considered the same narrow technology, in a fairly short span of history, in similar types of manufacturing concerns. Kelley attributes at least part of the different outcomes to the important role of management in designing and implementing technical change.

Webster (1986) reaches a similar conclusion for a different computerized technology—dedicated word processing—in a study of eight British firms. The occupations studied included clerical jobs, and the range of firms studied included a university setting, a mail order firm, a motor parts firm, and a building society. Similar to Kelley's major finding, in some cases secretarial/clerical roles were redivided and fractionated with the introduction of dedicated word processing: One job involved only entry, another involved setup, another correction and final production, another supervision, and so on. In other firms, secretarial and clerical roles expanded with the introduction of word processing systems, with the work becoming more complex (that is, a single role involved judgments as to setup, format, input, and correction) and more autonomy-control in deciding how the task would be done. Webster attributes some of the differences to varying managerial practice across the firms.

Thus, managerial discretion adds uncertainty to the relationship between technological changes and skill transformations. The uncertainty is not merely a matter of managers responding to an uncertain (market) environment according to a rational choice or decisionmaking model. It is additionally the uncertain response of managers to uncertain environments. Although there are substantial traditions in economics and management science that model and generate theories of managerial behavior, firms and managers often behave in ways that are often at variance with textbook images. A number of studies document the idiosyncrasies when firms and managers implement technological changes (for review, consult Berg et al. 1981; also see Jaikumar 1986). For example, frequently management does not calculate or estimate the relative labor costs or skill mixtures associated with technology options. A growing number of studies are even less complimentary about the motives, strategies, and tactics of U.S. managers—for example, in short- versus long-term maximizing strategies, in takeovers, buyouts, plant shutdowns and international relocations (Bowles and Gintis 1976; Edwards 1979; Bluestone and Harrison 1982).

Markets

Another source of uncertainty occurs in market phenomena, broadly defined (Standing 1984). For example, larger variations in supply and demand, productivity change, and growth in the economy can substantially modify or overshadow skill changes that more directly derive from technological changes, assuming these various factors are analytically separable.

Osterman's (1986) recent study illustrates how market phenomena can obfuscate judgments about how technological change affects the skill requirements of work. In the period from 1972–78, Osterman found that the net effect of computers (as estimated by quantities of main memory available in industries in the Standard Industrial Classification) on the demand for clerical and managerial labor was negative (ignoring content shifts). But underlying the larger pattern was a reversal. Over the short term (two to five years) computerization in the various industries depressed demand for labor. Over the longer term (five to seven years) the effects were positive, with increased demand for clerical and managerial labor. The time period in question approximately spanned one short-term business cycle. Osterman interprets the empirical data with a bureaucratic reorganization hypothesis: In the short term the technological change (computers) displaces labor because a substitution of technology for labor is more efficient in the production function of the firm, given the original purpose and scope of the innovation. Over the longer-term, computerization lowers unit costs, perhaps allowing expanding production, and induces structural reorganization in the firm, perhaps allowing the production of new products, movement into new market domains, and so on.

With Osterman's study as backdrop, consider two dimensions underlying the socioeconomic environment in an industry: (1) resource-rich versus resource-lean environments and (2) the adoption-diffusion curve for a technological change within an industry, and the extent to which firms are staggered versus in synchrony in their rate of adoption of the technology. Further, let bureaucratic reorganization reflect a longer-term upgrading potential of a technological change that initially produces compositional downgrading. First, firms may be more or less likely to adopt an innovation, and second, to reorganize structurally depending on whether their local environments are resource rich or resource lean. Thus, in the cross-section, the location of an industry in the upgrading-downgrading cycle will depend on the resource richness or leanness of the environment, which is largely defined by larger market factors. Further, a study of an industry in resource-rich versus resource-lean environment might reach very different conclusions about the effects of technological change on skill requirements. Additionally, consider a situation in which the adoption-diffusion of an innovation is in synchrony for firms in an industry: Collectively, the firms would show initial downgrading then upgrading via structural reorganization. In this situation, the student of skill transformations would "see" a strong initial downgrading followed by skill upgrading. Contrast this situation with a highly staggered adoption-diffusion curve: Many firms in an industry are at different points in downgrading-reorganization-upgrading cycle described by Osterman. In this situation, the student of skill transformation would see a very different picture: approximate aggregate stability or slight downgrading.

In short, there are many ways in which market factors define and condition the relationship between technological change and skill transformations. Some of these we understand and some we do not. Further, there are even intricate ways in which market factors can obscure what our research studies show to be the relationship between technological change and skill requirements, in the absence of a complete specification and a long time series of data. The result is greater uncertainty, both in the intrinsic relationship between technological change and skill transformations, and in our knowledge of the relationship.

Organizational Cultures

A final source of uncertainty in our knowledge and in the technology-skill transformation relationship occurs in organizational cultures. By *organizational culture* I mean the social and cultural system of the work environment in which technological changes take place. Organizational culture so defined would include standard features of organizations such as size, age, differentiation, and hierarchy, but also the cumulative belief structures of employees, their images of technology, the norms and sources of power of managerial and worker groups, and more generally the demography and social psychology of organizations. Across the disciplines of anthropology, psychology, economics, management science, and sociology there is nothing approaching a comprehensive theory of how organizational cultures operate, to say nothing of specific implications for skill transformations. Rather, there are many midlevel theories, concepts, and research programs (for examples, reviews and citations consult Ouchi and Wilkins 1985 or Pfeffer 1985). Here, I illustrate one or two ways in which organizational culture comes to bear on the relationship between technology and skill requirements.

No less than other social phenomena, *skill* is subject to social definition, change, and process (for example, see Jones 1980; Littler 1982; Adler 1983; or DiPrete 1986). For example, in the steel and automobile industries, what comes to be defined as "skilled" in relation to job titles, classifications and hierarchies is some complicated mixture of the technical features of tasks, the past and present of union-management negotiations over job classification systems, and even day-to-day politics of performance-norm definition, worker-supervisor interactions, and other group dynamics. What workers define as skill and what researchers see and designate as skill in part depends on these relationships.

Another example of the conditioning role of organizational culture can be found in the studies of computerized typesetting in the printing industry. One of the major studies of the industry by Wallace and Kalleberg (1982) (see Table 4–2 above) documents the deskilling of hand compositor and typesetter occupations in the industry over the last forty years, as indicated by indirect wage comparisons. At first reflection, one would conclude that the consequences for workers in this industry were tragic. Nonetheless, a longitudinal study by Hull, Friedman, and Rodgers (1982) showed a group of printers from three large New York City newspapers experienced substantial upgrading in their subsequent work careers, even in the face of deskilling of their earlier jobs. Several years after the technological change, a majority of printers reported more expanded and challenging jobs, along with higher income and job satisfaction. What occurred? The severance package obtained by the printers and their union contained substantial benefits, including resources for retraining and a relatively long time period to accomplish the transition. This is not to suggest the printers' fate is the modal outcome for U.S. workers; tragically it is not. However, a study focusing on the original technological change and the affected occupations in this case would reach one conclusion; a study that followed the affected workers would reach another conclusion. Relationships defined by organizational cultures differentiate the two. Thus, organizational cultures provide an additional source of uncertainty, particularly if the focus shifts from affected jobs or occupations to affected workers.

EDUCATION AND TRAINING POLICY IN THE FACE OF UNCERTAINTY

The central theme of this essay has been uncertainty: uncertainty because of inadequacies in the knowledge base, and in what we do know, an uncertain relationship between technology and the upgrading or downgrading of occupations. Many other things come into play that are difficult to predict. Just as there is no single, simple answer to the question of technology and skill requirements, there is no single, simple answer to the question of an optimal education and training policy in the face of uncertainty. And just as there are strong popular beliefs about technology and work, there are strong beliefs about the role of education and training in relation to skill change.

For some, education and training operate as leading institutions that anticipate and even modify the skill requirements of jobs (Levin and Rumberger 1987). For others, education operates as a "trailing" institution that follows developments from other corners of society and has little to do with the skills needed for work. In between are those who suggest education and training systems slowly adjust to the requirements of the economy, affecting demand little in the short-term but more substantially over the long-term (see Rubison and Ralph 1984 for review). In recent years, the policy counsel is frequent and varied on the topic of technology, and education and training. For example, some cite mathematics and science education in Japan as the chief factor fostering industrial robotics and suggest mathematics and science education as a top policy priority (Lynn 1983). I know of no direct tests of the causal link. Ferman (1983) suggests a restructuring of all levels of community education to accomplish a closer link between new technologies in factories and training for work. Groff (1983) and others counsel bringing state-of-the-art technology directly into the curriculum and into methods of instruction. Werneke (1983) suggests more general skills that enhance adaptability to change;

others counsel training in more specific skills linked to specific technologies. The calls for computer literacy at all levels of education abound (see King 1985). And as if a diversity of counsel were not sufficient, a succession of national panels in the last five years has resoundingly indited first, elementary education, second, high school or secondary education, and most recently postsecondary education, particularly colleges and universities. In each of these reports, there is both direct assertion and indirect suggestion that education and training—at the same time—are in a desperate failing situation, on the one hand, and on the other hand can or will play a crucial role in national productivity, technical innovation, adaptation to increasing rates of social change, and so on. These are but a sampling of the views.

Many of the policy recommendations assume some knowledge of the relationship between technological change and occupational skill transformations. Based on this review such recommendations are on uncertain ground. A difficult and complicated enterprise to begin with—public and private policy on education and training— becomes even more tenuous in the face of uncertainty. Rather than a hopeless case, though, education and training policy can attempt to take the uncertainty into account (versus ignoring it or assuming its nonexistence). It is here that the scientific literature can sharpen some of the policy issues and options.

Limitations and Possibilities

Education and training operate in different ways according to upgrading and downgrading perspectives. In the upgrading tradition, the relationship between education and jobs is a functional one: Schooling and training impart the general and specific skills required by the economy in general and by jobs in particular. Further, schooling and training are forms of human capital, subject to appreciation and reward in the workplace. Proponents would cite the broad range of relationships between work outcomes and schooling levels (Becker 1964; Cain 1976). In downgrading perspectives, schooling bears little instrumental relation to job performance or productivity. Proponents would cite weak relationships between schooling and employee performance within occupations, and the massive growth of education and training in the face of stable skill requirements, fluctuations in productivity, and related phenomena (overeducation, underemployment) (Berg 1970; for review consult Rubinson and Ralph 1984). Rather, ours is a credential society where education and training provide the keys that open employment doors (Collins 1979). Further, education socializes workers into the values of the U.S. workplace and more often reinforces existing inequalities rather than alleviating them (Bowles and Gintis 1976). Most downgrading theorists would assign a lesser role to on-the-job training, vocational training, and other supply-side features, instead directing our attention to larger demand-side operations at the level of firms and the economy.

The evidence suggests schooling and training operate somewhat in line with both perspectives. That is, workers with more education and training are advantaged in the labor force, types of jobs, earnings, stable work histories, job satisfaction, and so on. In part, the advantages flow from the general and specific skills that students acquire in schools and training programs; in part the advantages flow from credentialing effects, the values that students are assumed to acquire or hold with higher education and training, and locations in job access queues, where education gives people greater access to jobs but does not greatly affect job performances (Thurow 1975). Education and training are cited as cures for many contemporary social ills: to arrest productivity declines, to provide the skills for high technology, to provide for job satisfaction, income, and personal growth, to remedy the dislocations of structural unemployment, and to assure equality of opportunity and the optimal use of human resources, to cite a few. Yet the research on education and training suggests important limits on its role as a solution to these and other problems.

First, if education and training modify the relationship between technology and skill requirements it is over the longer term. The ability of a well-educated population to innovate and solve its problems with technology is constrained in the short-term and subject to a host of limiting economic relationships, time-lags in adoption and diffusion, an existing physical plant structure and capital base. *Changes* in education and training systems that affect a small fraction of the population might take an even longer time to exhibit even modest effects on technological innovation in the economy.

Second, education and various forms of training bear an imperfect relationship to job acquisition and performance (Granovetter 1981; Thurow 1975). If anything, pre–labor force education and training better predict access to jobs than job performance. There is little evidence to suggest education and training differentiate workers in the same occupation in terms of productivity (Berg 1970).

Third, the quality of the match between workers' capacities and skills and job requirements is modest at best (Berg et al. 1981). The available pool of workers with sufficient skills is far less problematic than the process that matches workers with given mixtures and levels of skills to jobs with given skill requirements. Investments in better matching of workers to jobs might yield far greater return than across-the-board increases in schooling and training. For example, many workers are unwilling to migrate or are unaware of available jobs, how to find jobs, or how to interview for a job; many jobs, perhaps one-half or more, are obtained in particularistic fashion (relatives, friends, inside recruiting, and tips) (Granovetter 1974; U.S. Department of Labor 1976).

Finally, managers and organizations operate in a far less rational manner than suggested by textbook images. For example, the skill capacities of existing employees of a firm often guide decisions about technical innovations—versus skills estimated to be available from a larger local or regional labor pool or schools. The rational acquisition of schooling and training are of limited value if the labor market and organizations do not provide for its rational use.

Prelabor Force Education and Training

Education and training bear modest relationships to job acquisition and work outcomes. They will continue to be part of the skill acquisition and employment equations, if not in reality then in the minds of students and employers. There is nothing in this review thus far that would suggest a dramatically increased or decreased role for education and training. The modest levels of skill change, the mixtures of skill change, the forms of uncertainty including the role of modifying factors (managers, markets, and organizational cultures) provide the reasons for this conclusion. If a case for dramatic change in the general or vocational education enterprises were to be made, then the arguments and evidence must come from a domain broader than technology and skill requirements. There are perhaps two exceptions.

The first exception involves literacy. The national investment in bringing all of the population to a minimal level of literacy, with basic high school levels of language, reasoning, and mathematical skill, is worthwhile because those without these skills will be left out of the labor market entirely or restricted to the most menial jobs. If technology makes increasing or changing demands on workers, for example, in terms of added specific skills, then these workers will least be able to compete and survive in their jobs in the absence of basic literacy skills. At this most elementary level, workers of today need basic literacy skills, such as the ability to read, whereas a large fraction of the labor force of eighty to a hundred years ago (that is, those in agricultural settings) could easily survive without such skills.

The strongest case for more postsecondary and vocational education comes from those studies that show more educated workers are better able to adapt to technological change (in their current or next jobs) and are less likely to experience as many or as serious adverse consequences of technological change (Hull et al. 1982; Jaffe and Froomkin 1968; Mueller et al. 1969; Office of Technology Assessment 1986). In terms of policy, this means more schooling for the lowest one-third or one-half of the schooling distribution, versus the top half of the distribution, which might increase this form of inequality. The adaptability advantage seems to derive in part from having additional job-relevant skills that can be translated into a new job and from the more general skills such as cognitive ability, intellectual flexibility, and problem-solving capacities because these are obtained in general schooling (versus vocational) and might apply to rapid adjustment to new technology or a greater capacity to quickly switch careers.

Some of the policy questions center on specific versus general skills. In the face of uncertainty, it makes no sense to stress one to the exclusion of the other. Further, reduced reliance on a single vocational skill makes sense for several reasons. U.S. workers have more complicated careers now compared with fifty or a hundred years ago. Where a single vocational skill might suffice then, it may not now. For example, high school students of the mid- and late 1960s averaged over four different full-time jobs for men, and over three for women, before age thirty (Spenner, Otto, and Call 1982).

European societies provide one possible model, where vocational students often prepare in the skills of more than one field. In this sense, given the uncertainties, more specific skills seem a reasonable alternative, particularly for certain sectors of the economy.

Finally, the review of changing skill requirements should temper some of the enthusiasm for specific skills in computer literacy. The effects of computers on work, particularly clerical work, involve mixtures of upgrading and downgrading (Menzies 1981; Werneke 1983; Webster 1986; Hartmann, Kraut, and Tilly 1986). Many jobs do not directly use or require computer literacy skills; many jobs that do require them frequently change systems so that the ability to learn a new system far outweighs knowledge of any single system, particularly at the lower levels likely to be taught in elementary and secondary education. A large fraction of jobs that use video display terminals and related systems involve less complicated skills that are rapidly acquired with on-the-job training. The impacts of the computer on our lives and our jobs are substantial but substantially indirect. Moreover, the direct impacts on our jobs and lives have been overemphasized, particularly in the short term (Brooks 1983).

National versus Local Policies

In some ways, a coherent, clear national policy on education and training seems wise—for example, in standards for training or in uniform policies for notifying workers of displacement and providing retraining benefits. One way in which a national policy does not make sense involves local and regional variations in technological changes that depart from the national average. The research evidence shows such variations—for example, in the location of robotics innovations. Planning and policy should allow substantial room for local and regional variations and should actively and periodically monitor such variations.

Structural Unemployment and Retraining

Recent studies document the tremendous diversity of layoff notification, severance packages, and retraining benefits for workers displaced by technological change, foreign competition, and so on (Office of Technology Assessment 1986). Some of the discussion has cast these issues in terms of education and training policy. The real issues are not whether to give workers advance notice of layoffs or whether to provide retraining. If research can inform these questions, the answers are clear: The more advance warning the better in terms of minimizing adverse consequences and maximizing positive adjustment (psychologically and in terms of work) (for example, see Hull et al. 1982); and workers who have the time and resources for retraining fare better than those who do not (Office of Technology Assessment 1986). Rather, the issues and debate are more moral, ethical, and economic than education and training. Can firms economically afford to provide advance warning of layoffs, as is custom and law in a number of European countries (and one or two states)? Can firms morally afford not to provide such notice? And who will pay for the education and retraining? Can the society (business, government) afford such? Can it afford not to? No less important is the case for research on technology where social considerations play a central role, for the consequences of technology policies need to be informed by both technical and social considerations.

Curriculum Planning in the Face of Uncertainty

Building a vocational curriculum around the skill requirements of a specific technology involves large uncertainties. The collective research evidence suggests that the effects of technology on skill requirements are governed as much or more by aggregate demand, macroeconomic policy, and managerial discretion compared with the intrinsic features of the technology. Curricula based solely on the technical side are subject to greater uncertainties than curricula based on the technical *and* broader social implications of a technology on skill requirements, however difficult that might be. Further, few vocational programs can use approximate aggregate stability as a basis for curriculum planning. The markets for vocational skills and graduates are typically local and regional, not national. A specific skill in low or declining demand nationally may be in high demand in the recruitment area of a local vocational program (or vice versa). For example, even under the worst-case predictions for manufacturing employment, this sector will still employ around one-fifth of the nation's workers for the foreseeable future (Etzioni and Jargowsky 1984). Wholesale and large-scale abandonment of training areas in perceived decline does not make sense because the replacement needs in some declining industries will provide more jobs than some of the high-growth industries that start with a much lower base.

Curriculum planning in the face of uncertainty should include several features. Overreliance on single curriculum programs, unless there is strong justification, seems unwise. Without spreading the curricula too thin, there is strength in diversification of vocational programs and curricula, if for no other reason than it cuts the loss in the event of a major plant dislocation or change in the fortunes of a particular industry.

Further, the uncertainty over skill requirements suggests periodic review by vocational and educational policy analysts and planners along several lines: (1) What are the occupational destinations of local and regional program graduates? (2) What is the quality of the match between graduates' skill training and the skill requirements of their jobs? (3) What are the likely local and regional changes in the demand for labor in specific occupations and possible major changes in the quality of the local/regional job structure? Given uncertainty, some policies will be better and others worse; but given knowledge of uncertainty, there is no excuse not to diversify and periodically monitor and adjust to change. In short, the best strategies in the face of uncertainty and lack of knowledge about skill transformations include curriculum diversification and periodic review aimed at eliminating programs no longer in demand and creating ones where demand arises.

SUMMARY

The debate over the relationship between technology and work is long-standing and comprises a substantial number of studies in several scientific disciplines. The major positions in the debate—the upgrading, downgrading, and mixed-change or conditional arguments—have failed to receive a clear verdict from the collective body of evidence. There are a number of reasons for the uncertainty.

Empirical studies rely on aggregate and case study designs to assess the consequences of technological change for skill variations in work. Each type of design has strengths and weaknesses and tends to reach different conclusions. Comprehensive judgments require assessments of skill change in terms of compositional shifts in jobs and content changes in the nature of work. Different theoretical approaches and different types of designs have afforded uneven coverage to changes in skill requirements along these two tracks. The empirical studies do suggest mixtures of upgrading and downgrading, offsetting and contradictory trends for some occupations, industries, technologies, and sectors of the economy compared with others. The available data base is far short of a complete temporal and spatial sampling for technologies or jobs in the U.S. economy, even for the twentieth century.

The concepts and measures of skill in the research literature are uneven and poor. An optimistic reading of the empirical literature suggests two basic dimensions of skill: substantive complexity and autonomy-control. One interpretation of the evidence suggests the possibility of contradictory skill shifts over the course of this century: In the aggregate, a slow and gradual upgrading of the skill requirements of jobs in terms of their substantive complexity is counterbalanced by a small decline in the autonomy-control of jobs. This hypothesis awaits comprehensive test. A pessimistic reading of the empirical studies simply concludes that no conclusion is possible because of inadequate samples, concepts, and measures, threats to the validity of inferences based on the analysis and design, and other forms of noncomparability across studies.

Given the limitations, a review of select aggregate and case studies shows several provisional conclusions. The skill requirements of jobs in the U.S. economy are changing. The rate of change over the course of this century is slow for the labor force taken as a whole but particular occupations, industries, and sectors of the economy have experienced more substantial shifts in skill requirements. Compositional shifts appear to account for relatively more of the upgrading of skill requirements, particularly for substantive complexity; content shifts appear to account for more of the downgrading, particularly changes in the autonomy-control of jobs. Of far greater importance than these trends are the roles played by managers, market forces, and organizational cultures in conditioning the effects of technology on the number and quality of jobs. The empirical literature provides a number of illustrations of the same technology affecting similar jobs in opposite ways conditional on market forces, managerial prerogatives in how or when to implement a technology, and the organizational cultures in which technological changes occur. Several recent arguments suggest a changing relationship between technological change and the skill requirements of work, postulating new ways in which technology affects work, particularly current and future effects on the mental parts of work. An evaluation shows this thesis is plausible but has not yet been rigorously tested.

Perhaps more important than the intrinsic effects of a technology on skill requirements are the modifying role of managers (that is, in how and when to implement a technological change), markets (modifying or offsetting the effects of a technological change), and organizational cultures (also modifying or offsetting a technological change). Future research and policy deliberations should devote much more attention to the role of these modifying factors.

Finally, just as uncertainty characterizes the relationship between technological change and what we know about the relationship, education and training policy in relation to technological change faces considerable uncertainty. Education and training as policy vehicles are limited in many ways—for example, in the lag between schooling and greater productivity, the loose linkage between skills of workers and skill demands of jobs, and the uneven and uncertain responses of managers and firms to uncertainty. Arguments for increased education hinge primarily on bringing all of the population to minimal levels of literacy and additional schooling for the less well educated because education and training appear to minimize the adverse consequences of technological change for workers, and workers with such training and education adapt to change better and quicker. The research evidence suggests the need for education and training policies that are sensitive to local and regional variations in labor demand and technological change and for periodic monitoring of the available supply and demand for education and training levels, along with monitoring the quality of match of local graduates to available jobs and changing job conditions. Curriculum planning in the face of uncertainty should rely on curriculum diversification and periodic review aimed at adjusting available programs to demand. Thus, within modest limits, the research on technological change and skill requirements informs education and training policy, and within modest limits, education and training can be expected to inform and solve human problems associated with technological change.

NOTES

Support for this research was provided by the Panel on Technology and Employment, National Academy of Sciences, and by a grant from the University Research Council, Duke University. I gratefully acknowledge the assistance of Anita Hume, Michele Wurtzel, Claudette Parker, and Amby Boatright.

1. Portions and earlier versions of this section were originally presented in Spenner (1983, 1985).
2. For arguments and evidence, consult Baron and Bielby (1982), Braverman (1974), Bright (1958), Cain and Treiman (1981), Field (1980), Gottfredson (1984), Hunter and Manley (1982), Karasek, Schwartz, and Pieper (1982), Kohn and Schooler (1983), Littler (1982), Rumberger (1984), Spaeth (1979), and Spenner (1979, 1980, 1983, 1985).

REFERENCES

Adler, P. 1983. "Rethinking the Skill Requirements of New Technologies." Working Paper HBS 84–27. Cambridge, Mass.: Graduate School of Business Administration, Harvard University.

———. Forthcoming. "Does Automation Raise Skill Requirements? What Again?" In *Microelectronics and the Future*, edited by R. Gordon. Norwood, N.J.: Ablex Publishing.

Albin, P. S., F. Z. Hormozi, S. L. Mourgos, and A. Weinberg. 1984. "An Information System Approach to the Analysis of Job Design." In *Management and Office Information Systems*, edited by S. Chang, pp. 385–400. New York: Plenum.

Baron, J. N., and W. T. Bielby. 1982. "Workers and Machines: Dimensions and Determinants of Technical Relations in the Workplace." *American Sociological Review* 47:175–188.

Becker, G. 1964. *Human Capital.* New York: National Bureau of Economic Research.

Bell, D. 1973. *The Coming of Post-Industrial Society.* New York: Basic Books.

Berg, I. 1970. *Education and Jobs: The Great Training Robbery.* New York: Praeger.

SKILLS AND THE DISTRIBUTION OF EARNINGS AND INCOME

Berg, I., R. Bibb, T. A. Finegan, and M. Swafford. 1981. "Toward Model
    Specification in the Structural Unemployment Thesis: Issues and
    Prospects." In *Sociological Perspectives on Labor Markets*, edited by I.
    Berg, pp. 347–367. New York: Academic Press.
Berg, I., M. Freedman, and M. Freedman. 1978. *Managers and Work Re-
    form*. New York: Free Press.
Blauner, R. 1964. *Alienation and Freedom: The Factory Worker and His Indus-
    try*. Chicago: University of Chicago Press.
Bluestone, B., and B. Harrison. 1982. *The Deindustrialization of America*.
    New York: Basic Books.
Bowles, S., and H. Gintis. 1976. *Schooling in Capitalist America*. New York:
    Basic Books.
Braverman, H. 1974. *Labor and Monopoly Capital: The Degradation of Work
    in the Twentieth Century*. New York: Monthly Review Press.
Bright, J. R. 1958. "Does Automation Raise Skill Requirements?" *Har-
    vard Business Review* 36:84–98.
————. 1966. "The Relationship of Increasing Automation and Skill Re-
    quirements." In National Commission on Technology, Automation,
    and Economic Progress, *The Employment Impact of Technological Change*.
    Appendix Vol. 2 of *Technology and the American Economy*, pp. 203–222.
    Washington, D.C.: U.S. G.P.O.
Brooks, H. 1983. "Technology, Competition, and Employment." Annals
    of the American Economy. *Annals of the American Academy of Political
    and Social Science* 470:115–122.
Burawoy, M. 1979. *Manufacturing Consent*. Chicago: University of Chi-
    cago Press.
Cain, G. G. 1976. "The Challenge of Segmented Labor Market Theories
    to Orthodox Theory." *Journal of Economic Literature* 14:1215–1257.
Cain, P., and D. J. Treiman. 1981. "The Dictionary of Occupational
    Titles as a Source of Occupational Data." *American Sociological Review*
    46:253–278.
Chirot, D. 1986. *Social Change in the Modern Era*. San Diego, Calif.: Har-
    court Brace Javanovich.
Clogg, C. C. 1979. *Measuring Underemployment*. New York: Academic
    Press.
Clogg, C. C., and J. W. Shockey. 1984. "Mismatch Between Occupation
    and Schooling: A Prevalence Measure, Recent Trends and Demo-
    graphic Analysis." *Demography* 2:235–257.
Collins, R. 1979. *The Credential Society: An Historical Sociology of Education
    and Stratification*. New York: Academic Press.
Crossman, E. R. F. 1960. *Automation and Skill*. London: HMSO.
Davis, L. E., and J. C. Taylor. 1976. "Technology, Organization, and Job
    Structure." In *Handbook of Work, Organization and Society*, edited by R.
    Dubin, pp. 379–419. Chicago: Rand-McNally.

DiPrete, T. A. 1986. "The Upgrading and Downgrading of Occupations: Status Redefinition vs. Deskilling as Alternative Theories of Change." Unpublished manuscript, Department of Sociology, The University of Chicago.

Dubnoff, S. 1978. "Inter-occupational Shifts and Changes in the Quality of Work in the American Economy, 1900–1970." Paper presented at the annual meeting of the Society for the Study of Social Problems, San Francisco, August 17–20.

Eckaus, R. 1964. "The Economic Criteria for Education and Training." *Review of Economics and Statistics* 41:181–190.

Edwards, R. C. 1979. *Contested Terrain: The Transformation of the Workplace in the Twentieth Century.* New York: Basic Books.

Etzioni, A., and P. Jargowsky. 1984. "High Tech, Basic Industry and the Future of the American Economy." *Human Resource Management* 23:-229–240.

Faunce, W. A. 1958. "Automation in the Automobile Industry." *American Sociological Review* 23:401–407.

Ferman, L. A. 1983. "The Unmanned Factory and the Community." *Annals of the American Academy of Political and Social Science* 470:136–145.

Field, A. J. 1980. "Industrialization and Skill Intensity: The Case of Massachusetts." *Journal of Human Resources* 15:149–175.

Fuchs, V. R. 1968. *The Service Economy.* New York: National Bureau of Economic Research.

Gallie, D. 1978. *In Search of the New Working Class: Automation and Social Integration Within the Capitalist Enterprise.* Cambridge, England: Cambridge University Press.

Ginzberg, E. 1982. "The Mechanization of Work." *Scientific American* 247:67–75.

Glenn, E. N., and R. L. Feldberg. 1979. "Proletarianizing Clerical Work: Technology and Organizational Control in the Office." In *Case Studies on the Labor Process,* edited by A. Zimbalist, pp. 51–72. New York: Monthly Review Press.

Gottfredson, L. S. 1984. "The Role of Intelligence and Education in the Division of Labor." Report No. 355. Baltimore, Md.: The Johns Hopkins University, Center for Social Organization of the Schools.

Granovetter, M. S. 1974. *Getting a Job: A Study of Contacts and Careers.* Cambridge, Mass.: Harvard University Press.

———. 1981. "Toward a Sociological Theory of Income Differences." In *Sociological Perspectives on Labor Markets,* edited by I. Berg, pp. 11–47. New York: Academic Press.

Groff, W. H. 1983. "Impacts of the High Technologies on Vocational and Technical Education." *Annals of the American Academy of Political and Social Science* 470:81–94.

Hartmann, H. I., R. E. Kraut, and L. A. Tilly. 1986. *Computer Chips and*

Paper Clips: Technology and Women's Employment, Volume 1. Washington, D.C.: National Academy Press.

Hirschhorn, L. 1984. *Beyond Mechanization: Work and Technology in a Postindustrial Age.* Cambridge, Mass.: MIT Press.

Horowitz, M., and I. Herrnstadt. 1966. "Changes in Skill Requirements of Occupations in Selected Industries." In National Commission on Technology, Automation, and Economic Progress, *The Employment Impact of Technological Change.* Appendix Vol. 2 of *Technology and the American Economy* Washington, D.C.: U.S. G.P.O.

Hull, F. M., N. S. Friedman, and T. F. Rodgers. 1982. "The Effect of Technology on Alienation from Work." *Sociology of Work and Occupations* 9:31–57.

Hunter, A., and M. Manley. 1982. "The Task Requirements of Occupations." Paper presented at the annual meeting of the Canadian Sociology and Anthropology Association, Ottawa, Canada,

Jaffe, A. J., and J. Froomkin. 1968. *Technology and Jobs: Automation in Perspective.* New York: Praeger.

Jaikumar, R. 1986. "Postindustrial Manufacturing." *Harvard Business Review* November–December:69–76.

Jones, F. F. 1980. "Skill as a Dimension of Occupational Classification." *Canadian Review of Sociology and Anthropology* 17:176–183.

Kalleberg, A. L., and K. T. Leicht. Forthcoming. Jobs and skills: A multivariate structural approach. *Social Science Research.*

Karasek, R., J. Schwartz, and C. Pieper. 1982. "A Job Characteristics Scoring System for Occupational Analysis." Unpublished manuscript, Center for the Social Sciences, Columbia University.

Kelley, M. R. 1986. "Programmable Automation and the Skill Question: A Reinterpretation of the Cross-National Evidence." *Human Systems Management* 6:223–241.

Kerr, C., J. T. Dunlop, C. Harbison, and C. A. Myers. 1964. *Industrialism and Industrial Man.* New York: Oxford University Press.

King, K. M. 1985. "Evolution of the Concept of Computer Literary." *EduCom Bulletin* 20:18–21.

Kohn, M. A. 1971. "Bureaucratic Man: A Portrait and an Interpretation." *American Sociological Review* 36:461–474. .

Kohn, M. L., and C. Schooler. 1983. *Work and Personality: An Inquiry into the Impact of Social Stratification.* Norwood, N.J.: Ablex Publishing.

Kraft, P. 1977. *Programmers and Managers: The Routinization of Computer Programming in the United States.* New York: Springer-Verlag.

Levin, H., and R. Rumberger. 1983. "The Low-Skill Future of High-Tech." *Technology Review,* 86:18–21.

———. Forthcoming. "Educational Requirements for New Technologies: Visions, Possibilities and Current Realities." *Educational Policy.*

Littler, C. R. 1982. *The Development of the Labour Process in Capitalist Societies: A Comparative Study of the Transformation of Work Organization in Britain, Japan and the U.S.A.* London: Heinemann.

Lynn, L., 1983. "Japanese Robotics: Challenge and—Limited—Exemplar." *Annals of the American Academy of Political and Social Science* 470:-16–27.

McCormick, E. J., R. C. Mecham, and P. R. Jeanneret. 1977. *Technical Manual for the Position Analysis Questionnaire.* Lafayette, Ind.: Purdue University.

McLaughlin, D. B. 1983. "Electronics and the Future of Work: The Impact on Pink and White Collar Workers." *Annals of the American Academy of Political and Social Science* 470:152–162.

Menzies, H. 1981. *Women and the Chip: Case Studies on the Effects of Informatics on Employment in Canada.* Montreal: Institute for Research on Public Policy.

Miller, A., D. J. Treiman, P. S. Cain, and P. A. Roos. 1980. *Work, Jobs, and Occupations: A Critical Review of the "Dictionary of Occupational Titles."* Washington, D. C.: National Academy Press.

Miller, R. J., ed. 1983. "Robotics: Future Factories, Future Workers." Special issue of *Annals of the American Academy of Political and Social Science* (November).

Mueller, E., J. Hybels, J. Schmiedeskamp, J. Sonquist, and C. Staelin. 1968. *Technological Advance in an Expanding Economy: Its Impact on a Cross-Section of the Labor Force.* Ann Arbor, Mich.: Survey Research Center.

National Commission on Technology, Automation, and Economic Progress. 1966. *Technology and the American Economy.* 2 vols. Washington, D.C.: U.S. G.P.O.

National Resources Committee. 1937. *Technological Trends and National Policy.* Washington, D.C.: U.S. G.P.O.

Oakley, K. P. 1954. "Skill as a Human Possession." In *A History of Technology,* Vol. 1, edited by C. Singer, E. J. Holmyard, and A. R. Hall, pp. 2–3. New York: Oxford University Press.

Office of Technology Assessment. 1986. *Technology and Structural Unemployment: Reemploying Displaced Adults.* Washington, D.C.: U.S. G.P.O.

Osterman, P. 1986. "The Impact of Computers on the Employment of Clerks and Managers." *Industrial and Labor Relations Review* 39:175–186.

Ouichi, W. G., and A. L. Wilkins. 1985. "Organizational Culture." In *Annual Review of Sociology,* Vol. 11, edited by R. Turner and J. Short, pp. 457–483. Palo Alto, Calif.: Annual Reviews, Inc.

Pfeffer, J. 1985. "Organizations and Organization Theory." In *The Handbook of Social Psychology,* Vol. 1, 3d edition, edited by G. Lindzey and F. Aronson, pp. 370–440. New York: Random House.

Piore, M. J., and C. F. Sabel. 1984. *The Second Industrial Divide*. New York: Basic Books.

Rawlins, V., and L. Ulman. 1974. "The Utilization of College Trained Manpower in the United States." In *Higher Education and the Labor Market*, edited by M. Gordon, pp. 195–235. New York: McGraw-Hill.

Rubison, R., and J. Ralph. 1984. "Technical Change and the Expansion of Schooling in the United States, 1890–1970." *Sociology of Education* 57:134–152.

Rumberger, R. W. 1981. *Overeducation in the U.S. Labor Market*. New York: Praeger.

———. 1983. "A Conceptual Framework for Analyzing Work Skills." Project Report No. 83–A8. Stanford, Calif.: Institute for Research on Educational Finance and Governance.

———. 1984. "High Technology and Job Loss." *Technology in Society* 6:263–284.

———. Forthcoming. "The Potential Impact of Technology on the Skill Requirements of Future Jobs." In *The Future Impact of Technology on Work and Education*, edited by G. Burke and R. Rumberger. Philadelphia: Falmer Press.

Rumberger, R. W., and M. Levin. 1985. "Forecasting the Impact of New Technologies on the Future Job Market." *Technological Forecasting and Social Change* 27:399–417.

Scott, J. W. 1983. "The Mechanization of Women's Work." *Scientific American* 247:166–187.

Shaiken, H. 1984. *Work Transformed: Automation and Labor in the Computer Age*. Lexington, Mass.: Lexington Books.

Singlemann, J., and M. Tienda. 1985. "The Process of Occupational Change in a Service Society: The Case of the United States, 1960–1980." In *New Approaches to Economic Life: Economic Restructuring, Unemployment and the Social Division of Labor*, edited by B. Roberts, F. Finnegan, and D. Gallie, pp.   . Manchester, England: University of Manchester.

Smith, H. L. 1986. "Overeducation and Underemployment: An Agnostic Review." *Sociology of Education* 59:85–99.

Sobel, R. 1982. "White Collar Structure and Class: Educated Labor Re-Evaluated." Ph.D. dissertation, School of Education, University of Massachusetts, Amherst.

Spaeth, J. 1979. "Vertical Differentiation Among Occupations." *American Sociological Review* 44:746–762.

Spenner, K. I. 1979. "Temporal Changes in Work Content." *American Sociological Review* 44:968–975.

———. 1980. "Occupational Characteristics and Classification Systems:

New Uses of the *Dictionary of Occupational Titles* in Social Research." *Sociological Methods and Research* 9:239–264.

———. 1982. "Temporal Changes in the Skill Levels of Work: Issues of Concept, Method, and Comparison." Paper presented at the 10th World Congress of Sociology, Mexico City, Mexico, August 10–17.

———. 1983. "Deciphering Prometheus: Temporal Change in the Skill Level of Work." *American Sociological Review* 48:824–837.

———. 1985. "The Upgrading and Downgrading of Occupations: Issues, Evidence, and Implications for Education." *Review of Educational Research* 55:125–154.

———. Forthcoming. "Occupations, Work Settings and the Course of Adult Development: Tracing the Implications of Select Historical Changes." In *Life-Span Development and Behavior*, Vol. 8, edited by P. Baltes, D. Featherman, and R. Lerner. Hillsdale, N.J.: Lawrence Earlbaum Associates.

Spenner, K. I., L. B. Otto, and V. R. A. Call. 1982. *Career Lines and Careers.* Lexington, Mass.: Lexington Books.

Standing, G. 1984. "The Notion of Technological Unemployment." *International Labour Review* 123:127–147.

Stark, D. 1974. "Class Struggle and the Transformation of the Labor Process: A Relational Approach." *Theory and Society* 9:89–130.

Stone, K. 1974. "The Origins of Job Structures in the Steel Industry." *Review of Radical Political Economics* 6:61–97.

Thurow, L. C. 1975. *Generating Inequality: Mechanisms of Distribution in the U.S. Economy.* New York: Basic Books.

United States Department of Labor. 1949. *Dictionary of Occupational Titles.* Washington, D.C.: U.S. G.P.O.

———. 1965. *Dictionary of Occupational Titles,* 3d edition. Washington, D.C.: U.S. G.P.O.

———. 1972. *Handbook for Analyzing Jobs.* Washington, D.C.: U.S. G.P.O.

———. 1976. *Recruitment, Job Search, and the United States Employment Service.* Research and Development Monograph 43. Washington, D.C.: U.S. G.P.O.

———. 1977. *Dictionary of Occupational Titles,* 4th edition. Washington, D.C.: U.S. G.P.O.

Vallas, S. Forthcoming. "The Implications of New Technology for Work and Alienation: The Communications Industry as a Critical Case." *Work and Occupations.*

Wallace, M., and A. L. Kalleberg. 1982. "Industrial Transformation and the Decline of Craft: The Decomposition of Skill in the Printing Industry, 1931–1978." *American Sociological Review* 47:07–324.

Webster, J. 1986. "The Impact of Dedicated Word Processors on Office

Labour in the UK." Paper presented to the 11th World Congress of Sociology, New Delhi, India, August 16–22.

Werneke, D. 1983. *Microelectronics and Office Jobs: The Impact of the Chip on Women's Employment.* Geneva, Switzerland: International Labour Office.

Wright, E. O., and J. Singlemann. 1982. "Proletarianization in the Changing American Class Structure." In *Marxist Inquiries,* edited by M. Burawoy and T. Skoopol, pp. 176–209. Chicago: University of Chicago Press.

5 TECHNOLOGY AND SKILLS
Lessons from the Military

Martin Binkin

Advances in technology have had a dramatic influence on all sectors of U.S. society.[1] Through developments in such diverse fields as materials and processing, biomedicine, energy conversion, and especially information and communications, technology has become the linchpin of contemporary U.S. commercial and industrial activity. Yet the implications of these changes for the nature of work in general and, more specifically, for the skill requirements of affected occupations remain a subject of debate. Important questions regarding the desirable attributes of jobholders in a high-tech environment have gone unresolved, with obvious ramifications for educational researchers and policymakers faced with the task of devising curricula to meet the needs of a changing occupational structure.[2]

At the heart of the debate is the issue of whether changes in technology foster the need for a broader or a narrower variety of skills or for a higher or a lower average skill. The purpose here is not to settle the question, but rather to contribute to a better understanding of the issues involved by drawing on the relevant experiences of one of the nation's principal—and earliest—consumers of technology: the armed forces.

TRENDS IN MILITARY TECHNOLOGY

Technological change since World War II has profoundly affected the nation's military establishment. Indeed, during the postwar period research and development for military purposes often have been forerunners of U.S. commercial and industrial applications, at times serving to accelerate the overall pace of technological substitution. Today, the armed forces of the United States undeniably enjoy a technological edge over the military forces of any other world power. As the weapons of war have become more sophisticated, so too has the military's roster of jobs.

Past Trends

World War II marked the first extensive uses of armor and tactical aircraft in land warfare, which expanded the fields of battle and the scope of military maneuvers. Aircraft and submarines replaced the ships of the line as key naval weapons. Tactical aircraft were employed in a variety of roles: as interceptors, fighter bombers, and long-range fighter escorts to accompany bombers. These new weapon systems had a substantial effect on the military's occupational structure. By the end of the war only 39 percent of U.S. Army and 34 percent of U.S. Marine Corps enlisted jobs were classified as "ground combat," and fewer than half of these were infantry billets. The remaining personnel were assigned to a wide range of support jobs, which in addition to the traditional services and craft occupations included specialists associated with communication, radar, and fire control equipment; medical and dental technicians; surveyors and draftsmen, photographers, weather observers, intelligence analysts, and a host of other specialized personnel. Thus the military establishment that emerged from World War II was substantially more industrialized than its prewar counterpart, and the distinction between many military and civilian jobs began to blur.[3]

Along with aeronautics and submarines, the major technologies to emerge from World War II were nuclear weapons and modern electronics. Electronics technology initially provided better means for target detection and tracking through radar, sonar, and infrared techniques and later, combined with parallel advances in aero-dynamics and propulsion and supported by developments in materials, chemistry, and structures technologies, gave the United States a sizable lead in guided weapons. More recently, dramatic advances in computer-based command, control, and communications systems and finally, the introduction of space technology have influenced the structure of the armed forces and the skills that the military requires.

The evolution in the mix of military jobs in the postwar period can be traced in Table 5–1, which presents the distribution of trained enlisted personnel by occupational category for selected

Table 5–1.   Distribution of Trained Military Enlisted Personnel, by Occupational Area, Selected Years, 1945–85 (percentage).

Occupational area[a]	1945	1957	1963	1973	1985
White-collar	28%	40%	42%	46%	47%
Technical workers[b]	13	21	22	25	29
Electronics	6	13	14	18	21
Other	7	8	8	7	8
Administrative and clerical workers[c]	15	19	20	20	18
Blue-collar	72[d]	60	58	54	53
Craftsmen[e]	29	32	32	28	27
Service and supply workers	17	13	12	13	10
General military skills, including combat	24	15	14	13	16

Sources: Data for 1945, 1957, and 1963 from Harold Wool, *The Military Specialist: Skilled Manpower for the Armed Forces* (Baltimore, Md.: Johns Hopkins Press, 1968), p. 42. Data for 1973 and 1985 provided by Office of the Assistant Secretary of Defense for Manpower, Reserve Affairs, and Logistics. Percentages are rounded.

a. Categories are based on the Department of Defense occupational classification system.

b. Percentages before 1973 consist of "electronics" and "other technical" categories. Percentages for 1973 and 1985 consist of "electronic equipment repairers," "communications and intelligence specialists," "medical and dental specialists," and "other technical and allied specialists" categories.

c. Percentages before 1973 are the "administrative and clerical" category. Percentages for 1973 and 1985 are for the "functional support and administration" category.

d. Includes 2 percent classified as miscellaneous.

e. Percentages before 1973 consist of "mechanics and repairmen" and "craftsmen" categories. Percentages for 1973 and 1985 consist of "electrical/mechanical equipment repairers" and "craftsmen" categories.

years. In this table military occupational areas are classified to correspond roughly to civilian occupational categories.[4] The effects of technological change are particularly evident in the shift between white- and blue-collar occupations, which reflects the move away from work requiring general military skills toward that requiring special skills. White-collar workers in the military now make up 47 percent of the total versus 28 percent in 1945, closely paralleling the growth of white-collar employment in the economy as a whole. Although the steepest increase occurred between 1945 and 1957, the trend toward a more technical workforce has since continued, but at a slower pace.

The most conspicuous change during the period, as the table shows, has been the increase in electronics-related occupations, which now account for more than one out of five enlisted jobs compared with about one out of twenty at the end of World War II. Moreover, in recent years, the expansion of the electronics occupational field has remained brisk and, in fact, accounts for all of the growth in the technical occupations since 1957. Not surprisingly, the increase in the proportion of electronics technicians in the armed forces closely parallels changes in the electronics content of military equipment (as measured by cost), which has grown from an estimated 10 to 20 percent in the 1950s, to 20 to 30 percent in the late 1960s/early 1970s, and to nearly 40 percent by 1983.[5]

It should be noted, too, that the growth in the proportion of technical jobs has been accompanied by an increase in the technical complexity of *specific* jobs. For example, although the percentage of clerical positions has remained relatively stable over the years, special skill requirements are now imposed even on these workers, many of whom must be familiar with the use and operation of data processing systems. The tasks of today's basic infantrymen, some equipped with laser designators to "light-up" enemy targets, are also more involved than those of the foot-slogging "grunts" of earlier eras. The duties of shipboard propulsion specialists have likewise changed over the years. As naval steam and diesel systems have been outfitted with automatic control systems they have become more complicated and more demanding of technically proficient specialists. This trend will be further reinforced as older steam-driven ships are replaced by newer vessels powered by gas turbines, which will require propulsion maintenance personnel with an even higher order of technical ability.

Technological Prospects

Advances in technology are expected to continue to have a marked influence on the armed forces of the future. Planned changes in the nation's military posture over the next decade emphasize improvements in (1) survivability of U.S. forces; (2) detection and tracking of enemy targets; (3) accuracy and lethality of U.S. weapons; and (4) responsiveness of battle management systems. These improvements, it is anticipated, will be made possible by the exploitation of a wide variety of technological opportunities. Advances in aerodynamics and propulsion, which will accompany developments in materials and chemistry, are expected to enhance the capabilities of weapon platforms; higher thrust-to-weight ratios will allow future aircraft to fly faster and higher and to carry heavier payloads; advances in propulsion will enable ships to "steam" for longer periods; and stronger new materials will make tanks and armored fighting vehicles much tougher to destroy.

Emerging electronics technologies, however, are expected to have the most dramatic and pervasive effects. Many knowledgeable observers believe that the U.S. military is on the threshold of a major breakthrough in electronics that, like nuclear weapons and jet propulsion in earlier eras, will literally revolutionize warfare. Even allowing for exaggerations inherent in projections of breakthroughs, by the mid-1990s the concept of the "electronic battlefield"—on land, on sea, and in the air—could well move from the realm of science fiction several steps closer to reality.

The individual services will not be affected equally by these changes. The air force and the navy, both of which have a long association with complex weapon systems, should be the least affected during the transition from one generation of technology to the next. On the other hand, the army and, to a lesser extent, the Marine Corps are likely to encounter substantial challenges as they convert from systems that are largely electromechanical to systems incorporating advanced integrated electronics.

Nevertheless, a key issue for all the services is the extent to which the introduction of new systems will affect the character of military jobs. Will the manpower skill mix change? Will recruits of the future need to be more or less qualified?

SKILL REQUIREMENTS AND WORKER QUALIFICATIONS

Intuitively, the shift toward a higher concentration of technical occupations should have implications for the skill content of military jobs and for the qualitative profile of the military workforce. As jobs have become more technical, it would seem the tasks should have become more complex and the people performing them should be more highly qualified. Yet this connection is not easy to make because changes in job skills are difficult to assess, the concept of manpower "quality" is elusive, and precious little is known about the relationship between them.

Problems in Assessing Skill Requirements

The implications of new technologies for manpower requirements can best be analyzed by looking at the effects that these technologies will have on the operations and maintenance of military systems and their supporting infrastructures. Practically, however, such analyses have been hampered in the past by a lack of reliable data as well as by the low priority afforded by the services to manpower research in general and analysis of occupational requirements in particular.

The military services do have in various states of development tools and techniques for predicting the occupational effects of new weapons. Because of its dependence on leading-edge technology, the air force has led the services in devising models for assessing these implications. The U.S. Air Force Management Engineering Program (MEP), for example, is designed to provide a scientifically based, systematic process to determine work center manpower requirements and an ability to project aggregated manpower requirements for use in its planning, programming, and budgeting process. Management engineering teams develop manpower standards based, when practical, on industrial engineering techniques (such as time study and work sampling) or, when the tasks are not suitable to work measurement, on less statistically rigorous methods involving previous staffing patterns and historical performance. For new systems, however, these approaches are generally infeasible, and the air force relies extensively on manpower "guides," which are developed through surveys or evaluations of planning and programming data, staff estimates, or contractor estimates.

The air force also employs a computer model for the specialized purpose of projecting aircraft maintenance manpower requirements. The Logistics Composite Model (LCOM) simulates aircraft operations and supporting functions under expected operational scenarios and calculates direct labor requirements for aircraft maintenance performed in applicable shops. Task analysis data and design engineer assessments are supplemented where possible by comparability analysis, which uses an existing system similar in use, design, and operating environment to project the maintenance requirements for the new system.

In 1976 the navy initiated its Military Manpower and Hardware Integration Program (called Hardman), designed to "ensure that the Navy explicitly identifies the manpower and training resource implications of its weapon systems acquisition decisions early enough . . . to influence the magnitude and skill characteristics of the resources required."[6] This program includes analytical tools and techniques to determine manpower, personnel, and training needs for new systems and to integrate these with existing systems, thereby projecting aggregate demands for individual skills.

The keystone of the Hardman approach, as in the air force methodology, is "comparability analysis," wherein a new system is compared to an existing system providing the closest match of requirements, concepts, functions, and performance standards. Unlike the air force models, Hardman goes beyond estimates of organizational maintenance skill requirements to higher-order effects on the navy supporting structure.

The army, for its part, recently adopted the Hardman concept as a part of a comprehensive plan, called Manprint (Manpower and Personnel Integration), to ensure that new weapon systems are designed to the capabilities of the average soldier. This program, which currently enjoys a high priority, is intended to provide a systematic means for considering manpower issues during the systems acquisition process.

As promising as these efforts appear, in 1987 the army's program was still embryonic, the navy was in the early stages of implementing the Hardman program fleetwide, and while the air force program had been in existence for two decades, the extent to which its results were incorporated into projections of future manpower requirements was limited.

In 1984 the U.S. Army Science Board found that "the current personnel system components of the Army are not well integrated. . . . Manpower quality and quantity requirements cannot be assessed in an integrated way."[7] Similarly, an analysis done for the air force in 1983 concluded that "the Air Force currently has no method for aggregating and assessing the total demand for acquisition-related MPT [manpower, personnel, and training] requirements during the out years. . . . There is no capability for assessing Air Force requirements against projected resources in order to develop the necessary plans and programs for personnel training that would ensure that future force structure will be capable of supporting Air Force–wide system requirements."[8]

Thus despite attempts by the manpower, personnel, and training communities in the military services to project skill requirements more accurately, current capabilities do not appear to be much advanced beyond those that prompted the initiation of the Hardman project in 1976: "A not uncommon piece of folklore, or perhaps reality, is that if all systems currently under development are deployed, there will not be sufficient manpower to support the weapon systems. *In fact, no one seems to really know.*"[9]

In the absence of accepted methods for projecting skill requirements, the tendency has been to underestimate manpower needs for future systems, often by considerable amounts. For example, the DD-963 Spruance-class destroyer experienced a 32 percent growth in manning requirements during its first five years in operation, jumping from an initial estimate of 224 enlisted sailors to a typical crew complement of 295 by 1980.[10]

Similarly, initial estimates of the maintenance manpower needed for the army's Black Hawk (UH-60) helicopter were understated. Although "the Army had a lot of information, gained through years of experience with the UH-1" helicopter, for projecting Black Hawk requirements, "manpower predictions varied widely for some MOS [military occupational specialties]."[11] For example, the number of tactical transport helicopter repairers needed for a typical platoon in a combat support aviation company grew sixfold (from four to twenty-four) between 1976 and 1983.[12] Although UH-1 usage data indicated that nineteen maintenance positions would be required, the smallest of the available estimates was used in the early stages of the decision process. It has been suggested that underestimates of manpower needs may be intentional, motivated by a desire to hold down estimated life-cycle costs.[13]

These tendencies and shortcomings should be kept in mind when assessing the military services' projections of changes in skill requirements. Each service, it should be noted, uses a different scheme to categorize the technical content of its job roster. The categories used by the army and the changes that are projected for 1984–90 are shown in Table 5–2. The consistency in the skill mix over the period is striking, especially in view of the growing emphasis on electronic warfare and the proliferation of high-tech hardware programmed to enter the army's inventory in the 1990s.

A similar pattern emerges from projections made by the navy. Under its arrangement of skills in the petty officer grades by level of technical difficulty, Table 5–3 shows that the percentage distribution is projected to change only modestly by the early 1990s. But unlike the army, whose overall strength is expected to remain constant over the period, the total number of navy jobs is programmed to increase as the goal of a 600-ship fleet is reached. Thus roughly 7,000 "highly technical" and 10,000 "technical" positions are expected to be added to the navy personnel inventory by 1993.

Finally, the air force's mix of skills is also expected to change little over the period (see Table 5–4). New jobs requiring electronics aptitude are projected to increase by 3.6 percent, just keeping pace with an overall rise of 3.7 percent in air force strength. The largest scheduled growth (5.5 percent) is in mechanical jobs. On balance, the differences are slight, and taken together, the mechanical and electronics categories amount to fewer than 10,000 additional jobs.

Table 5–2. Comparison of the U.S. Army's Enlisted Personnel Skill Distribution, Fiscal Years 1984 and 1990.

Occupational Category	1984		1990		Change, 1984–90	
	Number	Percentage	Number	Percentage	Number	Percentage
Combat	146,178	25.1%	147,270	25.1%	+1,092	+0.7%
Very technical	113,397	19.5	117,615	20.1	+4,218	+3.7
Technical	81,678	14.0	83,621	14.3	+1,943	+2.4
Administrative	186,153	31.9	184,077	31.4	−2,076	−1.1
Semi-skilled	55,410	9.5	53,506	9.1	−1,904	−3.4
Total	582,816	100.0	586,089	100.0	+3,273	+0.6

Source: Data provided by the Department of the Army (July 1985).

Table 5–3. Comparison of the U.S. Navy's Petty Officer Skill Distribution, Fiscal Years 1984 and 1990 (number in thousands).

Level of Technical Difficulty	1984		1990		Change, 1984–90	
	Number	Percentage	Number	Percentage	Number	Percentage
Highly technical	75,400	19.8%	82,500	20.5%	+7,100	+9.4%
Technical	207,000	54.3	217,000	54.1	+10,000	+4.8
Semi-technical	99,000	26.0	101,900	25.4	+2,900	+2.9
Total	381,400	100.0	401,400	100.0	+20,000	+5.2

Source: Robert J. Murray, "Technology and Manpower: Navy Perspective," in William Bowman, Roger Little, and G. Thomas Sicilia, eds., *The All-Volunteer Force after a Decade: Retrospect and Prospect* (McLean, Va.: Pergamon-Brassey's, 1985), p. 141.

Table 5-4. Comparison of the U.S. Air Force's Enlisted Personnel Skill Distribution, Fiscal Years 1985 and 1990.

Occupational Category	1985		1990		Change, 1985–90	
	Number	Percentage	Number	Percentage	Number	Percentage
Mechanical	117,047	25.8%	123,517	26.3%	+6,470	+5.5%
Administrative	66,643	14.7	66,175	14.1	−468	−0.7
General	176,368	38.9	183,904	39.1	+7,536	+4.3
Electronic	87,680	19.3	90,858	19.3	+3,178	+3.6
Dual[a]	5,549	1.2	5,699	1.2	+150	+2.7
Total	453,287	100.0	470,153	100.0	+16,866	+3.7

Source: Data provided by the Department of the Air Force (September 1985).
a. Consists of twenty-four skills with dual aptitude requirements, such as supervisory skills fed by specialties from different aptitude categories.

Worker Qualifications

Not captured in the above statistics is the possibility that, within a group of jobs clustered by level of technical difficulty, advances in technology may require a more qualified workforce. Thus even if the number of electronics jobs remained fairly constant (say, because of capital-labor substitution), those jobs arguably could be more demanding and require individuals with a higher degree of proficiency.

Here again, however, the connection is difficult to make because specifications concerning qualitative manpower requirements are largely arbitrary since there are no hard-and-fast rules for judging how smart or how well-educated individuals must be to function effectively in the armed forces. This is a result mainly of difficulties in measuring military output in general and job performance in particular.

Job performance, after all, depends on many characteristics, such as mental ability, educational achievement, job aptitude, physical condition, experience, motivation, adaptability to change, and an ability to get along with co-workers. All are interrelated, and their relative importance varies by type of job and experience level within a given occupation. Further complicating the problem is the fact that job performance itself is often difficult to measure, especially in skills—such as combat infantryman—in which standards of measurement in a peacetime environment are not particularly well developed.

Because of the difficulty of constructing individual profiles and predictors of performance and because of the large number and wide diversity of military occupations, manpower "quality" is most often described in the easy-to-measure terms of educational level and standardized test attainment. Entry into the armed forces and into a specific occupation is restricted to those who meet minimum educational and aptitude standards.[14] Changes in these qualitative characteristics of military recruits can be traced with the aid of Table 5–5, which shows educational and aptitude test score attainment for selected periods since 1960 under both conscription and volunteer conditions.

Enlisted volunteers entering the armed forces since the end of the draft have been less likely than recruits during the draft era to

Table 5-5. Percentage Distribution of Recruits, by Aptitude Category and Level of Education, All Services, Selected Fiscal Years, 1960–85.

Aptitude Category and Educational level	Draft Era			Volunteer Era		
	1960–64	1965–69	1970–73	1974–76	1977–80	1981–85
Armed Forces Qualification Test category						
I and II (above average)	38%	38%	35%	35%	29%	38%
III (average)	49	41	45	55	43	51
IV (below average)	14	21	22	10	28	11
Educational level						
College degree	2	3	5	1	1	2
Some college	11	15	13	5	4	7
High school diploma	51	56	52	60	66	78
Total (having at least high school diploma)	64	74	70	66	71	87

Sources: Fiscal years 1960–69, Richard V.L. Cooper, *Military Manpower and the All-Volunteer Force* (Santa Monica: RAND Corp., 1977), p. 133; fiscal years 1970–73, U.S. Department of Defense, Office of the Assistant Secretary of Defense for Manpower, Reserve Affairs, and Logistics, *America's Volunteers: A Report on the All-Volunteer Armed Forces* (Washington, D.C.: DOD, 1978), pp. 193, 196, 197; fiscal years 1974–85, author's estimates based on data provided by Defense Manpower Data Center.

have acquired some college credits but more likely to have a high school diploma; on average, however, the educational attainment of volunteer-era recruits has not differed appreciably from that of their draft-era predecessors. The sizable proportion of entrants with college experience during the conscription era can be explained largely by the influence of the draft, while the growing percentage of high school graduates during the volunteer era can be attributed both to a growth in the proportion of young Americans completing high school and the higher priority placed by the services on attracting graduates.

The pattern of aptitude test scores has been more irregular. For the first several years under the volunteer concept (1974–76), the services attracted roughly the same proportion of high-aptitude recruits as they had during the draft years, but a substantially larger proportion of average recruits and, commensurately, a smaller percentage of below-average recruits. The latter part of the 1970s saw a reversal, however, as recruit quality plummeted. Of all volunteers who entered in 1977–80, for example, 28 percent were in the lowest acceptable category. In 1981 the recruitment climate improved dramatically, and since then a much larger proportion of enlistees have scored in the above-average range and a much smaller fraction below average.

These changes in the qualifications of the military workforce, by most accounts, are more attributable to changes in the recruitment market than to changes in the skill content of military jobs,[15] and thus yield little useful information for assessing the implications of technological changes. Therefore, despite the military's long association with the products of U.S. technological know-how, their ability to estimate the impact on job skills is too underdeveloped to serve as an effective model for civilian applications.

SKILL REQUIREMENTS AND TRAINING

Although educational background and standardized test scores are useful for determining general eligibility for military service, the great majority of recruits, it should be noted, are under the age of twenty-one, have little job experience, and are drawn from the least skilled segment of the population. And because many military skills do not have a civilian counterpart, few individuals enter service without needing additional training. Some skills can be learned on the job, but the military services feel that most are better learned through formal classroom training and thus they operate one of the nation's largest training institutions. In fiscal year 1987, for example, over 350,000 enlistees entered recruit training, over 450,000 service personnel underwent initial skill training courses, and roughly 230,000 entered advanced skill training.[16]

Skill Entry Standards

As a gauge of trainability, the individual services have developed composite sets of subtests, designed to predict training success in particular job clusters, such as electronics repair and mechanical maintenance. Success is defined in terms of either course grades or time spent to achieve a specified standard of performance.

Although scores on composite tests are used primarily to assess vocational aptitude, the armed forces also use them to supplement AFQT scores and educational attainment in determining basic eligibility for entry into certain services. To qualify for enlistment in the army in 1985, for example, a high school graduate was required not only to place above the fifteenth percentile on the AFQT but also to score eighty-five or above on one of ten army occupational-area aptitude composites. A nongraduate with a general equivalency diploma needed to score above the thirtieth percentile and at least eighty-five on any one composite, while a dropout without an equivalency certificate needed to be in the thirtieth percentile on the AFQT and to score at least eighty-five on *two* composites. The air force, navy, and Marine Corps use similar test formulas although cutoff points vary.

Data in Table 5–6 show how the combination of educational and test score requirements affect the supply of eligible people. Obviously, more stringent standards mean fewer qualified applicants. Thus 76 percent of young men could be expected to meet minimum entry standards for the army, compared with 63 percent for the air force. Among young women, fewer than half would be expected to qualify for the Marine Corps, while nearly four out of five would be eligible for the army.

Although this discussion has focused on entry qualifications expressed as "minimum" requirements, the armed forces cannot,

Table 5–6.   Percentage of U.S. Population Ages Eighteen to Twenty-three Eligible for Enlistment Based on Fiscal 1984 Educational and Aptitude Standards, by Service and Sex.

Sex	Army	Navy	Marine Corps	Air Force
Male	76.3%[a]	75.0%	68.3%	62.6%
Female	78.3	78.1	46.4	60.4
Total	77.3	76.5	57.5	61.5

Source: Adapted from data appearing in Mark J. Eitelberg and others, *Screening for Service: Aptitude and Education Criteria for Military Entry* (Alexandria, Va.: Human Resources Research Organization, 1984), pp. 177–80.

a. Percentage estimates calculated on the basis of results from the "Profile of American Youth" and the fiscal 1984 educational and aptitude test requirements used by the armed services. Excludes such factors as medical and moral requirements.

of course, be staffed completely with individuals who meet just the minimum standard. They seek a range of talent to fill a variety of jobs with differing demands, depending to a large extent on occupational needs. Those services whose job rosters contain large numbers of specialists and technicians, such as the air force and navy, would be expected to need recruits with higher qualifications, on the average, than would the army and Marine Corps, whose jobs cluster toward the low end of the skill spectrum.

This result is ensured in part through the use of composite tests to determine eligibility for skill training courses. Generally speaking, training courses for the technical occupations (those involving computer specialists, electronics technicians, medical technicians, and the like) require high aptitude scores. This is illustrated in Table 5–7, which shows for major army occupational categories the percentage of male youths likely to qualify, based on their test scores, for the most-selective job, the least-selective job, and an "average" job in each cluster. Certain electronics jobs demand a high degree of skill, but other billets in the electronics field are less technical and consequently have less stringent entry standards. Only one out of four male youths, for example would qualify for training as a calibration specialist (the most-selective electronics skill), while close to three out of four would qualify for training as a cable splicer (the least-selective skill). On balance, most electronics jobs in today's army tend to cluster toward the least-selec-

202 SKILLS AND THE DISTRIBUTION OF EARNINGS AND INCOME

Table 5–7. Percentage of U.S. Males Ages Eighteen to Twenty-three Eligible for Army Skill Training, by Aptitude Area and Level of Selectivity.[a]

Aptitude Area	Percentage of Total Army Positions	Percent Eligible for Training		
		Most Selective Skill in Area[b]	Least Selective Skill in Area[b]	Skill of Average Selectivity[c]
Electronics	7.7%	24.9%	72.8%	57.2%
Skilled technical	15.9	34.1	72.4	60.7
Clerical	16.6	29.0	68.9	63.0
Mechanical maintenance	12.0	54.2	70.3	63.1
Surveillance/communications	5.7	51.6	66.5	64.1
Field artillery	4.1	56.3	71.3	68.2
Operators/food	10.6	53.8	71.3	68.4
General maintenance	5.9	59.0	74.6	68.7
Combat	21.4	63.2	71.5	71.4

Source: Derived from data obtained from Defense Manpower Data Center and the Department of the Army.

a. Estimates of the percentage of males eligible for training were calculated on the basis of results from a nationwide administration of the ASVAB in 1980 and the fiscal 1984 standards used by the army.

b. Most and least selective skills are those for which the smallest and largest percentages, respectively, of the eighteen- to twenty-three-year-old male population would be likely to qualify based on composite scores.

c. Proportion of eighteen- to twenty-three-year-old male population qualified for each skill in the area, weighted by the percentage of jobs in each skill.

tive end of the scale; close to 50 percent of the electronics positions, for example, require a minimum score of only ninety-five on the electronics composite, which about 65 percent of the male population could be expected to attain.

In contrast, while air force electronics skills fall within a relatively narrow range of job difficulty, they tend to be more demanding, on average, than the army's. For example, 32 percent of male youths could qualify for the air force's most-selective electronics skill, aircraft control and warning radar specialist, and 62 percent for electric power line specialist, the least-selective. For the average air force electronics skill, about 50 percent would qualify, compared with 57 percent for the average army electronics skill.

Limitations of Skill Entry Standards

In recent years, the aptitude standards used by the armed forces for enlistment and classification have become a subject of increased controversy. A major issue is the extent to which these standards are related to job performance. The weakness of the relationship has been widely acknowledged, but efforts to strengthen it have been hampered by difficulties involved in measuring job performance for a wide range of military positions. Since 1980 the military services have pursued research programs addressing job-performance measurement and the linkage of job-performance data to enlistment standards. While some progress has been made, it is too early to predict future changes in current standards. In the meantime, imperfect though it may be, ASVAB (Armed Services Vocational Aptitude Battery) remains the best instrument available for selection and classification purposes and in all probability will continue to be used at least until the end of the decade, when the results of the ongoing research program are expected.[17]

Potential Improvements in Military Skill Training

The U.S. military's position at the forefront of technological applications notwithstanding, their skill training programs have been criticized for being outdated conceptually and for failing to exploit recent developments in training technology.

Training Concepts.   Advances in technology have contributed to the growing belief that the philosophy underlying military training programs needs to be reconsidered. One of the principal criticisms of technical training in the U.S. armed forces centers on its "front-loading"; that is, most formal training is provided to the service member early in the first enlistment period and often before reporting to an initial assignment. This approach has been challenged on both cost and effectiveness grounds. The returns to the upfront investment are often low, it is contended, because of high turnover after the first term of service; moreover, early training ill prepares the individual to acquire necessary job proficiency through on-the-job experience.[18]

There are signs, however, that the shortcomings of front-loaded training are beginning to be recognized. The navy, for example, in 1977 instituted a project to develop, as part of an integrated personnel system, a revised approach to training called Epics (Enlisted Personnel Individualized Career System), designed to "prepare personnel for early operational and technical contribution, minimize and defer the initial training investment, and utilize the available enlisted resource pool to the fullest."[19] The system integrates the use of job performance aids (JPAs), self-teaching packages for shipboard training, integrated shore-based training episodes, and most notably, the notion of deferred formal technical training and early at-sea experience with an individualized career advancement structure.[20] Under the deferred-training concept, on completion of basic training the recruit would be sent directly to sea for an indoctrination period of eight to twelve months, during which time he would receive "transition-to-shipboard-life" training and would use JPAs and self-paced instructional materials in his apprentice-technician duties. At the completion of this period, those who demonstrated satisfactory job performance to their supervisors would attend formal equipment-technician training and, following another period of sea duty, would return to shore for system-technician training (after serving roughly twenty-four months).[21]

Research is also under way with the aim of developing courses that will turn newcomers to electronics, regardless of their entering aptitudes or literacy, into technicians capable of entry-level work in a variety of technical jobs (assembly, radio/television repair, small-appliance repair) or will qualify them for more advanced training in basic electricity and electronics or related programs.[22] In contrast to traditional methods of training electronics technicians, which emphasize the theory of electricity, using this approach (the "functional-context" method)

> students start with a known piece of equipment and go from the whole, and an understanding of how it works, to a series of analyses of parts and subparts and how they work. When needed to explain how

something functions, basic electricity concepts are introduced. Additionally, mathematics is taught when it is needed to understand a phenomenon. Reading skills are developed by reading block diagrams that display functional relationships among inputs/components/outputs, and by reading functional stories which, in a very direct manner, with very little use of analogy, tell how the particular subsystem works.[23]

By working from the familiar, known equipment to the unfamiliar (such as volt-ohm meter) and unknown, the course organizers "hope to develop a more competent electronics technician out of people whose literacy and general technology knowledge is lower than is usually regarded as adequate for such training."[24]

Training Technologies.   Standing in sharp contrast to the growing influence of high technology on military weapon systems is its relatively meager influence on the support infrastructure in general and the training base in particular. Some observers are optimistic about the prospects: "Opportunities exist to turn technology itself from the driver of ever increasing quantitative and qualitative manpower requirements to the provider of new and innovative ways to provide trained personnel that are fully capable of using and maintaining our most advanced systems."[25] Advanced computer-based instruction in secondary schools and colleges, for example, seems to substantially increase the effectiveness of training (as measured by final exam scores). Comparisons of two groups of students enrolled in the same courses in the same secondary schools indicated that the average final exam score of those who received computer-based instruction was at the sixty-third percentile of those who received conventional instruction. A similar experiment at the college level found a slightly narrower but nevertheless substantial difference: those in computer-based classes were at the sixtieth percentile of those in conventional classes.[26]

Although the military, too, has adopted computer-based instruction in many of its programs, critics say these new techniques have generally been applied to reduce the length of training courses (and hence training costs) rather than to produce a more effective graduate. This has happened because the military has used computer-based instruction to produce an equally effective performer in less time, while secondary schools and colleges generally have used it to increase student achievement in a fixed period of time.

The difference stems from divergent philosophies of training. The armed forces, unlike most civilian educational institutions, train individuals to do a specific job. Accordingly, under the military's approach to computer-based training, students set their own pace and graduate when they demonstrate they have met the course requirements. Brighter individuals will graduate sooner than their slower counterparts, but presumably with the same level of achievement. In civilian institutions, on the other hand, the length of the school term is fixed, and because *all* students are trained for the same amount of time, brighter individuals learn more than slower ones. "Changing the way of using computer-based instruction in military training from saving time to increasing student achievement at school," according to a noted expert in military training technology, "can make a significant contribution to the increasing need of the military services to improve the performance of their personnel."[27]

Research also suggests that computer-based instruction could succeed where conventional methods have failed in training individuals who typically have not fared well in a text-and-lecture setting, but who have been "exposed to a culture with greatly enhanced receptivity to computer training (e.g., great exposure to television, personal calculators and computers, arcade games, smart games, etc.)."[28] Computers afford an advantage, for example, by being better able to convey the necessary course material in a mode of presentation more closely matched to the learning skills and styles of many of today's youth. The purpose of military training is largely to teach personnel how to operate and maintain equipment, skills that experts believe can be honed more effectively through the use of visual aids in an interactive setting rather than by the use of the written word in a passive classroom environment.[29]

The above are only a few of many technological opportunities that the military could exploit to improve the capabilities of its training institutions. Mechanisms to improve language skills, for example, include language-training cassettes and speech recognition devices. Because the private sector is apparently well ahead of the military in many training technologies, it has been suggested that "the military should 'piggyback' on this technological revolution, finding its own special applications and creating only the software and courseware peculiar to its needs."[30]

TECHNOLOGY AND SKILLS: THEORY AND PRACTICE

By most accounts, the armed forces have failed to realize a full return on their substantial investment in technology by allowing the sophistication of many of their new weapon systems to outstrip the capabilities of their workforce. In theory, new technologies were supposed to produce user-friendly systems that would require fewer and less qualified people to operate and maintain; in practice, that promise has gone unfulfilled as the armed forces have fielded systems that are more complex, less reliable, and harder to maintain than had been envisioned.

The issue is discussed here in a historical context, concentrating on experience gathered over the past decade with the application of solid state electronics to avionics systems. In many ways, avionics systems—fire control radars, weapons delivery equipment, navigation computers, and electronic countermeasures—represent the cutting edge of military technology, and since they incorporate one of the first widespread uses of microelectronics, they provide an excellent medium for examining the broader issues involved with technology and manpower.

"Transparent" Complexity: The Theory

Devotees of high technology argue that its application to weaponry not only gives a critical performance edge to U.S. armed forces in the face of the numerical superiority of the Soviet Union but should also reduce the number and requisite qualifications of people to operate and maintain it.[31] This contention rests on the assumptions that technology is applied in a manner that improves reliability and maintainability, that makes testing and servicing equipment less labor intensive, and, for electronics systems, exploits "the exploding technology of microprocessors in automatic test equipment, built-in test, and in simulators and training devices."[32]

According to this argument, the evolution of electronics technology—from vacuum tubes to transistors to integrated circuits—has been accompanied by dramatic improvements in reliability and efficiency. Today's hand-held electronic calculator, for example, is much faster, far less likely to break down, and less costly by a factor of a hundred than its bulkier electromechanical forerunner.[33] Similar comparisons are made between mechanical and electronic watches, and between tube and transistor radios. And, despite more complicated circuitry, today's calculators are easier to use since the operator requires no knowledge of the technology embedded in them.

User-Friendly Weapons.   In the military setting, this "transparent complexity," to use the technologist's jargon, should permit major advances in equipment capabilities without placing greater demands on the operators. Experience with the M-1 Abrams, the army's newest tank, supports that argument. Equipped with a variety of high-tech features (such as laser range finder, ballistic computer, thermal-imaging night gunsight, full stabilization, a muzzle reference system to measure gun-tube distortion, and a wind sensor), the M-1 is clearly more complicated than its predecessors. By all indications, however, it is a far simpler tank to operate as demonstrated in Table 5–8, which compares range firing results by crews manning M-1 tanks with those manning the

Table 5–8.   Comparison of M-60 and M-1 Tank Crew Performance.

AFQT Category Gunner/Tank Commander	Tank Equivalent Kills[a]		
	M-60	M-1	Percentage Improvement
I (above average)	10.23	12.75	25%
II (above average)	9.51	12.47	31
IIIA (average)	8.52	12.05	41
IIIB (average)	7.47	11.57	55
IV (below average)	5.84	10.72	84

Source: Barry L. Scribner and others, "Are Smart Tankers Better Tankers: AFQT and Military Productivity," Office of Economic and Manpower Analysis, Department of Social Sciences, U.S. Military Academy, West Point, N.Y.

a. Because the test included a variety of targets (tanks, troops, and so forth) fired at both with the main gun and with machine guns, "tank equivalent kills" is the measure of effectiveness. The maximum possible "tank equivalent kills" was 24.5.

older M-60. M-1 crews consistently scored more tank "kills" than M-60 crews with similar aptitudes. More interestingly, the number of kills by M-1 crews was less influenced by their aptitudes, suggesting that the M-1 is easier to employ than its predecessor and capable of being operated by crewmembers with lower aptitude scores for a given level of effectiveness.[34]

Black-Box Maintenance. Technologists likewise contend that modern "user-friendly" electronics makes the maintenance task easier. With the advent of modular systems architecture incorporating smaller, replaceable subassemblies ("black boxes") accompanied by improved diagnostic capabilities, the maintenance burden—and hence the requirement for highly skilled technicians—can be reduced or at least shifted from the operational unit rearward to major depots.

The "remove and replace" maintenance concept for the advanced avionics system in F-15 aircraft is illustrative. The F-15 "was designed from the beginning to recognize the decreasing availability of both pilots and maintenance personnel. The goal was to reduce flight line test sets, provide built-in test equipment, and simplify support tasks, thereby requiring fewer personnel and less training. For example, there was to be no soldering at the intermediate level. All components were designed to plug into either the aircraft or 'Black Boxes.' "[35]

The F-15 avionics system consists of some forty-five black boxes, each containing a number of solid-state electronic circuit boards. Built-in test equipment (BITE) aboard the aircraft monitors the avionic components and indicates when a failure has occurred. Under the "remove and replace" concept the entire black box is removed and replaced with a functioning unit. No attempt is made to repair the faulty component on the flight line; rather it is sent to an Avionics Intermediate Shop (AIS), which is equipped with automatic test stations able to isolate the problem to the malfunctioning circuit card by checking internal circuitry against predetermined tolerances. The AIS technician replaces the defective card, checks out the component, and returns it to supply to be used as a serviceable replacement; the defective card is returned to a depot for repair.

Compared with earlier avionic systems incorporating bulkier vacuum tube technology, analog design, and traditional manual maintenance methods using meters, oscilloscopes, and other standard test instruments, the on-board fault diagnostic capabilities of the F-15 ostensibly reduce the need for external flight-line support equipment and avionics equipment repairmen and, by transferring more of the maintenance burden from the flight line to the depot, allow greater dependence on civilian rather than military technicians. Moreover, while the operators and maintainers of the Automatic Test Equipment must be proficient in the use of the specialized test station, it is contended that "the skill level required is not as great as that required if the maintenance personnel were employing individual test instruments."[36] Overall, "remove and replace" maintenance, made practical by solid state electronic technology, presumably renders advanced systems more maintainable and diminishes the requirements for skilled personnel, especially those in deployed units. Accordingly, only three avionics repair occupations are involved in F-15 maintenance, down from twelve for the F-4. Similarly, training time for avionics technicians was shortened from thirty-eight to twenty-five weeks as the electronics theory portion of the curricula was substantially reduced.[37]

If a technological bonus has not yet been realized, according to proponents of this approach, it is only because the systems now in the inventory do not reflect state-of-the-art technologies. "The problems [that the military services] are having are largely with equipment which was designed in the 1950s and 1960s and built in the 1960s and 1970s."[38] Dramatic improvements in reliability and maintainability are anticipated as the products of the recent modernization program reach the field over the next decade. Moreover, so the argument goes, "the methods and machinery of the logistics system have yet to adapt to the new requirements posed by advanced technology."[39]

Technology and "Oversophistication": The Reality

Theory aside, however, the historical experience with high-performance systems indicates that the manpower dividends expected to flow from technological substitution have rarely materialized. "Consider the fact," writes one inside observer, "that virtually every new weapon system proposed for development promises to 'save people.' . . . My hunch is that this promise is seldom, if ever, realized once the system is fielded and operational."[40] The problem has been ascribed to "an inbred tendency to build exotic weapons that promise high performance but invariably prove unreliable and difficult to repair and maintain."[41]

The underlying problem, according to a Defense Science Board panel of military and defense contractor representatives, is "that the maintenance concepts employed on our high performance systems have been force fit into maintenance and repair structures which have preceded the advent of such systems, and that these structures are not well-matched to today's needs."[42] Among the problems found to be associated with high performance systems were skilled personnel shortages, unnecessary maintenance, insufficient spare components, incompatible test equipment, and inflexible maintenance practices.[43]

The Automated Diagnostics "Disaster."   Many of the maintenance difficulties being encountered with high-tech weapon systems are being attributed to the "remove and replace" maintenance philosophy that depends so heavily on built-in diagnostic and automatic checkout equipment . . . [which] increase the burden on the skills of the maintenance personnel."[44]

In an unusually harsh critique of the concept, the Defense Science Board's panel concluded that

> In the early 1970s, the proliferation of complexity within avionics equipment, coupled with the advancing state of digital technology, led to the birth of the "Built in Test (BIT)" cult. Concerns relative to the maintainability of equipments with a multiplicity of removable assemblies were quieted with the promise of automatic fault detection and isolation capabilities that stretched into the high ninety percentile range. While these promises looked good on paper and were incorporated into almost all specifications, the actual field performance has been nothing short of a disaster.[45]

A senior air force general officer put it even more succinctly: "We swallowed a story hook, line and sinker that said 'We've gotten so good with test equipment that we've got dumb man-smart machines.' It turned out to be an incomplete promise; matter of fact, it turned out to be complete bull."[46]

In effect, these automatic systems for detecting and isolating faults have not performed as expected and indeed, many have turned out to be as complicated to operate and maintain as the prime mission equipment itself. The challenge, which the electronics industry appears to be having difficulty meeting, is to design fault detection systems sensitive enough to detect failures without also having high false alarm rates.[47] Available data indicate generally low fault detection and fault isolation rates and generally high false alarm rates, which cause unnecessary removals. These put additional burdens on the maintenance process, tying up automatic checkout equipment and escalating the requirement for spare parts.

By adopting this concept, which relies heavily on advanced technologies, the armed forces have virtually "painted themselves into a corner." Automated test equipment, it turns out, "almost completely cuts off communication between the levels of maintenance, and it places total reliance on the fault-isolation capability of ATE. The flexibility of manpower is reduced while the skill levels required to maintain and use the test equipment have increased."[48] One of the most challenging aspects of the electronic technician's job, then, is to find the faulty component when the automatic diagnostic system fails to do so. The task requires "the informed use of combinations of the external indicators provided by the system . . . [which] in turn, requires that the airmen know how the system is integrated, what functions are performed in what boxes, and how a failure in a particular box affects the system."[49] Since the adoption of the modular-design philosophy has generally led to reduced skill requirements and shorter training courses, many military technicians are ill-equipped to handle such troubleshooting tasks.[50]

The air force's experience with F-15 aircraft systems provides a case in point. The complicated computer test stations in the Avionics Intermediate Shop (AIS) have been critical bottlenecks and have proved difficult to keep in commission owing to their sheer complexity

the automatic test stations contain 220,000 parts that have to be fault isolated and replaced. This is more than double the total number of electronic parts in the F-15. There are over 280 different technical orders and 100 different computer programs on 530 reels of tape used in troubleshooting the avionics boxes. Hookup on the station requires up to 85 interface device connections. The maintenance technicians must also know how to break down, ship and reassemble their test stations within 12 hours on deployment.[51]

Although much of this discussion has been confined to avionics, reliability and maintainability problems are common to other systems incorporating complex electronics. The army's "fix-forward" concept for maintaining tactical weapons, for example, which emphasizes the performance of repairs as near as possible to the FEBA (forward edge of the battle area) to minimize system downtime, depends on the ability of organizational maintenance personnel to diagnose the causes of failures quickly and accurately. Accordingly, many of the newer army weapons are incorporating automatic test sets and other sophisticated diagnostic equipment to facilitate troubleshooting.[52] Thus yesterday's problems with avionics could well be tomorrow's problems with land warfare systems. The problems, in fact, are already in evidence. The M-1 tank, "probably the first major armored ground system to emphasize and incorporate advanced technology equipment," has experienced reliability, availability, and maintainability problems, some of which have been attributed to automated test equipment.[53]

The Computer Software "Nightmare." The tendency of the armed forces to understate skill requirements has not been confined to the operations and maintenance of hardware. Indeed, there is strong evidence that the military services have underestimated manpower needs—both quantitatively and qualitatively—associated with computer software requirements.

The situation has become particularly acute with the recent proliferation of microcomputers, the exponential growth in distributive processing applications, and their attendant software demands.[54] As more and more microcomputers have been put into service, the armed forces have run into software problems similar to those confronted by many organizations dealing with computer applications: The software may not work as advertised (that is, it "bombs"); it may not meet design specifications regarding speed, cost, or ease of maintenance; or it may require modifications as situational factors change.[55]

By its very nature, software maintenance is a manpower-intensive task but, because of its complexity, workload requirements are extremely difficult to forecast, especially in a distributed processing environment. Thus it is likely that military manpower planners have underestimated the magnitude of the task involved and they might well experience what has been termed the software maintenance "nightmare": the possibility that "in ten years, vital functions may be relying on a hodgepodge of poorly written programs that have never been adequately validated or tested."[56]

ORGANIZATIONAL PROBLEMS

The foregoing discussion suggests that despite promises that U.S. technological know-how would produce a more efficient and labor-saving military establishment, the weight of the evidence is that new systems demand ever-more skillful operators and maintainers, especially if the capabilities of these systems are to be fully exploited.

This situation is largely a product of the defense acquisition process, which traditionally has emphasized system capability rather than reliability and maintainability and hardware rather than people. These priorities have been described by former Defense Secretary Harold Brown, who maintains that the armed forces have tended "to achieve the best possible performance (speed, payload, range) in systems and to take full advantage of the newest technology only for that purpose." In faulting the armed forces for their failure to consider possible tradeoffs, Brown contends that

> the operating commands have often insisted, for example, on the highest possible speed for a given aircraft design, without asking what value the last 100 knots provides and what is sacrificed, to achieve that capability, in other desirable performance characteristics or in reliability. In other cases, fleet air-defense missiles have been given ranges considerably beyond those at which the radar associated with them could provide reliable target information. This situation is reversed in the new Aegis fleet air-defense system: There the radar outperforms the missile. Almost always, these unnecessary increments of performance have been paid for in unreliability, demonstrated in either more frequent equipment failure or more frequent maintenance requirements.[57]

Accepting performance 5 percent or 10 percent lower than the peak that could be obtained from new technology and using the design freedom thus achieved to operate engines at lower temperatures, structures at lower stresses, or circuits at higher redundancy, according to the former secretary, would pay "rich dividends in reliability."[58]

If the armed forces are to realize the full labor-saving potential of technology, reliability and maintainability will have to be given higher priority in the systems acquisition process. Reliable systems reduce life-cycle costs, result in higher output rates, and require less skilled manpower. Maintainable systems reduce out-of-commission time and require fewer people and lower skill levels.

Just as reliability and maintainability of weapon systems have received less emphasis than performance, schedule, and cost in the acquisition process, so too has training of military personnel. According to a former navy training official,

> *There is a traditional lack of appreciation for the role of training.* At least since the advent of sophisticated weapon systems into naval warfare, naval officers have been driven by an insatiable compulsion to use their resources for "better" and more technically complex hardware systems, be they platforms, propulsion plants, sensors, or weapons. . . . Because the emphasis has always been on hardware, and the resources are finite, people problems have taken a backseat in the Navy, especially concerning their proper training to maintain and operate the ever-changing hardware systems.[59]

Similarly, a panel established by the Defense Science Board in 1982 concluded that "currently, training aspects of systems development are too often sacrificed first when funds run short," and urged the implementation of its recommendations "if necessary, at the expense of hardware [or] force structure."[60]

If changes are to be made, however, it appears that the impetus will have to come from the services themselves. There is little cause to believe that the defense industry will take the initiative, for reasons explained by Walter B. LaBerge, technical director and vice president of technology at the Lockheed Missiles and Space Company:

> Lockheed would like to work on [and] . . . is competent at working on training and training devices. But there is in fact no market for that. That's the same for Hughes and TRW and all the major people in the business. From the major equipment supplier's standpoint, working

on the technology of training people, retraining of skills, and the application of these skills in the field by the presentation of information is, in fact, not related to the winning of any major program. If you can concentrate on a new high-powered laser, or how to point that laser, or how to correct its wavefront, you can in fact win the next generation space age laser. If you describe the training and training technology supporting it, it will make no difference at all. And hence, we put our money on how you point and track. Thus training and training technology are not important to the winning of contracts because they're not related to profits.[61]

This attitude, it should be noted, is reinforced by budget priorities within the Department of Defense. In fiscal year 1987, for example, out of a total budget of $36.7 billion for research and development, about $361 million (or 1 percent) was earmarked for manpower, personnel, and training.[62]

LESSONS FOR THE CIVILIAN SECTOR

Although the discussion in this chapter has centered on the link between military technology and defense manpower, several lessons drawn from the military's experience appear relevant to the civilian sector:

1. The armed forces' ability to project the implications of technology for job skill requirements is at best rudimentary, impeded both by methodological problems and by a lack of interest on the part of higher management.
2. The size and skill distribution of the military workforce depends to a large extent on the complexity of its systems. Generally speaking, more complicated equipment (as measured by the number of components) breaks down more often and, when it does, takes longer to repair.
3. Promises that technology will render complexity "transparent" and thereby diminish the requirement for skilled personnel should be viewed with skepticism. Although some newer military systems are more "user-friendly" than their less complicated predecessors, thus far they have proved to be more difficult to maintain.
4. The selection of individuals to attend skill training courses is

based on standards that, at best, are loosely related to job performance, which is difficult to measure in a military setting.

5. Military training methods have failed to keep pace with changes in technology; indeed, the private sector appears well ahead of the military in the application of advanced training techniques and hardware.

6. The principal culprit impeding the military's fuller exploitation of the United States' technological edge is its preoccupation with system performance at the expense of reliability and maintainability. Inattention to manpower, personnel, and training considerations in the weapons acquisition process has resulted in a mismatch between weapons and skills.

7. Affording manpower, personnel, and training considerations equal billing with performance, schedule, and cost in the acquisition process would brighten the prospects that the systems fielded by the armed forces in the future will be within the capabilities of their personnel and that the nation will realize a fuller return on its investment in military technology.

NOTES

1. This paper draws on my *Military Technology and Defense Manpower* (Washington, D.C.: Brookings, 1986).

2. For a comprehensive and even-handed treatment of the debate, see Kenneth I. Spenner, "The Upgrading and Downgrading of Occupations: Issues, Evidence, and Implications for Education," *Review of Educational Research*, vol. 5, no. 2 (Summer 1985): 125–54.

3. For an excellent historical analysis of the effects of military technology on the occupational structure of the armed forces, see Harold Wool, *The Military Specialist: Skilled Manpower for the Armed Forces* (Johns Hopkins Press, 1968), Chaps. 2, 3.

4. Although civilian and military classifications differ in some respects, for many purposes—and those of this chapter in particular—the two systems yield comparative information.

5. Frans Nauta, "Maintenance Training Technology Research: Future Technology Trends," Working Note NA 202–5 (Bethesda, Md.; Logistics Management Institute, July 1984), p. 4–2.

6. Commander George S. Council, Jr., and Allan Akman, "Planning

for Manpower and Training Needs on New Navy Weapon Systems," *Defense Management Journal* (First Quarter 1984): 24.

7. U.S. Army Science Board, *Final Report of 1984 Summer Study: Leading and Manning Army 21* (Washington, D.C.: Department of the Army, 1984), p. 28.

8. Akman Associates, *Enhancing Manpower, Personnel and Training Planning in the USAF Acquisition Process* (Silver Spring, Md.: Akman, 1983), p. VIII–21.

9. U.S. Department of the Navy, Chief of Naval Operations, *Military Manpower versus Hardware Procurement Study* (HARDMAN) (Washington, D.C.: Department of the Navy, 1977), p. I–2 (emphasis added).

10. Frans Nauta and Thomas A. White, *Manning of Recently Fielded Systems: Case Study of the DD-963 (Spruance) Class Destroyer*, prepared for the Office of the Assistant Secretary of Defense for Manpower, Reserve Affairs, and Logistics (Bethesda, Md.: Logistics Management Institute, 1981), p. 5–1.

11. Christine R. Hartel and Jonathan Kaplan, *Reverse Engineering of the Black Hawk (UH-60A) Helicopter: Human Factors, Manpower Personnel, and Training in the Weapons System Acquisition Process*, Research Note 84–100 (Alexandria, Va.: U.S. Army Research Institute for the Behavioral and Social Sciences, 1984), p. 38.

12. *Ibid.*, p. 39.

13. *Ibid.*, p. 40.

14. In addition, applicants must satisfy certain physical and moral requirements. The physical examination is designed to select individuals who are fit for the rigors of military life while moral standards are designed to screen out those who are likely to cause disciplinary problems. For a discussion of physical requirements, see Martin Binkin and Shirley J. Bach, *Women and the Military* (Washington, D.C.: Brookings Institution, 1977), pp. 787–81. A comprehensive treatment of moral standards can be found in Barbara Means, *Moral Standards for Military Enlistment: Screening Procedures and Impact* (Alexandria, Va., Human Resources Research Organization, November 1983).

15. For example, it was concluded in a report prepared for the Pentagon that "the formal minimum standards instituted by the Armed Services during any particular period may not be a true reflection of the actual 'operational standards' used for determining who gets in and who stays out. The minimum 'cutting score' on the military's enlistment test may be adjusted up or down (but not below the formal minimum requirements) in response to periodic changes in manpower retention and the recruiting market. These temporary

modifications are intended to increase recruiting efficiency by regulating the flow of applicants and skimming only the cream of the crop of otherwise qualified recruits. The impermanence and changeable nature of operational criteria have usually functioned to keep them from general public view." See Mark J. Eitelberg and others, *Screening for Service: Aptitude and Education Criteria for Military Entry*, prepared for the office of the Assistant Secretary of Defense for Manpower Installations, and Logistics (Alexandria, Va.: Human Resources Research Organization, 1984), p. 122.

16. Department of Defense, *Military Manpower Training Report, FY 1987*, Vol. 4 (Washington, D.C.: Department of Defense March 1986), pp. III–3, V–4, and V–8. The number entering initial skill training exceeds the number in recruit training because it includes personnel who entered the services in previous years but had not attended formal skill training and some who are cross-training from one skill to another.

17. For a description of the issues and a discussion of the Pentagon's research efforts, see Office of the Assistant Secretary of Defense for Manpower, Installations, and Logistics, *Joint-Service Efforts to Link Enlistment Standards to Job Performance*, Third Annual Report to the House Committee on Appropriations (Washington, D.C.: (Department of Defense, 1984).

18. Frans Nauta, *Alleviating Ship Maintenance Problems through Maintenance Training and Aiding Research*, prepared for Naval Training Equipment Center (Bethesda, Md.: Logistics Management Institute, 1985), pp. 78–79.

19. A.M. Megrditchian, *Enlisted Personnel Individualized Career Systems (EPICS) and Conventional Personnel System (CPS): Preliminary Comparison of Training and Ancillary Costs* NPRDC SR 83–23 (San Diego, Calif.: Navy Personnel Research and Development Center, 1983), p. vii.

20. Robert E. Blanchard, Robert J. Smillie, and Harry B. Conner, *Enlisted Personnel Individualized Career System (EPICS): Design, Development and Implementation*, NPRDC TR 84–15 (San Diego, Calif.: Navy Personnel Research and Development Center, 1984), p. vii.

21. *Ibid.*

22. T. Sticht, "Literacy, Cognitive Robotics, and General Technology Training for Marginally Literate Adults," Paper Presented at a Conference on the Future of Literacy in a Changing World: Syntheses from the Industrialized and Developing Nations, University of Pennsylvania, Philadelphia, May 9–12, 1985, p. 16.

23. *Ibid.*, pp. 19–20.

24. *Ibid.*, p. 20.

25. Bernard Rostker, "Manpower, Personnel, and Training: An Agenda for the Secretary of Defense," Paper Prepared for a conference on U.S. conventional force structure, Georgetown Center for Strategic and International Studies, January 29–31, 1985, p. 15.

26. Jesse Orlansky, "Commentary" in William Bowman, Roger Little, and G. Thomas Sicilia, eds., *The All-Volunteer Force after a Decade: Retrospect and Prospect* (McLean, Va.: Pergamon-Brassey's, 1986), p. 170.

27. *Ibid.*, p. 172.

28. Defense Science Board, *1982 Summer Study Panel on Technology Base,* (Washington, D.C.: Department of Defense, p. D–5.

29. Defense Science Board, *1982 Summer Study on Training and Training Technology, Supplementary Report* (Washington, D.C.: Department of Defense) p. B–9.

30. *Ibid.*, p. C–4.

31. One of the strongest proponents of technology is William J. Perry, former Undersecretary of Defense for Research and Engineering during the Carter administration. His views are set forth in a number of publications. See his article (coauthored with Cynthia A. Roberts), "Winning through Sophistication: How to Meet the Soviet Threat," *Technology Review*, vol. 85, no. 5 (July 1982): 27–35. "Defense Reform and the Quantity-Quality Quandary," in Asa A. Clark IV and others, eds., *The Defense Reform Debate: Issues and Analysis* (Baltimore, Md.: Johns Hopkins University Press, 1984); and especially for the relationship to manpower issues, "Impact of Technology on Military," in *Impact of Technology on Military Manpower Requirements, Readiness, and Operations,* Hearings before the Subcommittee on Manpower and Personnel, Senate Armed Services Committee, 96 Cong., 2 Sess., December 4, 1980, pp. 4–8.

32. Perry, "Impact of Technology on Military Forces," note 31 above, pp. 7–8.

33. Perry estimates that over the past twenty years, the cost of performing a given electronic function has been halved every two years. At that rate, ten years hence the cost will be one-thirtieth what it is today, and in fourteen years, one one-hundredth. *Ibid.*, p. 15.

34. Barry L. Scribner and others, "Are Smart Tankers Better Tankers? AFQT and Military Productivity," Office of Economic and Manpower Analysis, Department of Social Sciences, United States Military Academy, West Point, N.Y., December 5, 1984, p. 12.

35. Perry, "Impact of Technology on Military Manpower Requirements," note 31 above, p. 50.

36. *Ibid.*

37. Walter B. Bergmann, II, "Manpower and Logistics—Anything

New for the Eighties?" Paper Prepared for the Forty-fifth Military Operations Research Symposium, Annapolis, Md., June 1980.

38. Perry, "Impact of Technology on Military Manpower Requirements," note 31 above, p. 10.

39. Seymour J. Deitchman, "Weapons Platforms, and the New Armed Services," *Issues in Science and Technology*, vol. 1, no. 3 (Spring 1985): 10.

40. Colonel Warner D. Stanley III, "MANPRINT: The Leverage for Excellence," *Army Research, Development and Acquisition* (March–April 1985): 6.

41. William Rosenau, "Will R&M 2000 Make a Difference?," *Military Logistics Forum* (May 1985): 44.

42. U.S. Department of Defense, "Operational Readiness with High Performance Systems," Report of the Defense Science Board 1981 Summer Study Panel, Office of the Under Secretary of Defense for Research and Engineering, April 1982, p. 6-6.

43. *Ibid.*

44. *Ibid.*, p. 6–8.

45. *Ibid.*, p. 6–10.

46. Comments attributed to Lieutenant General Leo Marquez, Air Force Deputy Chief of Staff for Logistics and Engineering in Rick Atkinson and Fred Hiatt, "Military in a Fix," *Washington Post*, August 18, 1985.

47. Design engineers face a tradeoff between "type 1" and "type 2" errors. Type 1 errors are those in which a fault is indicated in a component that is operating properly. Type 2 errors are those in which the detection system fails to recognize a malfunctioning component. Emphasis has generally been placed on minimizing type 2 errors since they more directly affect safety or mission performance.

48. *Defense Resource Management Study: Case Studies of Logistical Support Alternatives* (Washington, D.C.: U.S.G.P.O., 1979), p. 39.

49. Polly Carpenter Huffman and Bernard Rostker, *The Relevance of Training for the Maintenance of Advanced Avionics* (Santa Monica, Calif.: Rand, 1976), p. vii.

50. In one survey of knowledge and skills required by graduates of entry-level navy electronics training courses, it was concluded that "(a) only 5 to 10 percent are able to use common basic test equipment; (b) basic electronic theory is not understood by 95 percent; and (c) most are able to run the built-in diagnostic checks associated with combat system preventive maintenance, but 85 percent have not the slightest idea what to do when a failure is detected and the automated diagnostics fail to isolate the failure to a single replace-

able module." See Frans Nauta, *Alleviating Fleet Maintenance Problems through Maintenance Training and Aiding Research* (Orlando, Fla.; U.S. Naval Training Equipment Center, May 1985), p. 26.

51. Bergmann, "Manpower and Logistics—Anything New for the Eighties?," note 37 above, p. 4.

52. Frans Nauta, *Fix-Forward: A Comparison of the Army's Requirements and Capabilities for Forward Support Maintenance* (Bethesda, Md.: Logistics Management Institute, April 1983).

53. Arthur Marcus and Jonathan Kaplan, *Reverse Engineering of the M1 Fault Detection and Isolation Subsystem: Human Factors, Manpower, Personnel, and Training, in the Weapons System Acquisition Process* (U.S. Army Research Institute for the Behavioral and Social Sciences, June 1984), pp. 36 and 88.

54. The air force, for example, has acquired over 27,000 Zenith Z–100 microcomputers over the last two years. James R. Van Scotter and Aaron R. De Wispelare, "The Microcomputer Software Maintenance Dilemma," *Air Force Journal of Logistics*, vol. 10, no. 3 (Summer 1986): 27.

55. *Ibid.*

56. *Ibid.*, p. 28.

57. Harold Brown, *Thinking About National Security* (Boulder, CO: Westview Press, 1984).

58. *Ibid.*

59. Captain Worth Scanland, "Training Sailormen: Why Isn't It Working?," *U.S. Naval Institute Proceedings*, vol. 109 (October 1983): 54.

60. Defense Science Board, *1982 Summer Study Panel on Training and Training Technology*, note 29 above, pp. v–vi.

61. W.B. LaBerge, "Commentary," in Bowman and others, eds., *The All-Volunteer Force after a Decade*, note 26 above, p. 176.

62. Data provided by U.S. Department of Defense, April 1987.

6 THE EFFECTS OF TECHNOLOGICAL CHANGE ON EARNINGS AND INCOME INEQUALITY IN THE UNITED STATES

McKinley L. Blackburn and David E. Bloom

Although technological progress has always been an important feature of the U.S. economy, the introduction and diffusion of new technologies has proceeded at an especially rapid pace during the past two decades. Production technologies are being powerfully affected by the development of microprocessors and microcomputers, automated production processes, lasers and satellite communications equipment, and data handling and information systems. These and other changes are fundamentally transforming the nature of traditional workplaces, as well as leading to the emergence of new work environments. Important academic and policy issues relating to the impact of technological progress on employment and unemployment, on labor productivity, and on earnings levels are being raised as a by-product of the changing nature of work. The impact of recent technological changes on the distribution of income has also been the focus of much recent discussion.

This chapter analyzes trends in both income and earnings inequality over the past two decades and explores their association with technological progress. This task is complicated by several factors. First, technological progress is a complex notion that potentially subsumes an extraordinarily wide variety of specific advances. It makes little sense to think of technological progress as a single variable. Different technological changes can have different effects on labor markets, with correspondingly big differences

223

in their effects on labor productivity, labor income, and income dispersion. Second, there appears to be no set of easily measured and easily interpreted variables that satisfactorily reflect the nature and importance of particular technological changes, and certainly not of aggregate technological change. Third, technological change is only one of many factors that lead to changes in income inequality. Our ability to isolate the influence of technological change on income inequality is quite limited by well-known difficulties involved in building a statistical model that controls for the "correct variables" in the "correct way."

Given these problems, we confine ourselves in this chapter largely to a descriptive analysis in which we measure changes in income and earnings inequality over the past two decades, a period of time during which technological change has been so substantial that it has been termed a "second industrial revolution." Whether any changes in inequality that occurred over that period are the direct result of technological change is not the focal point of our analysis. For example, we will not attempt to determine how much income inequality there would be today if there had been more or less technological change during the past ten years. We will, however, lay out a simple economic framework that provides some clues about the nature of the inequality measures one might like to calculate and compare in order to learn about the relationships between technological change and income and earnings inequality. It is important to stress at the outset, however, that the effect of technological change on income and earnings inequality is indeterminate in a general theoretical sense, largely because technological change can affect the structure of labor demand and of labor supply in a variety of ways.

In brief, we find little empirical evidence of an association between technological change and *earnings* inequality since the late 1960s. In contrast, there is evidence of a positive association between technological change and *family income* inequality, which is likely reflective of the effect of technological change on the size, structure, and labor supply behavior of U.S. families.

PRELIMINARY EVIDENCE

Before discussing the theoretical linkages between technological development and the distribution of income, we present evidence concerning the extent and nature of changes in income inequality in the United States over the years 1967 to 1985. Our source of data is the public use samples of the March Current Population Surveys for the years 1968 to 1986. Each of these surveys contains data on a representative sample of U.S. households, with information on social, economic, and demographic characteristics of all household residents, including their income and earnings in the year preceding the survey. A more complete discussion of the data is provided later in the chapter.

A basic issue that arises in all empirical work on income inequality involves the formulation of an operational definition of the term *distribution of income*. We stress this point early because much of the interpretation of the evidence we present focuses on variations in the underlying concepts defining a distribution of income. Many of the conflicting conclusions reached by different studies in this area are explained by differences in the particular distribution of income that is analyzed.

In this chapter, we wish to draw a sharp distinction between two concepts of the income distribution. One is the *distribution of total family income across families*. For every family in a particular year's sample, we compute a single income statistic that is the sum of the incomes of all members of the family. The sources of income included are earnings, interest and dividend income, and government cash transfers (that is, this statistic does not account for noncash transfers or tax obligations). We then measure the cross-family dispersion of these incomes. Both Census families (two or more related persons living together) and unrelated individuals (individuals living alone or with other individuals to whom they are not related) are included in our definition of the family. This distribution is usually viewed as being closely related to a distribution of economic well-being, although it does not account for important differences across families such as the number of family members.[1]

The second distribution on which we focus is the *distribution of earnings across individuals*. Only income reported as being the direct result of work-related activity is included in this distribution. Also, only those individuals reporting that they worked at some time in the year preceding the survey are included in the population.[2] In contrast to the distribution of total family income, families with more than one earner will be represented more than once in the earnings distribution, while families with no earners will not be represented at all. This distribution most closely corresponds to a distribution of job/wage opportunities present in the labor market. It is related to a distribution of well-being because earnings is the main component of income, but this relationship is less close than the relationship between the total family income distribution and well-being, since the earnings distribution does not include nonlabor income and does not account for sharing of resources within families. As a result, the dispersion of these two distributions need not move in the same direction over time.

We measure dispersion of these two distributions in two ways. First, we compute a Gini coefficient, a standard measure of inequality that is a positive function of the degree of inequality. Second, we classify each income unit (that is, family or individual) into one of five income (or earnings) classes based on the relationship of each unit's income to the median level of income. The classes are defined as follows:

1. *Lower class* (LC): income less than or equal to 60 percent of the median income;
2. *Lower middle class* (LMC): income greater than 60 percent but less than or equal to 100 percent of the median;
3. *Middle class* (MC): income greater than 100 percent but less than or equal to 160 percent of the median;
4. *Upper middle class* (UMC): income greater than 160 percent but less than or equal to 225 percent of the median;
5. *Upper class* (UC): income greater than 225 percent of the median.

This class-based analysis is especially useful for pinpointing the location of changes in the income distribution; such information is not provided by a unidimensional inequality measure like the Gini coefficient.

The estimated inequality measures for the total family income distribution for the years 1967 through 1985 are presented in Table 6–1. The Gini coefficient follows an upward trend over the period, indicating that income inequality has increased over time. Changes in the percentages of the population falling in the five income classes reveal that the increase in income inequality is largely associated with a decline in the share of families in the

Table 6–1.   Inequality Measures for the Distribution of Total Family Income.[a]

Year	Gini	LC	LMC	MC	UMC	UC
1967	.395	29.7%	20.3%	27.5%	14.3%	8.3%
1968	.389	29.6	20.4	27.4	14.5	8.0
1969	.393	29.4	20.6	25.8	15.1	9.1
1970	.406	30.4	19.6	26.1	14.4	9.5
1971	.405	30.0	20.0	25.1	14.2	10.7
1972	.404	29.7	20.3	25.6	14.2	10.2
1973	.403	29.9	20.1	25.3	14.0	10.8
1974	.393	30.2	19.8	26.2	14.3	9.5
1975	.400	29.1	20.9	25.7	13.9	10.3
1976	.410	30.6	19.4	23.7	15.2	11.2
1977	.409	30.6	19.4	23.8	14.4	11.7
1978	.402	29.6	20.4	22.9	15.2	11.9
1979	.412	30.7	19.3	23.0	15.5	11.5
1980	.392	29.7	20.3	24.6	14.4	11.0
1981	.412	29.9	20.1	23.5	14.5	11.9
1982	.414	30.7	19.3	23.2	14.2	12.6
1983	.425	31.3	18.7	21.9	14.8	13.2
1984	.416	29.8	20.2	22.5	14.5	13.0
1985	.426	29.7	20.3	21.3	14.5	14.2

a. The population includes both Census families and unrelated individuals. The class measures are defined in the text. Total family income includes earned income, interest and dividend income, and government cash transfer income.

middle class and increases in the shares in the upper middle and upper classes. No secular change seems to have occurred at the lower end of the distribution.

The same inequality measures were computed for the earnings distribution; the results are reported in Table 6–2. It is clear from these statistics that there has been no upward trend in earnings inequality. The class percentages reveal only small changes in the distribution over time: The upper-class percentage seems to have increased, while the lower-class percentage has decreased.[3]

One other feature of the statistics in Tables 6–1 and 6–2 deserves mention. There is a remarkably low correlation (0.29) between the Gini coefficients for the distributions of total family income and individual earnings. This suggests that there were important

Table 6–2. Inequality Measures for the Distribution of Annual Earnings across Individuals.[a]

Year	Gini	LC	LMC	MC	UMC	UC
1967	.459	34.4%	15.6%	20.2%	15.5%	14.3%
1968	.462	35.3	14.7	20.3	16.4	13.3
1969	.466	34.8	15.2	19.5	16.3	14.2
1970	.466	34.5	15.5	20.8	15.7	13.5
1971	.472	34.8	15.2	19.5	16.2	14.3
1972	.472	33.6	14.4	19.7	16.2	14.1
1973	.474	35.3	14.7	19.1	14.9	16.0
1974	.466	34.7	15.3	18.8	16.0	15.2
1975	.468	34.5	15.5	19.8	14.8	15.5
1976	.469	34.9	15.1	20.8	13.5	15.6
1977	.468	34.3	15.7	18.5	15.2	16.3
1978	.461	33.6	16.4	19.5	14.0	16.5
1979	.464	33.5	16.5	20.0	15.4	14.6
1980	.454	32.8	17.2	21.2	14.9	13.8
1981	.460	33.5	16.5	21.1	14.4	14.5
1982	.470	34.5	15.5	19.4	14.2	16.4
1983	.464	35.1	14.9	21.0	13.7	15.4
1984	.468	33.1	16.9	18.9	13.5	17.6
1985	.467	33.8	16.2	20.1	13.1	16.8

a. The earnings measure includes both wage and salary income and self-employment income. All individuals with positive earnings in the year preceding the survey were included in the sample.

changes over time in the variables that differentiate those distributions.

It is tempting to jump from the observation that there has been no trend in earnings inequality to the conclusion that technological change since the late 1960s has had no effect on the dispersion of income in the United States. There are several reasons why we hesitate to do so. First, this conclusion would rest largely on the presumption that technological change affects the income distribution via its effect on the distribution of job/wage opportunities. As described below, the simple theoretical framework we use to consider the impact of technological change on income inequality identifies factors that would be expected to change family income inequality without affecting earnings inequality. Second, as we shall see below, focusing on the distribution of earnings among all workers masks significant differences in the trends in earnings inequality for particular subgroups of workers, such as males and females and workers in different industries. Finally, as noted earlier, simple time-series patterns can be quite misleading when data are potentially generated by complex multifactor models.

THE ECONOMIC FRAMEWORK

At a theoretical level, technological change can affect the distributions of income and earnings both through changes in the structure of labor demand as well as through supply-side variables relating to labor force participation, hours of work, and family size and structure. Consider an economy in which identical firms produce a single good using capital and two kinds of labor: skilled labor and unskilled labor. For the time being, we will assume that the supply curves for each type of labor are perfectly inelastic. In such an economy, technological change can have three basic effects on the demand for each type of labor: (1) a pure technology effect associated with the fact that a given amount of inputs can be used to produce the same or more output; (2) a scale effect due to the downward shift of cost curves leading to an increase in product demand and, concomitantly, in labor demand; and (3) a technical bias effect arising from the fact that new ratios of factor inputs may be optimal at the old ratio of factor prices.

Whether technological change leads to a change in earnings inequality depends on whether there are differences in these three effects for skilled and unskilled labor. Because all firms have the same production functions in this model (and therefore the same cost curves), the first two effects are, by definition, the same for both types of labor. However, unless technological change is neutral, signifying that the technical bias effect is zero, labor demand will shift differently for the two types of labor. This should alter the ratio of skilled to unskilled wages, which will cause earnings inequality to change. For example, if technological change leads to the substitution of skilled for unskilled labor, earnings inequality will increase.

If we relax the assumption that the supplies of skilled and unskilled labor are perfectly inelastic, the change in relative wages (and therefore the change in earnings inequality) will also depend on the relative slopes of the supply curves. In the context of the preceding example, if we allow the supply of unskilled labor to be responsive to the unskilled wage, the technical bias effect on earnings inequality will be muted. Of course, the pure technology effect and the scale effect can also affect earnings inequality when the supplies of the two types of labor are not perfectly inelastic. Assuming that the supply of unskilled labor is more elastic than the supply of skilled labor, the negative technology effect on labor demand will tend to decrease earnings inequality (that is, because wages of skilled workers will fall more than the wages of unskilled workers) while the positive scale effect will tend to increase earnings inequality. Insofar as the employment shares of skilled and unskilled labor also change when labor supplies are elastic, these effects can also change sign (see Robinson 1976). The bottom line of this part of our analysis is that the overall labor demand effect of technological change on *earnings* inequality is theoretically indeterminate. It is indeed remarkable that economic theory has so little to offer here, despite the strong simplifying assumptions we have made.

Changes in labor demand can affect the distribution of family income independently of their effects on earnings inequality. Suppose that all families have at most one earner and that technological change leads to an overall decrease in labor demand and in employment. Even if earnings inequality remains unchanged, income inequality will tend to increase. This difference arises because workers who become technologically unemployed drop out of the population of earners but remain in the population of families. On the other hand, if families can have more than one earner, income inequality can increase or decrease depending on the income position of the families whose members become unemployed. Thus, the labor demand effects of technological change on *family income* inequality are also indeterminate.

So far this discussion has focused on the long-run effects of technological change. In the short run, the effect of technological change will also depend on the ability of workers to adapt to changes in the structure of labor demand. For example, workers whose human capital loses value as a consequence of technological change can respond by investing in new human capital, by investing in job search, or by taking a lower-wage job. Their choice among these alternatives will have different implications for the dynamic pattern of earnings and income inequality.

Technological change can also affect income and earnings inequality via the supply side of labor markets—that is, through its effect on labor/leisure, market-work/home-work choices made within the family. This effect arises from the effect that technological change has on the nature of commodities consumed by the household and on the nature of production within the household.[4] The major source of such changes in recent years involves the expanding supply of commodities that reduce the time required to maintain a household of a given quality (such as child care or new products in food preparation, house cleaning, and home entertainment). Also important are innovations that allow greater choice as to the type of family to which an individual belongs (such as new methods of contraception).

Significant changes in the family have taken place in recent years. Here, we focus on the changes related to the increasing proportion of families headed by an unmarried individual and to the increasing labor market activity of married women. According to the March 1968 Current Population Survey, 8.0 percent of all family units were two- or more person families headed by an unmarried female, and 12.4 percent were females living alone. By March 1986, 12.8 percent of all families were in the former category, and 16.2 percent in the latter. The labor force participation rate for married women increased from 45.7 percent in 1968 to 57.9 percent in 1986. The overall labor force participation rate for women also increased because of the shift to female-headed families, since female heads tend to have higher labor force participation rates than married women. The percentage of married women working full-time, year-round also increased, from 20.3 percent of all married women in 1968 to 28.9 percent in 1986.[5] This led to an increase in the percentage of married-couple families where both spouses worked full-time, year-round (15.9 percent in 1968; 21.6 percent in 1986).

It is not clear whether technological progress related to household production has been mainly a cause or consequence of changes in the structure of the family and its economic activities. Nonetheless, to the extent that technological progress has facilitated these changes, it has played an important role in increasing the labor supply of women and changing the demographic composition of households. Insofar as changes in family income inequality reflect changes in the family *and* changes in income, it can be argued that technological progress has a supply-side relationship to income inequality. In addition, depending on how technological change affects the supplies of skilled and unskilled labor, it can also affect earnings inequality—in a variety of possible ways.

The main message of this section is that economic theory offers no unambiguous predictions regarding either the size or even the net direction of the effect of technological change on family income and earnings inequality. The issues at hand are completely empirical in nature.

PREVIOUS EMPIRICAL FINDINGS

The principal source of information on the distribution of total family income in the United States is the published data contained in the *Current Population Reports* P-60 series. These reports are compiled by the Bureau of Labor Statistics from the same March Current Population Survey data used in this chapter. The published statistics relating to family income inequality reflect the same finding as mentioned earlier: inequality seems to have increased in recent years. The BLS measures are also available for years prior to 1967 and show essentially no trend in inequality during most of the post–World War II period up to 1967. However, the definition of the family used by the BLS differs from the one used in this chapter because the BLS distributions refer to income inequality among Census families only (that is, our definition includes unrelated individuals as separate income units as well). The omission of individuals not living in Census families dismisses a large and growing segment of the U.S. population and eliminates one route through which technological change can affect income inequality. Nonetheless, the basic finding that there has been an upward trend in income inequality since the late 1960s is robust with respect to the definition of the family.

The claim that family income inequality has increased over the years 1968 to 1985 is not in dispute in the literature in this area. In contrast, the trend in earnings inequality is a source of much debate among academic researchers. In this section, we review research findings related to recent trends in earnings inequality in the United States. Our major goal is to reconcile the seemingly conflicting conclusions reached by different investigators.

In reviewing this literature, it should be kept in mind that three operational concepts underlie an "income distribution":

1. *The population.* Does it refer to all earners, males only, wage and salary workers only, and so forth?
2. *The income measure.* Is it all earnings (including self-employment income), or just wages and salaries?
3. *The unit of time.* Is it annual earnings, weekly earnings, or hourly earnings?

Careful attention should be paid to the conventions adopted by each researcher because different conventions seem to explain much of the cross-study variance in conclusions.[6]

Differences in findings across studies can also result from the use of different techniques to measure inequality, such as the variance of the logarithm of earnings or the Gini coefficient. Differences in the method used for measuring inequality is another potential explanation for differences in findings among studies.

We first focus on analyses performed for the distribution of annual earnings. In a widely quoted article, Henle and Rsycavage (1980) use grouped March CPS data to calculate Gini coefficients for all earnings (and for wages and salaries), separately for men and women, over the period 1958 through 1977. They also compute Gini coefficients when the sample is restricted to full-time, year-round[7] workers only.[8] For men, they find an overall upward trend in inequality for all earners (with some slowing for the 1970 through 1977 period), but no trend for full-time, year-round workers. For women, they find no trend for all workers and a downward trend for full-time, year-round workers. Results for a sample with both sexes combined are not provided. The fact that Henle and Rsycavage use grouped data means that the Gini coefficients are computed from information on the percentage of total income received by various income quintiles.[9] Plotnick (1982) used the same grouped data to compute variances of the logarithm of income for all earnings, but for men only. He finds an upward trend in the variance of logarithms for the years 1968 through 1977, the same result reported by Henle and Rsycavage using Gini coefficients.

Dooley and Gottschalk (1984) use March CPS data on individual male workers to calculate the variance of logarithms for wage and salary income. They do this for both annual and weekly earnings over the 1967 through 1978 period.[10] They use a sample that is restricted so that it is representative of civilian males between the ages of sixteen and sixty-two who were either year-round (but not necessarily full-time) workers in the previous year or were looking for work in those weeks they were not employed. They find a steep upward trend in the inequality of annual earnings over the period, with a less-pronounced increase for the inequality of weekly earnings. In their 1985 paper, they calculate the percentage of workers in their sample who fall below an arbitrarily chosen minimum earnings level, held constant in real terms over the period. This measure is more closely akin to the concept of absolute poverty than to the concept of inequality; however, the relatively flat profile of average earnings over much of this time period implies that this research can shed some light on earnings dispersion as well. The basic finding is that both the percentage of males with low annual earnings and the percentage with low weekly earnings increased over the years 1967 through 1978.

One problem plaguing both Dooley-Gottschalk papers is that their samples exclude all individuals who did not respond to the earnings questions in the CPS. The Census Bureau uses an imputation procedure for these nonrespondents that involves allocating to them the earnings level of an individual with similar characteristics who did respond to the earnings question.[11] The omission of such individuals would be no cause for concern if nonresponse was random; however, it is known that individuals with high actual earnings are more likely to be nonrespondents. Not using the imputed earnings values for these individuals results in an over-weighting of individuals with low incomes.[12] For the estimation of a regression model using CPS microdata, omitting imputed incomes may be an appropriate strategy (see Welch 1979); for estimating a population average, such as a variance of logarithms, it is not.

A recent paper by Harrison, Tilly, and Bluestone (1986) uses March CPS data to look at the inequality of annual wage and salary income, for all earners, over the years 1964 through 1983. Using individual-level data to calculate the variance of logarithms, they find evidence of a "U-turn" in inequality—the variance of logarithms fell gradually until the late 1970s, after which it began to rise sharply. The same pattern emerges when "all earnings" is the income measure. When the sample is separated by sex, they find decreasing dispersion for all women over the 1967 through 1977 period, followed by a sharp increase. This result for 1967 through 1977 conflicts with the findings of Henle-Rsycavage, who find decreasing inequality for females only when the sample is restricted to females who worked full-time, year-round. Harrison et al also find virtually no increase in inequality for men over the 1969 through 1977 period, contradicting the Henle and Rsycavage finding that it increased. The difference in findings between these studies is not easily explained. However, it does not appear to be due to the measure of inequality used, given Plotnick's finding that the variance of logarithms increased (consistent with Henle and Rsycavage); neither does it appear to reflect differences in the earnings variables because Henle and Rsycavage reach the same conclusion using both "all earnings" and "wage and salary income."[13] Harrison, Tilly, and Bluestone et al. argue that the shift in employment from the goods-producing sector to the service-producing sector is the primary reason for their finding of an increase in inequality in earnings inequality since the late 1970s.[14] However, they do not present any evidence on the inequality of annual earnings within industrial sectors.

Evidence related to inequality within industries, using weekly earnings as the income measure, is reported in Lawrence (1984). This study concludes that the employment shift from goods to services explains only a small part of the decline from 1969 to 1983 in the proportion of earners who are "middle-income." Lawrence uses a measure of "usual weekly earnings" from the CPS that includes only wage and salary income. The sample is restricted to full-time workers who were employed at the time of the survey. The income variable—usual weekly earnings—is conceptually distinct from another possible weekly earnings measure—average weekly earnings—which is calculated as the ratio of annual earnings to weeks worked over the course of the year. Because usual weekly earnings is available in the CPS only for individuals who are employed at the time of the survey, the sample Lawrence uses will have a higher percentage of year-round workers than the samples used in the analyses of annual earnings inequality.

In Lawrence's study, workers are classified as either low earners (defined as less than 66 percent of the median level of earnings among males), high earners (more than 132 percent of the median for males), or middle earners. He finds that the percentage of males with "middle earnings" fell from 56 percent in 1969 to 47 percent in 1983; for females, it increased from 39 percent to 44 percent; while for both sexes combined it went from 50 percent to 46 percent.[15] Lawrence also subdivides the sample according to whether the worker is employed in a goods-producing or a service-producing industry and finds that little of the change in the percentage of middle earners can be attributed to the increase in the proportion of service workers from 1969 to 1983. Rather, the middle-class decline occurred within the service sector and, especially, within the goods sector. No data for intervening years are analyzed by Lawrence.

Rosenthal (1985) examines the same hypothesis as Lawrence—that the middle of the earnings distribution has declined—but, unlike Lawrence, comes to the conclusion that no decline occurred. Although Rosenthal uses the same income measure as Lawrence (usual weekly earnings for full-time workers), his analysis varies from that of Lawrence in several ways. Rosenthal separates 416 three-digit occupations into thirds based on the median income among workers in the occupation in 1982 (that is, the "top third" contains the 33 percent highest-paying occupations). He then calculates the percentage of employees in the occupations that make up each of the thirds of the occupational ranking, for both 1973 and 1982. He finds that the fraction of workers in the "middle third" did not change over the period, while there was a decline in employment in the lower third and an increase in the top third. Thus, the highest-paying occupations also have the highest rate of employment growth. Including part-time workers does not change the basic conclusion.

Rosenthal's analysis provides incomplete information on the extent to which the overall change in inequality is due to a changing occupational structure because it ignores all variation of incomes within occupations. It also ignores any changes in the variation of incomes across occupations that fall within the bottom third, middle third, and top third of the occupational ranking. Contrary to the claims made in his conclusion, his study does not address the question of whether the earnings distribution has seen a decline in its "middle." However, Rosenthal's study nicely complements the Lawrence study, in that together they cast doubt on the hypotheses about broad industry and broad occupational shifts being responsible for any increase in weekly earnings inequality that may have occurred.[16]

The only recent analysis of hourly wages has been conducted by Medoff (1984). This study has sample restrictions that are similar to those of Lawrence and uses data on "usual weekly earnings" and "usual weekly hours worked" to compute an hourly earnings measure for employed individuals in the May CPS for various years.[17] Medoff finds that the variance of the logarithm of hourly earnings was at about the same level in 1984 as it was in 1973 and 1975. However, his results do show that earnings inequality increased from 1981 to 1984 (which he attributes to changes in the macroeconomic environment). He separates the samples for males and females and finds little evidence of a trend in hourly earnings inequality for either. He does, however, find evidence of increasing inequality within the manufacturing sector for the 1980s relative to 1973 through 1975, but a slight (though uneven) decrease in inequality for the nonmanufacturing sector.[18]

What is the bottom line on trends in earnings inequality? As should be clear from the foregoing discussion, different studies have reached widely varying conclusions. Unfortunately, because these studies typically use methods that also vary widely, it is difficult to infer much about the trend in earnings inequality. For example, Medoff finds no increase in inequality in wages and salaries from 1973 to 1983, while Harrison, Tilly, and Bluestone do find an increase. This conflict could be due to differences in the underlying populations analyzed (Medoff excludes public and agricultural workers, and any individual not employed at the time of the survey), different income measures ("usual" earnings versus earnings in the previous year) and different time periods (hourly versus annual). The only directly comparable results among the studies reviewed are from Plotnick and Harrison, Tilly, and Bluestone for male annual earnings inequality in the late 1960s to early 1970s period; yet even here Plotnick finds an increasing trend, while Harrison, Tilly, and Bluestone do not.[19]

The weight of the evidence does, nonetheless, seem to suggest an increase in male earnings inequality and a decrease in female earnings inequality over the years 1967 through 1975. It also appears that male earnings inequality increased from 1975 to 1983, while the results for females and all earners do not strongly support any conclusion.

TRENDS AND PATTERNS IN EARNINGS INEQUALITY

In this section, we analyze earnings inequality over the years 1968 to 1985. We extend the empirical analysis presented in the preliminary evidence above in several ways. First, we consider inequality for males and females separately. Then we attempt to control for variations in hours worked by restricting the population to full-time, year-round workers. Finally, we examine inequality within six industrial sectors to see if there are any cross-industry differences. Before presenting the results, we discuss the source of our data and also our approach to measuring inequality.

Data

The data we analyze are drawn from the March Current Population Survey public use samples for the nineteen years from 1968 to 1986. These data are commonly used in studies of income inequality in the United States. We use 10 percent samples of the original data. For our purposes, a major strength of the CPS data is that they are representative of the U.S. population. Observations with imputed incomes are included in our analysis, and sample weights are used in our computations.[20]

The March CPS data are not without their shortcomings. One undesirable characteristic of these data is a tendency for certain sources of income to be underreported by survey respondents. This is not a problem for the earnings measures because earnings, especially wage and salary income, tend to be well reported. However, both cash transfers and interest/dividend income are not well reported, and these sources show up in the family income measures used later. We assume that such underreporting is fairly stable over time and therefore does not bias inferences concerning the trend in inequality, although it does limit our ability to measure inequality accurately at a point in time. The fact that the share of income received as transfers has not grown since 1973 supports this assumption.

Another problem that arises for the family income distribution but not for the earnings distribution is the fact that the family is defined at the time of the survey, which is in March, although the reported income corresponds to the previous calendar year. If changes in the composition of a family occur between the time income is reported and the time the survey is taken, then the measure of total family income may not reflect the actual income received by the family. Burkhauser, Holden, and Myers (1986) have looked at the effect of this problem on measures of the transition into poverty among newly widowed women and have found that the bias for estimated transition probabilities can be large. However, they found the bias for the overall poverty rate to be small, suggesting that the problem may be relatively minor when calculating aggregate inequality measures.

There is also a "top-coding" problem with the earnings data that does not seem to have been fully appreciated in earlier research using the CPS. There are three sources of information on earnings in the March CPS data: wage and salary income, farm self-employment income, and nonfarm self-employment income. Prior to the 1981 survey, these three sources of earned income were never recorded as being above $50,000—in nominal dollars. All incomes greater than that amount were coded as equal to $50,000. Given the substantial inflation over the 1967 through 1980 period, the effect of holding the top-coded income level constant was to reduce the upper bound for the earnings measure—in real terms—over time. Narrowing the bounds within which income can be reported will bias most inequality measures downward. We deal with this problem by recoding earnings so that no figure above $50,000 in 1980 terms will be used—that is, we use a consistent real-dollar top-code over the 1967 through 1985 period.[21] Our analysis of changes in inequality does not, therefore, account for changes in the shape of the upper tail of the income distribution.

Measuring Inequality

The purpose of an income inequality index is to summarize the degree of income dispersion among N income-receiving units. There are many measures of inequality, each of which implicitly weights the sample data differently (see Atkinson 1970; Champernowne 1974). Because different inequality measures can sometimes lead to different results, we base our analysis on three single-number inequality indices: the Gini coefficient; the mean logarithmic deviation; and the coefficient of variation.[22] All three measures satisfy the main properties that are generally considered desirable for an inequality index. However, each index is particularly sensitive to changes in different parts of the income distribution: the mean logarithmic deviation to changes at lower levels of income; the coefficient of variation to changes at higher levels of income; and the Gini coefficient to changes around the middle of the income distribution.

We also continue to use the more descriptive class measures outlined in the preliminary evidence section above. A weakness of this measurement scheme is its insensitivity to changes that might occur within classes of the distribution. However, in practice, this group of measures does seem to highlight much of the change in the shape of the distribution. These measures also have an attractive characteristic that the three indices mentioned above do not: the class measures are not biased by the top-coding of incomes in the CPS. This is because the cutoff point for the upper class—that is, 225 percent of the median income—always lies below the level at which earned income was top-coded. The upper-class cutoff for the family income distributions discussed in the section on family income distribution below also lies beneath the top-code.

Results for Earnings

The Gini coefficient and the class percentages for the distribution of earnings among all individuals are reported in Table 6–2. Table 6–3 extends the results by presenting the mean logarithmic deviation and the coefficient of variation for the same distribution for the years 1967 through 1985. Table 6–3 also reports the three inequality measures for the earnings distribution when the population is restricted to full-time, year-round workers. Our earlier conclusion that earnings inequality has not changed significantly since the late 1960s is further supported by examination of these additional measures.

Each time series of a particular inequality index in Table 6–3 was regressed on a simple trend variable to provide a descriptive

Table 6–3. Other Measures of Inequality for the Distribution of Annual Earnings across Individuals.[a]

Year	All Earners		Full-Time, Year-Round Only		
	MLD	CV	Gini	MLD	CV
1967	.608	.856	.313	.207	.588
1968	.609	.857	.308	.206	.574
1969	.630	.865	.302	.186	.560
1970	.630	.865	.307	.194	.571
1971	.638	.880	.310	.191	.577
1972	.629	.878	.307	.202	.568
1973	.627	.882	.310	.200	.573
1974	.601	.870	.310	.205	.578
1975	.604	.875	.299	.170	.558
1976	.603	.879	.303	.187	.564
1977	.616	.870	.300	.180	.554
1978	.583	.859	.302	.187	.562
1979	.600	.868	.309	.193	.576
1980	.560	.851	.302	.184	.569
1981	.577	.858	.311	.201	.585
1982	.597	.891	.323	.222	.607
1983	.600	.869	.319	.232	.592
1984	.600	.881	.322	.219	.602
1985	.591	.874	.320	.213	.593
Trend coefficient	−.0021 (.0009)	.0005 (.0005)	.0005 (.0004)	.0008 (.0010)	.0009 (.0009)
Trend coefficient without cycle	−.0025 (.0013)	−.0005 (.0008)	.0006 (.0006)	.0010 (.0013)	.0010 (.0011)

a. Full-time, year-round workers are defined as individuals who worked thirty-five or more hours per week for at least fifty weeks over the course of the year for which earnings is reported. MLD is the mean logarithmic deviation, and CV is the coefficient of variation.

measure of the time trend in inequality. Regressions of each inequality index on a trend and the adult male unemployment rate were also fit in an attempt to describe the trend in inequality controlling, at least crudely, for business cycle effects on inequality.[23]

The estimated trend coefficients in the earnings inequality regressions are presented at the bottom of Table 6–3. The conclusion that earnings inequality has not increased over time is supported by the estimates. The trend coefficient for all workers is small and insignificant when the dependent variable is the coefficient of variation, while it is significant and negative for the mean logarithmic deviation. We noted earlier in the discussion of Table 6–2 that the lower-class percentage for the earnings distribution fell over the period, while there was a slight increase in the upper-class percentage. The difference in trend coefficients for the coefficient of variation and the mean logarithmic deviation reflects the relatively greater sensitivity of the mean logarithmic deviation to changes at the lower end of the distribution.[24]

Table 6–4 reports earnings inequality measures calculated separately for males and females. The statistics in this table reveal that there were widely different trends in earnings inequality for males and females who worked in the years 1967 through 1985.[25] Earnings inequality did not change for females who worked full-time, year-round, and actually fell for all women. In contrast, earnings inequality for males increases, both for the population of all workers, and for full-time, year-round workers only. Blackburn (1987) presents evidence that the increase in earnings inequality for males is the result of changes in the age composition of the male labor force and, to a lesser extent, industrial shifts.[26]

Restricting the population to full-time, year-round workers is an attempt to control for changes over time in hours worked. A slightly different way to control for hours worked is to examine the distribution of hourly wage rates.[27] We are able to compute wage rates using the March CPS data by dividing the annual earnings measure for each individual by the product of weeks worked and hours worked per week for that same individual. Unfortunately, the information on hours worked is available only beginning with the 1976 survey. Table 6–5 reports the mean logarithmic deviation and the coefficient of variation for hourly wages, reported separately for males and females, over the 1975 through 1985 period. There is much variance in the indices over time (especially for the coefficient of variation), making it difficult to pinpoint any trend. It would seem that there is no clear trend for either males or females, except perhaps an increasing trend for

Table 6–4. Earnings Inequality for Males and Females.

	All Workers				Full-Time, Year-round Only			
	Males		Females		Males		Females	
Year	Gini	MLD	Gini	MLD	Gini	MLD	Gini	MLD
1967	.389	.468	.477	.630	.281	.174	.283	.178
1968	.390	.460	.481	.636	.278	.171	.278	.186
1969	.398	.477	.488	.677	.272	.161	.272	.151
1970	.401	.499	.481	.649	.281	.178	.265	.139
1971	.406	.496	.485	.661	.284	.172	.265	.141
1972	.405	.483	.480	.649	.278	.181	.259	.142
1973	.403	.479	.481	.633	.278	.171	.267	.153
1974	.401	.468	.476	.617	.282	.181	.275	.164
1975	.406	.473	.478	.622	.276	.153	.255	.122
1976	.408	.481	.470	.600	.280	.171	.252	.135
1977	.405	.490	.475	.624	.274	.166	.256	.127
1978	.403	.460	.466	.603	.277	.161	.259	.157
1979	.408	.478	.456	.606	.285	.176	.253	.134
1980	.399	.436	.454	.584	.278	.153	.259	.165
1981	.411	.486	.452	.572	.293	.196	.263	.138
1982	.428	.509	.461	.593	.301	.203	.279	.177
1983	.423	.527	.460	.591	.301	.231	.282	.176
1984	.428	.519	.460	.593	.308	.206	.275	.178
1985	.424	.491	.468	.612	.305	.197	.288	.184
Trend coefficient (/100)	.18 (.04)	.15 (.12)	−.12 (.05)	−.36 (.10)	.14 (.05)	.19 (.09)	.02 (.09)	.06 (.12)
Trend coefficient without cycle (/100)	.13 (.05)	−.01 (.15)	−.14 (.06)	−.32 (.14)	.15 (.07)	.14 (.14)	.03 (.11)	.13 (.17)

Table 6–5. Inequality Measures for the
Distribution of Wages, for Males and Females.[a]

Year	Males		Females	
	MLD	CV	MLD	CV
1975	.253	.752	.235	.865
1976	.271	.781	.215	.752
1977	.253	.728	.241	.937
1978	.251	.767	.248	.947
1979	.257	.689	.261	.870
1980	.226	.654	.273	.967
1981	.273	.748	.228	.793
1982	.280	.748	.242	.752
1983	.271	.753	.256	.777
1984	.291	.767	.280	.950
1985	.277	.753	.290	.869

a. The wages were computed as annual earnings divided by the product
of hours worked per week and weeks worked over the year. Wages above
$99.99 an hour (in 1983 dollars) were top-coded at $99.99.

male wage inequality when the mean logarithmic deviation is the
index used.

Earnings by Industry

This section analyzes trends in earnings inequality within and
across industries. As mentioned earlier, there has been a shift over
time from goods-oriented to service-oriented employment. For instance, 41 percent of full-time, year-round workers were employed
in goods-producing industries in 1967, with the remaining 59 percent employed in the service-producing industries. By 1984, the
goods-producing share of employment had fallen to 31 percent,
while the service-producing share had risen to 69 percent. Because
inequality is higher within service-producing industries, this shift
would, other things equal, tend to increase overall earnings inequality.

Table 6–6 reports Gini coefficients for earnings within six broad
industrial groupings.[28] Only full-time, year-round workers are in-

Table 6-6. Gini Coefficients for Earnings within Industry Groups, Full-Time, Year-Round Workers.[a]

Year	Manufacturing	Other Goods	Trade Services	Trade	Services	Public Administration
1967	.266	.368	.269	.345	.363	.240
1968	.263	.377	.278	.329	.355	.226
1969	.266	.324	.266	.334	.349	.246
1970	.270	.370	.276	.324	.342	.240
1971	.257	.374	.293	.336	.338	.256
1972	.273	.352	.285	.339	.324	.249
1973	.279	.333	.277	.332	.348	.258
1974	.273	.360	.285	.341	.325	.282
1975	.265	.333	.270	.327	.321	.257
1976	.261	.366	.277	.325	.325	.258
1977	.265	.339	.269	.329	.322	.245
1978	.284	.315	.273	.323	.323	.226
1979	.267	.346	.287	.340	.328	.268
1980	.275	.323	.289	.321	.318	.250
1981	.270	.362	.290	.340	.328	.245
1982	.286	.355	.297	.341	.351	.246
1983	.286	.360	.293	.345	.330	.273
1984	.287	.351	.298	.342	.336	.271
1985	.285	.330	.294	.355	.343	.260
Trend coefficient	.0012 (.0003)	−.0010 (.0007)	.0013 (.0004)	.0005 (.0004)	−.0008 (.0009)	.0008 (.0006)
Trend coefficient without cycle	.0014 (.0005)	−.0034 (.0009)	.0011 (.0007)	.0009 (.0007)	−.0011 (.0012)	.0014 (.0011)

a. "Other goods" includes agriculture, construction, and mining. "Traditional Services" includes transportation, communications, public utilities, financial services, insurance, and real estate. "Services" includes personal, business and repair, entertainment and recreation, and professional and related services.

cluded in the samples analyzed here. It is apparent that there are
substantial differences in the level of earnings inequality within
these industry groups, with manufacturing, public administration,
and the "traditional" services groups having the lowest Gini coefficients, and the services, trade, and other goods sectors having the
highest levels of inequality.[29] The share of full-time, year-round
workers in the three industries with higher inequality increased
from 49.7 percent in 1967 to 54.9 percent in 1984. This employment shift contributed to increased earnings inequality, although
the magnitude of the effect is small. We can calculate this magnitude using a decomposition of the mean logarithmic deviation that
allows us to express changes in inequality as a simple function of
changes in industry employment shares, changes in industry mean
incomes, and changes in inequality within industries (see Bourguignon 1979). Using this property, we calculate that the industry
employment shifts can account for an increase in the mean logarithmic deviation of .005, or about 40 percent of the (small) total
increase from .207 in 1967 to .219 in 1984.[30] These results provide
little support for the Harrison, Tilly, and Bluestone argument that
sectoral shift has led to increased earnings inequality.

Earnings inequality did not move in the same direction for each
industrial sector between 1967 and 1985. For instance, inequality
increased in manufacturing, traditional services, and public administration, fell slightly in services and other goods, and held
steady in trade. As a result, there was less variation in the level of
inequality across industries in 1985 than in 1967.

THE FAMILY INCOME DISTRIBUTION

This section presents a more detailed examination of trends in the
distribution of income when the family is the unit of analysis. As
has been discussed above, technological change can influence supply-side behavior in a way that affects income inequality measured
across families without affecting earnings inequality measured
across individuals. Indeed, the empirical facts reported in the preliminary evidence section of this chapter are consistent with the
hypothesis that the effects of technological change on the distribution of well-being operate primarily through the supply side, and
not the demand side, of labor markets.

As we alluded to earlier, there are conceptual problems with the income data available from the Current Population Survey. Ideally, one would like to have a measure of income that closely reflects the level of economic well-being of the family unit. The CPS income measure falls short of this ideal in several ways. First, it does not include noncash transfers, nor does it include capital gains income. Second, there is no natural control for the fact that families with different compositions will derive different amounts of well-being from the same level of income. Third, CPS income data refer to pretax income. With regard to this last point, Pechman (1987) shows that there has been little change in the progressivity of the tax system from 1966 to 1985.[31] This finding makes the use of pretax income in studying the dispersion of incomes somewhat less objectionable.[32]

It has been shown that there was an increase in total family income inequality over the years 1967 to 1985. To investigate the sources of this trend, we now consider three related distributions. First, we examine the distribution of equivalent income, which lets us control, though imperfectly, for the effects of family composition. Second, we examine the distribution of total family earnings, which allows us to assess the importance of changes in the distribution of income that families receive from the labor market. This distribution is compared to the distribution of earnings among families' principal earners. This latter distribution allows us to examine inequality among families when the number of earners per family is held constant.

Equivalent Income

The income measure for the distribution of equivalent income is constructed by dividing the level of income for each family by the number of equivalent adults in the family, determined through a set of equivalence scales. Each person is assigned the equivalent income of his or her family, with inequality measured across persons. As pointed out by Danziger and Taussig (1979), this distribution relates more closely to well-being than the distribution of family income because it explicitly recognizes certain key differences among families (for example, that large families need more income to achieve a given level of welfare than small families). The equivalence scales used are those implicit in the BLS poverty lines developed by Orshansky (1965).

The results for the distribution of equivalent income are reported in Table 6–7. The Gini coefficient increases for this distribution, as it did for the distribution of total family income. However, unlike the increase in total family income inequality, most of the increase for equivalent income occurred in the last five years studied. Looking at the class percentages, the major change appears to be a movement from the lower-middle class to the lower class.

Total Family Earnings

The distribution of total family earnings uses the sum of the earnings of each member of a family as the measure of income for that

Table 6–7.   Inequality Measures for the Distribution of Equivalent Income across Persons in the United States

Year	Gini	LC	LMC	MC	UMC	UC
1967	.367	25.5%	24.5%	27.7%	12.9%	9.4%
1968	.360	23.7	26.3	28.7	12.6	8.7
1969	.364	24.6	25.4	27.6	13.1	9.4
1970	.367	24.2	25.8	27.3	13.9	8.8
1971	.371	23.9	26.1	27.7	12.8	9.5
1972	.362	24.5	25.5	29.0	12.5	8.4
1973	.363	23.3	26.7	26.9	14.3	8.8
1974	.362	24.3	25.7	28.4	13.0	8.6
1975	.364	25.4	24.6	26.7	14.4	8.9
1976	.367	25.5	24.5	29.1	12.5	8.4
1977	.358	25.1	24.9	27.6	14.0	8.3
1978	.361	24.3	25.7	27.3	14.0	8.7
1979	.363	25.2	24.8	27.4	14.1	8.6
1980	.357	25.7	24.3	28.3	13.8	7.9
1981	.380	27.1	22.9	27.0	13.9	9.1
1982	.390	27.0	23.0	25.3	14.5	10.2
1983	.395	28.0	22.0	25.9	13.5	10.6
1984	.391	28.0	22.0	27.2	12.8	10.0
1985	.394	27.3	22.7	26.0	13.7	10.3

family. Some families have zero total earnings over the year in question; these families are dropped from our sample. This makes the sample comparable to the sample used for the distribution of earnings among principal earners that we discuss below. The total family earnings distribution suffers from the inconsistent top-code problem mentioned earlier. We deal with this problem by using income-class shares to study changes in the shape of the distribution and to compare its shape with other distributions.

Table 6–8 contains class breakdowns for the family earnings distribution. The inequality of family earnings appears to have increased over the years 1967 to 1985. Changes in the upper end of the distribution are similar to those that occurred for the distribution of total family income. There was also a shift from the lower-middle to the lower class, which does not occur for the total family income distribution.

Table 6–8. Inequality Measures for the Distribution of Total Family Earnings.[a]

Year	LC	LMC	MC	UMC	UC
1967	26.3%	23.7%	29.3%	14.1%	6.6%
1968	25.6	24.4	30.8	12.9	6.3
1969	26.5	23.5	28.9	13.7	7.4
1970	27.5	22.5	29.7	12.9	7.4
1971	28.3	21.7	28.6	13.3	8.0
1972	27.0	23.0	28.2	13.5	8.3
1973	26.8	23.2	27.3	13.9	8.8
1974	28.1	21.9	28.6	13.6	7.8
1975	27.2	22.8	27.7	13.7	8.6
1976	28.1	21.9	27.1	13.8	9.1
1977	28.8	21.2	26.8	14.8	8.4
1978	28.5	21.5	26.6	14.4	9.0
1979	29.7	20.3	26.9	14.1	9.0
1980	27.9	22.1	26.5	14.8	8.7
1981	27.8	22.2	27.1	13.8	9.1
1982	29.2	20.8	25.2	14.0	10.8
1983	29.4	20.6	24.8	14.8	10.3
1984	28.7	21.3	24.9	15.1	10.0
1985	28.6	21.4	22.9	15.5	11.6

a. The relevant population includes only those families with positive earnings for the year in question.

The Principal Earner

For each family, we define the principal earner to be (roughly) the family member with the highest level of earnings in the previous year.[33] We use this construct as an alternative to the "head of the household" because the Census Bureau's definition of the household head changed over the 1967 through 1985 period.[34] Reported earnings were consistently top-coded at $50,000, in 1980 dollars, in the same manner as described earlier for the earnings of all individuals.

The class breakdown and the Gini coefficient for the distribution of earnings among principal earners are reported in Table 6–9. As with the distribution of total family earnings, we observe a rise in the lower-class and upper-class shares that coincides with a fall in the middle-class share. Table 6–10 reports the Gini coefficient and the mean logarithmic deviation for principal earners separately by sex and by full-time, year-round status. The trends are similar to those observed for the distributions of individual earnings, with inequality rising for males and falling for females. The fact that males constitute a larger fraction of the principal earner population than of the all-earners populations explains why earnings inequality for principal earners increases, whereas no increase is observed for the earnings distribution measured across all earners.

A change in the percentage of principal earners who work full-time, year-round might also be expected to affect earnings inequality among principal earners. Surprisingly, there was little change in this statistic from 1967 to 1985. Table 6–11 reports the percentage of principal earners who worked full-time, year-round for four types of families in both 1967 and 1985. The statistics reveal a shift from married-couple families to nontraditional families and a decline in the proportion of full-time year-round workers among female-headed single-parent families. These changes would have led to a decline in the full-time year-round percentage for all principal earners because married couples have the highest probability of having a full-time, year-round principal earner. However, their effect seems to have been offset by increases in the percentages of female unrelated individuals and male-headed single-parent families and unrelated individuals who work full-time, year-round.[35]

Table 6–9.   Inequality Measures for the Distribution of Earnings among Principal Earners.

Year	Gini	LC	LMC	MC	UMC	UC
1967	.346	25.2%	24.8%	31.4%	13.0%	5.6%
1968	.342	23.6	26.4	33.1	10.4	6.5
1969	.351	26.1	23.9	31.5	12.3	6.1
1970	.357	26.0	24.0	31.6	11.5	6.9
1971	.360	26.8	23.2	32.0	11.2	6.8
1972	.359	24.8	25.2	30.5	11.7	7.8
1973	.354	26.4	23.6	29.5	13.3	7.3
1974	.360	27.1	22.9	32.5	11.4	6.1
1975	.357	25.0	25.0	28.9	13.7	7.4
1976	.367	27.0	23.0	27.9	14.0	8.2
1977	.357	27.6	22.4	30.2	13.3	6.5
1978	.362	26.1	23.9	27.8	14.6	7.6
1979	.367	27.6	22.4	28.0	14.1	7.9
1980	.357	25.4	24.6	29.4	13.8	6.7
1981	.360	25.2	24.8	29.4	12.8	7.7
1982	.375	28.3	21.7	29.1	12.6	8.2
1983	.373	28.5	21.5	28.1	14.6	7.3
1984	.372	27.1	22.9	29.4	12.4	8.2
1985	.374	27.7	22.3	27.3	14.0	8.6

a. The principal earner is defined as the head of household for nonmarried couple families. For married couples, the principal earner is the spouse with the higher level of earnings in the year preceding the survey.

In Blackburn and Bloom (1987), we presented results suggesting that the growth in married females' earnings has not contributed to increasing inequality of total family income over the years 1967 through 1984. However, comparing the class percentages for total family earnings and earnings among principal earners suggests that the earnings of nonprincipal earners has had a positive impact on inequality. Both distributions have become more disperse over time, but the change is larger for the distribution of total family earnings. To describe these changes more precisely, we fit regressions of the lower-class, middle-class, and upper-class percentages for the two distributions on a time trend, and on a time trend and the adult male unemployment rate. The results are reported in Table 6–12. The trend coefficients are uniformly smaller (in abso-

Table 6–10. Earnings Inequality among Principal Earners, Males and Females.

Year	All Workers				Full-Time, Year-Round Only			
	Males		Females		Males		Females	
	Gini	MLD	Gini	MLD	Gini	MLD	Gini	MLD
1967	.303	.227	.416	.452	.264	.146	.295	.195
1968	.302	.217	.417	.439	.263	.140	.307	.218
1969	.309	.244	.417	.407	.260	.143	.297	.178
1970	.316	.254	.408	.441	.267	.150	.291	.159
1971	.319	.246	.420	.460	.268	.135	.272	.139
1972	.317	.258	.409	.487	.264	.153	.270	.161
1973	.312	.251	.402	.435	.262	.148	.261	.130
1974	.317	.260	.395	.447	.264	.148	.253	.139
1975	.320	.242	.393	.413	.262	.131	.249	.117
1976	.326	.264	.393	.411	.265	.145	.250	.134
1977	.312	.236	.385	.418	.255	.145	.257	.131
1978	.321	.254	.396	.418	.263	.134	.256	.127
1979	.323	.264	.385	.415	.265	.149	.255	.122
1980	.318	.241	.382	.405	.260	.129	.260	.138
1981	.322	.256	.381	.380	.272	.155	.266	.140
1982	.338	.281	.393	.379	.280	.163	.280	.144
1983	.335	.294	.397	.392	.277	.173	.287	.163
1984	.340	.289	.386	.414	.282	.160	.272	.161
1985	.334	.260	.402	.410	.277	.153	.277	.137
Trend coefficient (/100)	.17 (.03)	.24 (.07)	−.15 (.05)	−.39 (.09)	.08 (.04)	.08 (.06)	−.10 (.12)	−.26 (.16)
Trend coefficient without cycle (/100)	.10	.13	−.17	−.22	.06	.06	−.14	−.23 (.19)

Table 6–11. Percentage of Principal Earners Working
Full-Time, Year-Round, by Family Type, in 1967 and 1985.[a]

Family Type	Percentage Full-time, Year-Round		Percentage of Principal Earner Population	
	1967	1985	1967	1985
Married-couple	78.4%	79.1%	77.3%	65.3%
Male-headed single-parent family or unrelated individual	62.5	67.2	8.6	15.5
Female-headed single-parent family	63.3	51.8	5.9	10.6
Female unrelated individual	48.5	65.7	8.2	10.9
All families and unrelated individuals	73.7	73.0	100.0	100.0

a. The percentages reported in this table pertain to the population of families with principal earners who have positive earnings and do not apply to the population of all families.

lute value) for the principal earner distribution and are much smaller when looking at the upper-class percentages. This indicates that a significantly larger percentage of families were moved into the upper class when using all earnings—which includes the earnings of nonprincipal earners—than when only the earnings of principal earners are included. This implies that changing family behavior related to the labor force participation of its members has had a positive impact on the inequality of total family income.

CONCLUSION

This chapter has explored the relationship between technological change and inequality in the United States since the late 1960s. Because technological change is so difficult to characterize and measure at an aggregate level, our analysis has focused primarily on studying patterns and trends in the dispersion of various distributions of earnings and income during this recent period of rapid technological progress. If technological change is related to inequality, we would expect the inequality data for this period to

Table 6–12. Trend Regressions for the Distribution of Total Family Earnings and of the Earnings of the Principal Earner.[a]

	Total Family Earnings			Principal Earner		
	LC	MC	UC	LC	MC	UC
Trend coefficient	.0015	−.0031	.0022	.0014	−.0024	.0010
	(.0003)	(.0002)	(.0002)	(.0004)	(.0004)	(.0002)
Trend coefficient without cycle	.0013	−.0032	.0018	.0010	−.0025	.0009
	(.0006)	(.0005)	(.0004)	(.0007)	(.0008)	(.0004)

a. The dependent variable is the class percentage series for either the lower (LC), middle (MC), or upper (UC) class, for either the total family earnings or principal earner distributions.

reveal systematic patterns. Although economic theory has little to offer regarding the nature of such patterns, it does provide some useful suggestions about the type of income and earnings distributions one might study and compare to explore the linkage between technological change and inequality. Thus, under the assumption that the effect of technological change on inequality operates primarily through the demand side of labor markets—by altering the nature of jobs and therefore of wage opportunities—we would expect to see shifts in the inequality of earnings measured across individuals. In contrast, if the effect of technological change has affected inequality primarily as a supply-side phenomenon—through changes in decisions about family size, structure, and labor supply—we would expect to see limited changes in earnings inequality measured across individuals but sizable changes in the distribution of total family income measured across families.

On the basis of our review of relevant literature, and several empirical analyses we performed using microdata from the March Current Population Surveys from 1968 to 1986, we have four main sets of results to report.

1. The often contradictory conclusions reached by studies of recent trends in income and earnings inequality are largely explained by the reliance of different researchers on a remarkably wide range of data analytic conventions. For example, the list of important dimensions in which previous studies vary includes (a) the time period covered; (b) the way family units are defined; (c) the population to which the studies of individual earnings generalize (such as all earners, private nonagricultural workers, male earners, wage and salary workers, full-time year-round workers, and so forth); (d) the measures of earnings and income (such as total family income, equivalent family income, total family earnings, wage and salary income, and so forth); (e) the unit of time for the measurement of earnings (such as annual, weekly, or hourly); (f) the nature of the earnings measure (such as usual earnings, or average earnings); (g) measures of inequality (such as the Gini coefficient, income-class shares, variance of logarithms, coefficient of variation, mean logarithmic deviation); (h) the use of individual or grouped income/earnings data; (i) the treatment of sample

weights; (j) the treatment of observations with imputed incomes; (k) the handling of top-coded values of income and earnings; and (l) other criteria for including observations in the sample such as the age of the respondent and whether the respondent was working at the time of the survey or in the year preceding the survey.

2. The time profile of earnings inequality, measured across individual workers has been quite flat since the late 1960s. Among females, earnings inequality fell over time, although it was flat for women who worked full-time, year-round. In contrast, earnings inequality increased for males, both among the population of all workers and that of full-time, year-round workers. The upward trend in earnings inequality for males is less apparent if one focuses on the dispersion in hourly earnings, suggesting that some of the increase in the dispersion of annual earnings is due to increased dispersion in the supply of labor by males. In related work, Blackburn (1987) presents evidence that the increase in earnings inequality for males is closely related to changes in the age composition of the male labor force and somewhat related to changes in employment shares across industries.

3. Earnings inequality among full-time, year-round workers varies substantially across industries. Although high-inequality industries increased their share of total employment from 1967 to 1984, this change can account for a only small fraction of the small increase in earnings inequality over those years. Thus, our results provide little support for either part of the compound hypothesis that earnings inequality has increased and that the increase was primarily the result of sectoral shift in the U.S. economy.

4. Inequality of total family income and total family earnings increased from 1967 to 1985. Inequality of equivalent income (that is, total family income divided by the number of equivalent adults in the family) also tended to increase over this period, though most of the increase took place in the 1980s. Dispersion in the distribution of earnings among families' principal earners also increased since the late 1960s, although the overall increase reflects a combination of an increase for male principal earners and a decrease for female principal

earners. A comparison of the magnitude of changes in the inequality of total family earnings and of earnings among principal earners leads one to conclude further that the earnings of nonprincipal earners has had a positive effect on income inequality over the past two decades.

The main message of this chapter is that there is little empirical evidence that earnings inequality has increased since the late 1960s, and even less evidence to support the hypothesis that any changes that have occurred have resulted from the effect of technological change on the demand for labor. However, the fact that inequality of total family income increased since the late 1960s and that some of the increase appears to be due to changes in family composition and labor supply behavior is consistent with the hypothesis that technological change has had positive supply-side effects on income inequality in the United States. Unfortunately, the nebulous nature of technological change, the multiplicity of ways in which technological change can affect inequality, and the fact that inequality is influenced by many other economic and demographic forces as well, makes it impossible to know whether recent trends in inequality will continue into the future.

NOTES

This report was prepared as a background paper for the National Academy of Sciences, Panel on Technology and Employment. The authors wish to thank David Mowery for his helpful comments on an earlier draft of this paper.

1. The correspondence between the income of a family and its well-being is only approximate. Several pecuniary and nonpecuniary factors are omitted in the analysis, such as cross-family variations in wealth and variations in price levels across regions of the country. We also treat each family identically, though some otherwise-equivalent families may receive higher levels of utility from a given level of income than other families. Families also differ in their income needs, such as larger families tend to need more income than smaller families to enjoy the same standard of living. This latter factor is taken into account later in the chapter.

2. The actual restriction is that only individuals with positive earnings in the calendar year preceding the survey are included in the population. This excludes those individuals who only "worked without pay."

3. This conclusion is supported by a regression of the Gini coefficient on a constant and a time trend; after correcting for first-order serially correlated errors, the estimate of the trend coefficient for the earnings distribution was .00004, with standard error 0.00031. Including the adult male unemployment rate on the right-hand side as a proxy for business cycle effects, the estimate falls to −.0001, with standard error 0.0004. For the total family income distribution, without the unemployment rate the estimated trend coefficient is .0013 (.0003), while controlling for the cycle results in a coefficient estimate of .0005 (.0006).

4. For a discussion of the theory of household production, see Becker (1981).

5. These statistics refer to the labor force activity in the calendar year prior to the year in which marital status is measured.

6. Another important aspect of a researcher's analysis is the dataset he uses. However, for the studies reviewed in this section, and for the empirical work in this paper, either the March or the May Current Population Surveys served as the primary source of data. Since the CPS does not differ by month in its method of sampling, it is doubtful that much of the differences in conclusions drawn by different studies are the result of differences in the datasets analyzed.

7. A year-round worker is defined as an individual who was employed at least fifty weeks in the previous year; a full-time worker is defined as an individual who works at least thirty-five hours per week.

8. The sample used by Henle and Rsycavage for their wage and salary distributions only includes wage and salary workers who are employed at the time of the survey. These results are not discussed here.

9. Henle and Rsycavage also use the share of income received by the bottom 10 percent, the top 10 percent, and the top 5 percent, in calculating their inequality measures. Gastwirth (1972) shows that there is a problem in the Census Bureau's method of calculating Gini coefficients from grouped data (mainly due to the fact that their method does not incorporate information on the average level of income within the relevant groupings), which, for most conventional income distributions, causes their estimates to be biased upward (relative to the Gini coefficient one would calculate from individual-level data).

10. Prior to 1976, March CPS data contain information on weeks worked in the previous year only in a coded interval form, such as one-to-fifteen weeks. Dooley and Gottschalk do not discuss how they construct a weekly earnings variable from annual earnings and a coded weeks variable.

11. For more information on the Census Bureau's imputation procedure, known as the "hot deck" method, see David et al. (1986).

12. The Bureau of Labor Statistics also provides "population weights" that account for both the sampling scheme used and the tendency of the CPS to oversample certain demographic groups. The Census Bureau advises that these weights be used in calculating population averages involving incomes. None of the articles discussed in this section mention use of the weights (though the grouped data used by Henle-Rsycavage were most likely computed using a procedure that takes account of the weights.)

13. In a later study, Bluestone and Harrison (1986) compare the earnings distributions among all workers in 1978 and 1984 and find that most of the "job growth" over that six-year period occurred at the lower end of the distribution. No evidence is presented on the sensitivity of this result to Bluestone and Harrison's choice of years.

14. The fact that they find decreasing inequality for 1964 through 1978, a period during which employment was also shifting to the service sector, leads one to question why this explanation should be given such importance. Indeed, Urquhart (1984) shows that the shift from goods to services may have been more rapid from 1967 to 1972 than it was from 1977 to 1982.

15. Because the income cutoffs for the females depend on the male, not the female, median level of earnings, part of the increase for females reflects the rise of female-male wage ratios by 1983.

16. This conclusion is confirmed by McMahon and Tschetter (1986) who show that whatever changes there were in the inequality of weekly earnings were due largely to changes within occupations, and not to employment shifts toward occupations with relatively high and relatively low average levels of earnings.

17. Medoff excludes public-sector and agricultural employees and includes part-time workers.

18. However, in all six years analyzed, inequality in the nonmanufacturing sector is substantially higher than in the manufacturing sector.

19. Actually, these two analyses still differ on two accounts: (a) Harrison, Tilly, and Bluestone analyze wage and salary income, while

Plotnick uses all earned income; and (b) Harrison, Tilly, and Bluestone calculate the variance of logarithms using individual-level
data, while Plotnick uses grouped data.

20. Because the use of imputed incomes is assumed when the BLS
computes its weights (see note 12), omitting observations with imputed incomes is tantamount to changing the weights used in the
analysis.

21. The lowest real value for the top-coded level of earnings occurs in
the March 1981 CPS (pertaining to income in 1980). The following
year, the nominal value of the top-code was raised to $75,000; in 1985
it was raised to $99,999.

22. The mean logarithmic deviation, proposed by Theil (1967), is the
logarithm of the ratio of the arithmetic mean of income to the
geometric mean of income. These measures, and their properties,
are discussed more fully in Blackburn (1987).

23. The trend coefficients at the bottom of Table 6–3 are the estimated
values of b_1 and b_2 in the following equations:

$$\text{Trend: } I(t) = a_1 + b_1{}^*t + e_1(t)$$
$$\text{Trend without cycle: } I(t) = a_2 + b_2{}^*t + c^*U(t) + e_2(t)$$

where $I(t)$ is the level of inequality in year t, and the error terms
$e_1(t)$ and $e_2(t)$ are assumed to be normally distributed, and to follow
an $AR(1)$ process. $U(t)$ is the adult male unemployment rate in year
t.

24. It also implies that there has not been an outward shift of the
Lorenz curve over time, which would occur if and only if there had
been an unambiguous increase in inequality (see Rothschild and
Stiglitz 1973).

25. Our results are not directly comparable to those of Henle and
Rsycavage because we use individual, not grouped data, and because we use consistent top-codes on earnings. However, the trend
in our Gini coefficients mirrors the movements reported by Henle
and Rsycavage for the 1967 through 1977 period. Our findings do
not concur with those of Harrison, Tilly, and Bluestone.

26. Blackburn shows that an increase in the covariance of education
and age among males and a rise in the return to schooling were also
important factors in this rise.

27. To the extent that hourly wage rates depend on the number of
hours supplied by workers, the distribution of wage rates is not
completely purged of labor supply influences.

28. The industry employment shares in 1967 and 1984 were as follows:

Industry	1967	1984
Manufacturing	30.0%	22.3%
Other goods	10.7	8.9
Traditional services	13.2	16.4
Trade	16.9	17.6
Services	22.1	28.4
Public administration	7.0	6.4

For a definition of the industry categories, see the footnote to Table 6–6.

29. These results are consistent with those reported by Medoff, who analyzed hourly wage rates using a manufacturing/nonmanufacturing breakdown.

30. The mean logarithmic deviations for the individual industries are not presented here, although they exhibit patterns and trends that are qualitatively similar to those presented for the Gini coefficient. Unfortunately, the Gini coefficient does not possess the same decomposition property as the mean logarithmic deviation.

31. Pechman provides evidence that tax rates have declined for the top decile of the income distribution, due to the decreasing importance of the corporate income tax and the property tax. This finding strengthens our conclusion that the upper part of the distribution has become increasingly skewed over time.

32. Levy and Michel (1983) analyze the effects of tax system changes on after-tax income inequality for the years 1981 to 1984 and find the tax system to have changed so as to increase income inequality. However, unlike Pechman, their analysis fails to include corporate income taxes. They also do not explore the extent to which their results are sensitive to the particular assumptions they make concerning the incidence of various taxes.

33. The principal earner is defined as the head of household for those families in which this concept is not ambiguous—that is, for families not headed by a married couple. For married-couple families, the spouse with the higher earnings is defined as the principal earner.

34. Before the 1980 CPS, the head of household was always an adult male if there was one present in the household. After 1980, the designation of head of household was made by the survey respondent.

35. In 1985, the population of female unrelated individuals consisted of

more young, divorced females than it did in 1967 when a female living alone was more likely to be older and widowed.

REFERENCES

Atkinson, Anthony B. 1970. "On the Measurement of Inequality." *Journal of Economic Theory* 2: 244–63.

Becker, Gary S. 1981. *A Treatise on the Family.* Cambridge, Mass.: Harvard University Press.

Blackburn, McKinley. 1987. "An Analysis of Changes in the Distribution of Income among Families in the United States." Unpublished doctoral dissertation, Harvard University.

Blackburn, McKinley, and David Bloom. 1987. "Trends in Family Income Inequality in the United States: 1967–1984." *Proceedings of the Industrial Relations Research Association, 1986,* pp. 349–57. Madison, Wis: Industrial Relations Research Association.

Bluestone, Barry, and Bennett Harrison. 1986. "The Great American Job Machine: The Proliferation of Low Wage Employment in the U.S. Economy." Report to the Joint Economic Committee of the U.S. Congress. December. 99th Cong., 2d sess.

Bourguignon, Francois. 1979. "Decomposable Income Inequality Measures." *Econometrica* 47 (4): 901–20.

Burkhauser, Richard V., Karen C. Holden, and Daniel A. Myers. 1986. "Marital Disruption and Poverty: The Role of Survey Procedures in Artificially Creating Poverty." *Demography* 23 (4): 621–30.

Champernowne, D.G. 1974. "A Comparison of Measures of Inequality of Income Distribution." *Economic Journal* 84: 787–816.

Danziger, Sheldon, and Michael K. Taussig. 1979. "The Income Unit and the Anatomy of Income Distribution." *Review of Income and Wealth* 25: 365–79.

David, Martin, Roderick J. A. Little, Michael E. Samuhel, and Robert C. Triest. 1986. "Alternative Methods for *CPS* Income Imputation." *Journal of the American Statistical Association* 81: 29–41.

Dooley, Martin, and Peter Gottschalk. 1984. "Earnings Inequality among Males in the United States: Trends and the Effects of Labor Force Growth." *Journal of Political Economy* 92 (1): 59–89.

———. 1985. "The Increasing Proportion of Men with Low Earnings in the United States." *Demography* 22 (1): 25–34.

Gastwirth, Joseph. 1972. "The Estimation of the Lorenz Curve and Gini Index." *Review of Economics and Statistics* 54: 302–16.

Harrison, Bennett, Chris Tilly, and Barry Bluestone. 1986. "The Great

U-Turn: Increasing Inequality in Wage and Salary Income in the U.S." Unpublished manuscript, Massachusetts Institute of Technology, Cambridge, Mass.

Henle, Peter, and Paul Rsycavage. 1980. "The Distribution of Earned Income among Men and Women." *Monthly Labor Review* 103 (4): 3–10.

Kuznets, Simon. 1955. "Economic Growth and Income Inequality." *American Economic Review* 65 (1): 1–28.

Lawrence, Robert. 1984. "Sectoral Shifts and the Size of the Middle Class." *Brookings Review* (Fall): 3–11.

Leontief, Wassily. 1982. "The Distribution of Work and Income." *Scientific American* 247 (3): 188–204.

———. 1983. "Technological Advance, Economic Growth, and the Distribution of Income." *Population and Development Review* 9 (3): 403–10.

Levy, Frank, and Richard C. Michael. 1983. "The Way We'll Be in 1984: Recent Changes in the Level and Distribution of Disposable Income." Unpublished manuscript, Urban Institute, Washington, D.C. November.

McMahon, Patrick J., and John H. Tschetter. 1986. "The Declining Middle Class: A Further Analysis." *Monthly Labor Review* 109 (9): 22–27.

Medoff, James. 1984. "The Structure of Hourly Earnings among U.S. Private Sector Employees: 1973–1984." Unpublished paper, Harvard University, Cambridge, Mass. December.

Orshansky, Mollie. 1965. "Counting the Poor: Another Look at the Poverty Profile." *Social Security Bulletin* 28 (1): 1–26.

Pechman, Joseph A. 1987. "Pechman's Tax Incidence Study: A Response." *American Economic Review* 77 (1): 232–34.

Plotnick, Robert D. 1982. "Trends in Male Earnings Inequality." *Southern Economic Journal* 48 (3): 724–32.

Robinson, Sherman. 1976. "A Note on the U Hypothesis Relating Income Inequality and Economic Development." *American Economic Review* 66 (3): 437–40.

Rosenthal, Neal. 1985. "The Shrinking Middle Class: Myth or Reality?" *Monthly Labor Review* 108 (3): 3–10.

Rothschild, M., and J.E. Stiglitz. 1973. "Some Further Results on the Measurement of Inequality." *Journal of Economic Theory* 5: 188–204.

Theil, Henri. 1967. *Economics and Information Theory*. Amsterdam: North-Holland.

Urquhart, Michael. 1984. "The Employment Shift to Services: Where Did It Come From?" *Monthly Labor Review* 107 (4): 15–22.

Welch, Finis. 1979. "Effects of Cohort Size on Earnings: The Baby Boom Babies' Financial Bust." *Journal of Political Economy* 87 (5, pt.2): S65–S97.

III SECTORAL PATTERNS OF TECHNOLOGY ADOPTION

7 THE CHANGING PATTERN OF INDUSTRIAL ROBOT USE

Kenneth Flamm

Robots are a symbol of mechanization in the industrialized society of the late twentieth century.[1] Technologically, they sit at the crossroads of the first industrial revolution—the mechanical energy, machinery technologies developed in the eighteenth and nineteenth centuries—and the information-processing revolution of the latter half of this century. A modern industrial robot blends the capacity to use mechanical energy to apply force to objects, and perform physical work, with the information-processing capability ("intelligence") of the computer.

Because a robot acts on the external world with a manipulator—an "arm"—that is an analogue of a human limb, it would seem that the actions of such a machine are limited only by the intelligence that is used to direct the motion of the arm. In principle, given sufficient "intelligence" in its computer controller, a suitably designed robot ought to be capable of doing just about any form of physical labor that a human worker can perform. A robot can also perform reliably and consistently in harsh or constrained environments in which a human worker cannot function satisfactorily. Robots therefore represent about the most advanced and flexible form of industrial automation that can be envisioned.

Steep declines in the cost of computing power—which amounted to perhaps 20 to 25 percent per year (in nominal terms)

267

over the last thirty years—show every sign of continuing.[2] One might hope that the dramatically lower cost of computing hardware, as well as continued advances in software that mimics selected aspects of human cognition, which together direct and control the actions of robot manipulators, would create a situation in which cheap and powerful robots move quickly into factories.

Such has not been the case, however. In the United States, robots have remained a very marginal factor in U.S. manufacturing, overall. Despite a history of predictions of a forthcoming boom in the use of sophisticated industrial robots, actual usage in U.S. industry has consistently lagged well behind even moderately optimistic predictions.[3]

Ironically, though robotics technology originated in the United States, the technology has been introduced into manufacturing plants at a much more aggressive pace by our foreign competitors, particularly in Japan. This is all the more baffling because the economic logic arguing for their use would seem as applicable to the United States as to other nations. Most any argument for the adoption of robotics technology in manufacturing, if correct, should be at least as compelling in the United States as abroad.

Equally puzzling differences in perspective come into view when the underlying logic justifying the use of industrial robots is examined. For example, in the United States, manufacturing engineers and others appear to be struggling to overcome management resistance to installing such machinery. In the face of seemingly low rates of return on these investment projects, they argue that the full benefit of such advanced manufacturing technology cannot be realized through incremental, piecemeal installations but rather that an entirely new "factory of the future" must be built from the ground up to realize the full benefits of advanced manufacturing automation. A typical summary of this position is that

> a division may propose and undertake a series of small improvements in its production process; to alleviate bottlenecks, to add capacity where needed, or to introduce islands of automation based on immediately and easily quantified labor savings. Each of these projects, taken by itself, can have a positive net present value. But by investing on a piecemeal basis, without a long-term investment or technology strategy, the company never gets the full benefit from a complete redesign and rebuilding of the plant which can exploit the latest organization and technology of manufacturing operations.[4]

It is therefore all the more striking that robot installations in Japan frequently seem to adhere precisely to the pattern against which U.S. manufacturers are being admonished: piecemeal automation added in an incremental fashion to existing plants.[5]

In the view of some observers, this lag in the adoption of robotics technology is symptomatic of a central problem troubling U.S. industry—a decline in the technological sophistication of its manufacturing capacity relative to that of its international competitors, which in turn has played a crucial role in the deteriorating trade balances of key sectors of U.S. industry.[6] This now seems to be a widely held analysis of mature U.S. manufacturing industries like steel, shipbuilding, and autos. Similar problems, however, now appear to have cropped up at the technological leading edge where U.S. firms once flourished with virtually no serious foreign competition.

Numerous binational comparisons of U.S. and Japanese technology have established that in high-technology industrial sectors where U.S. firms pioneered, and once had a significant technological edge over their foreign competitors, that lead has since been seriously eroded and with some frequency even transformed into a lag. Examples of such sectors include semiconductor manufacturing equipment, many types of semiconductor devices, optoelectronics, certain types of computer hardware, robotics and factory automation, selected telecommunications hardware and components, and certain types of advanced materials.[7] A typical theme of all these reports is that the United States continues to lead in the basic science that underlies these technologies but lags in the timely transformation of new concepts into manufacturable commercial products and in the ability to manufacture these items at low cost. To the extent that these reports offer a clear vision of the problem, a deterioration in the sophistication of U.S. production technology relative to its foreign competition is a serious root cause of a decline in the competitive position of key industries, areas that have in recent times stood out as a particularly significant strength of the U.S. economy.

This chapter explores the extent to which this perception is correct by providing a detailed look at a key capital good—the industrial robot—that sits at the very center of most visions of the "factory of the future." Because they are at the leading edge of innovation in manufacturing technology, a focused analysis of the use of robots may clarify two significant issues: the degree to which manufacturing technology has thus far been transformed by computer-based capital goods; and the speed at which this process is taking place in the United States relative to the rest of the world.

THE CONTEXT: INDUSTRIAL ROBOTS AND MANUFACTURING AUTOMATION

An industrial robot is essentially defined to be a multifunctional, programmable manipulator. In the real world, however, machines—such as manipulators—are endowed with varying degrees of functionality and programmability, and deciding where the boundary line between "hard" and "flexible" automation ought to lie is a difficult task. In attempting to analyze the growing use of industrial robots, one immediately faces large and bewildering international differences in the scope of categories to which such flexible manipulators are assigned.

Industrial robots, moreover, are not the only machines sharing the characteristics of programmability and multifunctionality. There is a whole spectrum of programmable machinery found in automated manufacturing systems, and it is important to locate programmable manipulators within these categories. The atom of manufacturing systems is the *workcell,* or work station, where a set of related operations involved in the manufacturing process are situated. Programmable machinery found in the individual workcell includes not only industrial robots but also computer numerically controlled (CNC) machine tools,[8] programmable transfer machines of various kinds (not necessarily using manipulators), and programmable automated guided vehicles (AGVs) to shuttle materials between workcells, often under some form of central computer control.

With individual workcells (containing industrial robots, NC machine tools, programmable transfer lines, or automated guided vehicles) increasingly operating under direct computer control, the next obvious step in automating the factory has been to link the controllers supervising individual workcells together on a computer network. In this fashion, it is possible to write complex manufacturing control software and construct an automated manufacturing system that is capable of taking raw materials and processed inputs from an automated warehouse complex, then feed them through a variety of processing steps under central computer control, with minimal human intervention, to produce a finished product. Numerous variants on this idea of an automated, computer-controlled manufacturing system have now been built and are generally known as a flexible manufacturing system (FMS). If product design is also engineered on a computer-assisted design (CAD) system, and the design data downloaded to the computer-assisted manufacturing (CAM) system over a computer link, the set of technologies involved is often referred to as CAD/CAM.

The next step up in the automation of manufacturing is to link all the functions necessary for a complex manufacturing business together: manufacturing resource planning (often labeled MRP— and pronounced "murp") software used to plan materials use and schedule the detailed sequence of production operations, sales and corporate planning programs and databases, accounting and financial software, all woven together into a seamless, computer-controlled system bearing the magical acronym CIM (computer-integrated manufacturing). Carried to its logical end, an order for an item entered into a portable terminal by a salesperson automatically generates the planning and scheduling of all operations required for its manufacture, the coordination of its processing and distribution, the updating of all relevant company databases, and notifications for relevant action distributed electronically to all human decisionmakers in the control loop for the system.

No company yet claims to have achieved the nirvana of CIM, but many firms are proceeding toward the promised land in an attempt to link together the many "islands of automation" that can increasingly be found in manufacturing in advanced industrial economies. The move toward CIM is of special relevance to the use of robots because it is often argued in the United States that the full economic benefit of robot use can be perceived only when isolated islands of robot workcells are linked together in a fully integrated, computer-controlled manufacturing system.

As of the mid-1980s, robot workcells remained a rather minor star in the overall firmament of manufacturing automation in the United States (see Figure 7–1). In 1985 robots were about 2 percent of the value of shipments of manufacturing automation products in the United States, compared to a fifth of the market each for machine tools and material handling automation, respectively, and 12 to 15 percent market shares for inspection and test, process

Figure 7–1. U.S. Manufacturing Automation Market ($25.6 Billion U.S., 1985).

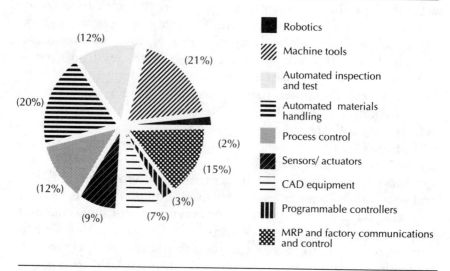

Source: John Kerr, "Made In The U.S.A.," *Business Computer Systems* (December 1985), P. 35. Note: Maintenance/ repair equipment and Rugged PCs each represent less than 1 percent and thus are not shown on this chart. Percentages may not total 100 due to rounding.

control, and sensor equipment. However, because robots are arguably the most flexible and most sophisticated of all automation machinery, they provide an excellent vehicle for studying how and why leading edge manufacturing technology has penetrated the commercial marketplace.

THE DIFFUSION OF THE INDUSTRIAL ROBOT

The first question on our agenda is how robots first entered the factory and became a significant factor in particular industrial niches. The early years of robotics are the story of complex machinery far too expensive to be economically viable in any but the most special of applications.

The industrial robot originated in the United States, built on a base of remote manipulator and servocontrol technology developed for defense users during and after World War II.[9] The first modern industrial robot—the Unimate, produced by Unimation (for *uni*versal auto*mation*)—was not sold until the 1960s, and the visionaries who brought it to market conceived of it as a highly programmable transfer machine. For over a decade, the Unimate faced only one significant competitor—the VERSATRAN (*versa*tile *tran*sfer machine)—in the U.S. manufacturing automation market.

Yet over the decade of the 1960s, only 200 Unimates were sold, along with a handful of VERSATRAN machines.[10] It was scarcely a lucrative market. Unimation did not finish a year with a profit until 1975, in spite of the fact that it still commanded roughly 80 percent of the U.S. robot market as late as the mid-1970s.[11]

There was relatively little further development in the basic design of commercial industrial robots (though improved components were used) until the early 1970s. In 1975 U.S. machine tool producer Cincinnati Milacron began to sell the first robot that interfaced a manipulator to a general purpose minicomputer. From this point on, industrial robot designs have basically regarded a robot manipulator as a sophisticated and specialized computer peripheral. Perhaps more important, the continued improvements in computer price and performance were directly reflected in the capabilities of industrial robot systems.

Japan's involvement in the industry dates back to 1967, when a VERSATRAN was imported and installed at Toyoda Automatic Loom, parent company of the Toyota Automobile Group.[12] In that very same year, Toyota's arch-rival, Nissan Motors, installed a variable sequence manipulator for materials handling that had been developed in-house (Nissan continues to develop and manufacture robots solely for in-house use).[13] Production of more sophisticated playback robots in Japan did not commence until some years later. Kawasaki Heavy Industries signed a licensing agreement with Unimation in 1968 but did not begin shipping Unimates it had manufactured (rather than imported) until 1970. As in the United States, business was initially slow—between 1968 and 1975, only 200 robots were sold by Kawasaki.[14]

THE GROWTH OF ROBOT USE

Industrial robots were not used in significant numbers until the late 1970s, and available estimates of robot use for earlier years should be approached with utmost caution. They are complicated by the many differences in definition as to precisely what constitutes an industrial robot. Japanese statistics, for example, show a stock of 16,000 industrial robots in place in 1975, but these are mainly manual and fixed-sequence manipulators. When the definition is restricted to variable-sequence manipulators (programmable "pick and place" machines) and more sophisticated models, that figure drops to 1,000 industrial robots in 1975.[15] U.S. statistics on robot production and use appear to be the worst among the larger industrialized countries—quasi-official estimates from the RIA do not start until 1981, and market studies for earlier years use wildly variable definitions.[16] Even the RIA statistics from 1981 on, which cover many countries and attempt to use only the U.S. definition, contain definitional errors in the estimates of robot populations in countries other than the United States.[17]

Given these reservations about the quality of robot statistics, Table 7–1 presents available and roughly consistent estimates of the robot population in a number of industrial countries from 1970 on. In several cases, these estimates are the product of calculations detailed elsewhere.[18] In particular, an effort has been made to exclude variable sequence manipulators from the count for all countries. Figure 7–2 makes relative growth rates more evident by plotting the natural logarithms of these estimates against time.

The figures show that Japan overtook the United States in robot use some time in the early to middle 1970s and has since been by far the greatest user of robots in the world, by any definition. Growth rates in U.S. robot use seem to have been somewhat lower than those in other countries in the 1970s, rising closer to the international norm in the early 1980s. In most countries, there was a sharp upturn in robot installations over the 1979 through 1981 period, followed by some slackening and resumption of a growth pattern closer to the long-run trend.

These economies have considerably different sizes, and in terms of relative intensity of robot use, a different pattern emerges. Figure 7–3 displays the stock of robots per 1,000 civilian workers in manufacturing.[19] Until 1983 (when a surge of robot installations pushed Japan into the number one slot) Sweden had the largest number of robots per manufacturing worker (over 2 per 1,000 workers). Germany and Belgium trail far behind Sweden and Japan (a little over one-half robot per 1,000 workers), and slightly

Table 7-1. Estimated Industrial Robot Populations, Selected OECD Nations.

	Japan[a]	Japan[b]	United States	Germany	Sweden	Britain	France	Italy	Belgium	Spain
1970	—	—	200	—	55	—	—	—	—	—
1971	—	—	—	—	—	—	—	—	—	—
1972	—	—	—	—	—	—	—	41	—	—
1973	—	—	—	—	135	—	30	45	—	—
1974	—	—	—	—	—	—	—	55	—	—
1975	—	1,000	—	—	—	—	—	—	—	—
1976	—	—	2,000	—	—	—	—	64	—	—
1977	2,900	—	—	541	490	80	—	255	12	—
1978	3,600	—	—	—	—	125	300	300	21	14
1979	4,700	9,000	—	—	940	—	—	321	30	40
1980	7,600	14,000	—	1,255	—	371	580	353	58	56
1981	12,600	21,000	4,700	2,300	1,250	713	790	—	242	118
1982	20,800	32,000	6,300	3,500	1,650	1,152	1,385	—	361	284
1983	31,700	—	9,400	4,800	1,850	1,753	1,920	1,800	514	416
1984	48,700	—	14,500	6,600	2,400	2,623	2,750	2,600	860	518

Source: Flamm (July 1986); "The State of Robotics in Italy at the Start of the 80s," *Industrial Robot* (September 1981): 176; "Industrial Robot in France," *Robot* (in Japanese), No. 48 (June 1985): 29–31.

a. Excludes variable sequence manipulators.
b. Includes variable sequence manipulators.

Figure 7–2. Growth in Robot Use, Selected OECD Nations.

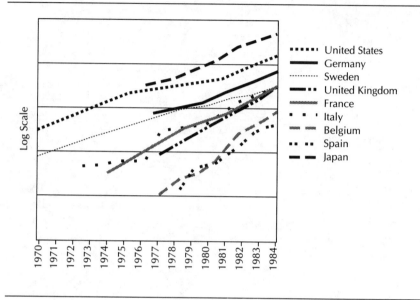

Figure 7–3. Robots per 1,000 Manufacturing Workers, Selected OECD Nations.

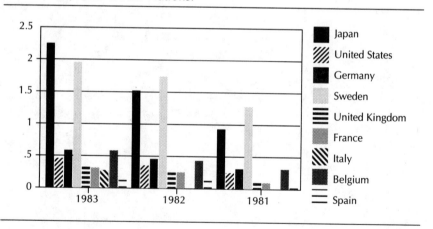

further back in the pack, one can find the United States (just under one-half robot). Still further back (at about one-third robot per 1,000 manufacturing employees), Britain and France poke along, followed closely by Italy.

The Pattern of Robot Use

In virtually all countries, use of robots is dominated by just a few applications. Historically, the first significant installations of industrial robots were in metal processing and fabrication—in foundries, casting, painting, and other hazardous and uncomfortable tasks. The first robot shipped by Unimation in 1961, in fact, was used in a die casting application at the Ford Motor Company.[20] In the 1970s spot welding emerged as a second major use for industrial robots.[21] The first robot welding line was installed in General Motors' Lordstown assembly plant for the ill-fated Vega automobile in 1969. In 1972 robot welding lines were first installed in Europe by Fiat and in Japan by Nissan.[22] And most recently, electronics assembly has emerged as the third major focus for growth.

Since the mid-1970s Japanese firms have moved most quickly and consistently to commercialize new applications. The pattern of industrial robot use by U.S. companies generally has resembled that of Japan, but after a lag of several years. Robot applications in Europe (with the important exceptions of particular specialized niches—like arc welding and spray painting systems, developed in Scandinavia, and precision assembly and forging, in Italy)—seem to have further lagged the pattern set in Japan and the United States by several additional years.

This pattern is most evident in Table 7–2 in the area of robotic assembly, currently the fastest growing application for industrial robots throughout the world. The bulk of these systems are being used in the electronics industry. Though the first specialized commercial assembly robot appears to have been marketed by Italian electronics producer Olivetti in 1975,[23] large-scale adoption of the technology was pioneered by Japanese industry. At the end of 1982 such applications accounted for over a fifth of the robots used in Japan,[24] compared to about 1.5 percent of the U.S. robot stock. The U.S. electronics industry did not begin to invest heavily in these systems until 1983, when the share of assembly robots shot up to 16 percent of the stock of U.S. applications. Use of robotic assembly systems continued to lag in Europe (except in Italy, where early adoption of the technology led to a stable share of assembly applications in the robot stock), though it was the fastest growing

Table 7–2.   Distribution of Robot Use by Application, Selected OECD Nations (percentage).

		Application			
	Distribution Year	Spot Welding	Arc Welding	Painting /Coating	Fini- shing
Belgium	1982	——— 71.80% ———		——— 7.21% ———	
	1983	46.69%	8.75%	3.70%	.97
	1984	60.00	7.33	N/A	.70
France	1983	27.86	12.49	7.31	2.24
Germany	1982	——— 44.56 ———		——— 9.70 ———	
	1983	32.50	17.83	12.21	.46
	1984	28.70	20.21	11.02	.33
Italy	1982	——— 44.73 ———		——— 8.82 ———	
	1983	——— 35.00 ———		——— 10.00 ———	
Japan	1982	——— 25.24 ———		——— 3.36 ———	
	1983	14.07	14.63	——— 3.53 ———	
Sweden	1982	——— 17.93 ———		——— 16.90 ———	
United Kingdom	1982	——— 40.94 ———		——— 15.86 ———	
	1983	19.91	13.35	9.53	1.54
	1984	19.37	14.02	7.28	1.77
United States	1982	——— 38.94 ———		——— 7.78 ———	
	1983	24.14	10.66	2.82	2.16

Source: RIA, *Worldwide Robotics Survey and Directory* (various years); BRA, *Robot Facts* (various years).

application in most of these countries in 1984. Assembly robots can be viewed as the third, and most significant wave to date, in a slow-paced robotic invasion of the factory.

The distribution of industrial robot use across industries, in different countries, varies considerably and is determined by two factors. The first is the industrial specialization of a nation's manufacturing sector. Robot use is concentrated in a handful of industries and applications. Until recently, countries that had no motor vehicle industry were unlikely to have large stocks of robots, particularly welding robots. In the near future, given current growth rates, countries with relatively large electronics sectors may be expected to have relatively more robots used in their manufacturing industry.

The second factor affecting the size of the robot stock is the

	Application				
Assembly	Loading /Unloading	Material Handling	Casting	Other	Total
1.31%	11.15%	3.28%	.66%	4.59%	100.00%
.78	11.67	2.53	.39	24.51	100.00
.47	8.37	2.44	N/A	N/A	100.00
6.97	——— 34.53 ———			8.61	100.00
2.84	4.49	13.65	2.79	21.98	100.00
5.17	6.67	4.04	2.75	18.38	100.00
6.85	7.06	3.18	2.23	20.42	100.00
10.09	9.27	——— 9.09 ———		18.00	100.00
10.00	——— 25.00 ———			20.00	100.00
19.12	8.08	21.31	1.75	21.15	100.00
26.02	8.94	24.76	2.52	5.53	100.00
1.03	40.00	8.97	13.10	2.07	100.00
1.84	9.62	8.70	5.53	17.50	100.00
5.88	9.41	23.50	2.97	13.92	100.00
8.18	8.76	23.97	2.22	14.43	100.00
1.14	16.83	20.63	13.89	.79	100.00
16.22	8.33	——— 26.23 ———		9.44	100.00

intensity of robot use (that is, input per unit output) within a sector. Intensity of use is clearly affected by the economic variables that make use of robots financially attractive, the technical characteristics of the product and manufacturing process, and social factors and national policies that may affect the speed of diffusion within national industry of the most advanced robotics applications. The auto industry, for example, is an industry with considerable amounts of metalworking and assembly tasks, relatively high wage rates in most countries, and therefore fertile ground for labor-saving automation. In countries where robots have been embraced aggressively, and their applications encouraged and promoted as national policy, relatively more robots may be found.

After controlling for the size of robot-using industries, the intensity of robot use in countries with roughly similar costs for

Table 7–3. Use of Robots by Industry in Selected OECD Nations (percentage distribution based on number of units).

Industry	Percentage
Canada (1981)	
Automobiles	63%
Plumbing fixtures	9
Electrical engineering	6
Metalworking	6
Appliances	5
Germany (1981)	
Transportation	46
Electrical engineering	14
Mechanical engineering	12
Metalworking	6
Plastic and materials	6
Other	16
Italy (1979)	
Automobiles	28
Household appliances	8
Metal	8
Electrical	6
Rubber	1
Exports	49
Netherlands (1982)	
Metalworking	64
Mechanical engineering	12
Electrical engineering	9
Transport equipment	5
Construction materials	3
Rubber and plastics	1
Others	6
Sweden (1979)	
Metalworking	51
Mechanical engineering	15
Transportation	22
Electrical engineering	9

Sources: OECD (1983: Table III.7). Canada: National Research Council Canada, *op. cit.;* Germany: KUKA, quoted in *Wirtschaftswoche* No. 15 (April 8, 1983); Italy: *Le Progrès Scientifique* (March-April 1981); Netherlands: Stichting Toekomstbeeld der Techniek; Sweden: Swedish Electronics Commission.

other manufacturing inputs may give us some clue to the speed of diffusion of robotics technology. Table 7–3 shows available data on the industrial distribution of robots for a selection of European countries. There is considerable diversity evident. Figure 7–4 shows cumulative shipments of robots to industry in Japan, and the successive shifts from metal processing, to autos, to electronics on the leading edge of robot use (note that these figures refer to the broadest Japanese definition of a robot).

Because industries come in very different sizes, looking at total stocks of machines may indicate little about their relative importance as an input to production. In general, however, the largest numbers of robots are also found in industries that make the greatest use of the machines per unit of output. Table 7–4 shows that the largest customers in Japan—autos, electrical machinery, and plastics—also are among the sectors making the most intensive use of robots, as measured by value of robots installed per worker, per million yen of value added, or per million yen of tangible fixed assets. However, there are some surprises: By the latter two measures, the bicycle industry is the most robot-intensive sector in Japan, and the construction and metal processing machinery industries are also major customers relative to their size. Yet none of these industries show up as first-rank users when ranked by the absolute numbers of robots purchased.

Figure 7–4. Cumulative Japanese Robot Shipments (1976–1984 by destination).

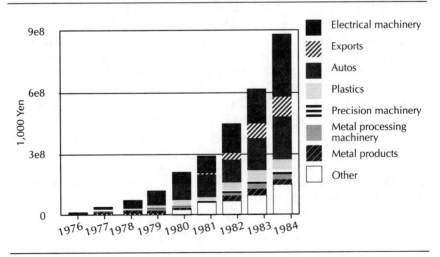

Table 7-4. Measures of Intensity of Robot Use by Industry: Japan, 1983.

Industry	Value of Installed Robots (1,000 Yen) per Worker	Value of Installed Robots (Yen) per Million Yen Value Added	Value of Installed Robots (Yen) per Million Yen Tangible Fixed Assets
Food processing	2,294	350	652
Textiles	6,377	1,550	3,503
Lumber products	2,812	634	1,270
Pulp and paper	1,602	221	231
Chemicals	7,651	429	669
Oil and coal products	18,128	633	427
Rubber products	5,442	835	1,664
Ceramic and stone products	5,451	723	1,041
Steel	29,740	3,002	1,829
Nonferrous metals	55,588	6,430	6,247
Metal products	36,924	6,386	14,435
Electrical machinery	111,648	15,104	41,618
Precision machinery	40,031	6,824	16,059
Boilers and motors	19,910	1,751	3,752
Construction machinery	112,020	11,765	25,410
Metal processing machinery	132,768	18,624	43,324
Automobiles	233,529	24,974	40,970
Bicycles	200,044	26,852	66,681
Shipbuilding	9,432	1,210	1,970
Plastics	141,603	21,835	42,837

Source: Author's calculations, using data described in Flamm (1986). Value of stock of installed robots is defined as estimated cumulative shipments 1976–83 of robots (broad Japanese definition).

Unfortunately, reliable data breaking out robot usage by industry in the United States is unavailable.[25] Table 7–5 shows available (and often contradictory) assessments by experts, which generally resemble the European and Japanese patterns. One exceptional feature of the U.S. use of robots is the relatively large (though small in absolute terms) share of the aerospace industry. This is almost certainly a direct consequence of the historical primacy of the U.S. Air Force in sponsoring government-funded research on manufacturing automation in general, and robotics in particular, and aggressive programs to subsidize the installation of such systems in defense contractor facilities. The consensus in the industry is that the auto industry now accounts for roughly half of the U.S. robot stock, with electronics gaining rapidly but still considerably behind autos.[26]

Table 7–5. Distribution of U.S. Industrial Robots, by User Industry (percentage).

	Stocks: End 1982 (Number)	1983 (Number)	End 1985 (Number)	Shipments: 1979 (Number)	1985 (Value)
Autos	40%	50%	50%	18%	51%
Foundry	20			21	13[a]
Nonmetals light manufacturing	17			37	9[b]
Electrical, electronics	11			11	8[c]
Heavy equipment	10			10	10[d]
Aerospace	2			1	6
Other	—			2	3

a. Includes primary metals and fabricated metal products.
b. Includes consumer nondurables, nonmetal primary commodities, and nonmetal fabricated commodities.
c. Includes electronics and precision equipment.
d. Includes machinery and other transport equipment.
Sources: Column 1—Tech Tran, as reported in OTA (1984: 291); column 2—DHR, Inc. (1984: 306); column 3—RIA, *Robotics News* (May 1986: 1); column 4—Smith and Wilson, *Industrial Robots: A Delphi Forecast of Markets and Technology* (1982); column 5—Smith and Heytler, *Industrial Robots,* 2nd ed. (1985).

The best indication that this is approximately correct may be found in statistics on the use of robots within large U.S. companies. Table 7–6 gives recent estimates of robot use within the U.S. auto industry and IBM—together, robots used in just four firms amounted to almost 60 percent of the U.S. robot stock at the end of 1984. This calculation overestimates the concentration of U.S. robots within these four large users because all are multinationals with extensive foreign production (and some of their robots, therefore, located overseas).

Nonetheless, given that their most automated plants are located within the United States, along with the lion's share of their manufacturing capacity, it seems safe to conclude that robot use in U.S. manufacturing is highly concentrated in a few large firms, primarily in the motor vehicle industry, and secondarily, in electronics. This conclusion is also born out by a 1983 survey by the U.S. International Trade Commission, which found that some 37 percent of the robots shipped domestically by U.S. producers over the years 1979 through 1983 went to the automobile industry.[27]

When an attempt is made to measure robot use relative to industrial value added on a more disaggregated basis, by industry, international comparisons become considerably more complex, as Table 7–7 shows. There is no consistent international ranking that holds across all industrial sectors, though Japan and Sweden generally lead in most activities. In chemicals, for example, Sweden is far ahead of any other country, followed by Britain and Japan. But if

Table 7–6.   Robot Use in Selected U.S. Companies.

Firm	Number of Robots in Use		As Percentage of Stock of U.S. Robots, December 1984 (=14,550)
General Motors	4,250	(February 1985)	29%
Ford	2,508	(February 1985)	17
Chrysler	640	(February 1985)	4
IBM	1,000	(Spring 1985)	7
Total	8398		57

Sources: GM, Ford, Chrysler—from Frank DiPietro, *Robotics: State of the Art Applications* (Warren, Mich.: GM Training Center, April 1985); IBM—from "Robots Getting Smarter, but Slowly, Say Experts," *IEEE Institute* (July 1985: 8). U.S. installed robot stock from RIA, *Robotics News* (June 1985).

Table 7-7. Number of Robots per Billion U.S. Dollars of Value Added by Industry and Country, Selected OECD Nations.

ISIC Industry/ Country	1980	1981	1982	1983
35 Chemical products				
Austria			3.20	3.51
Belgium	.71	3.09	4.56	5.43
France	2.90	4.11	7.04	9.62
Germany	1.32	2.46	3.79	5.14
Japan	2.45	4.84	8.60	13.19
Sweden		42.53	56.23	60.46
United Kingdom	3.13	6.46	10.10	14.92
United States				
356 Plastic products (not elsewhere classified)				
Austria				
Belgium	3.32	14.46	21.34	25.40
France	28.56	40.53	69.37	94.81
Germany	4.19	7.81	12.07	16.34
Japan	8.78	17.32	30.79	47.20
Sweden				
United Kingdom				
United States				
37 Basic metal industry				
Austria				
Belgium	1.47	6.39	9.43	11.22
France	3.27	4.64	7.95	10.86
Germany	4.19	7.81	12.07	16.34
Japan	1.69	3.32	5.91	9.06
Sweden				
United Kingdom	.19	.40	.62	.91
United States		10.37	15.23	21.87
302 Machinery (not elsewhere classified)				
Austria				
Belgium				
France	7.43	10.55	18.05	24.67
Germany	7.24	13.51	20.87	28.25
Japan	10.54	20.79	36.95	56.65
Sweden		75.03	99.20	106.68
United Kingdom	2.96	6.12	9.58	14.14
United States		4.68	6.87	9.87

Table 7–7. (*Continued*)

ISIC Industry/ Country	1980	1981	1982	1983
383 Electrical machinery				
Austria			14.99	16.45
Belgium	.96	4.17	6.15	7.32
France	5.58	7.91	13.54	18.51
Germany	6.94	12.95	20.02	27.09
Japan	40.22	79.35	141.03	216.20
Sweden		59.79	79.05	85.01
United Kingdom	2.13	4.41	6.89	10.17
United States		7.26	10.66	15.31
384 Transport equipment				
Austria			16.26	17.85
Belgium	12.24	53.27	78.60	93.55
France	13.73	19.48	33.34	45.56
Germany	22.77	42.48	65.66	88.95
Japan	53.57	105.68	187.83	287.94
Sweden		103.77	137.21	147.55
United Kingdom	13.52	27.93	43.68	64.49
United States		25.72	37.75	54.22
3843 Motor vehicles				
Austria			29.34	32.20
Belgium				
France				
Germany				
Japan	64.21	126.68	225.15	345.15
Sweden				
United Kingdom	21.92	45.30	70.84	104.58
United States		54.80	80.44	115.53
3000 Total manufacturing				
Austria			3.91	4.30
Belgium	2.91	12.68	18.70	22.26
France	4.21	5.98	10.23	13.98
Germany	5.82	10.87	16.79	22.73
Japan	14.78	29.16	51.83	79.46
Sweden		69.60	92.02	98.96
United Kingdom	3.61	7.47	11.68	17.24
United States		6.45	9.47	13.60

Table 7–7. (Continued)

Note: Number of robots per dollar value added calculated by multiplying sectoral share of robots divided by sectoral share of value added, times aggregate robots per value added for all manufacturing. Robot shares by sector are generally available for only one year over the 1980–83 period and are assumed the same in other years.

Sources for share of robots by manufacturing sector: Figures for Austria, Belgium, and the Netherlands are based on data in Economic Commission for Europe, *Production and Use of Industrial Robots* (New York: United Nations, 1985), pp. 57–58, 65. Figures for Sweden are based on data in *ibid.,* annex 6, p. 14, and Jan Carlsson. "Swedish Industries' Experience with Robots," *Industrial Robot* (June 1982), p. 89. Figures for Germany and the United Kingdom are based on data in Organization for Economic Cooperation and Development, *Industrial Robots: Their Role in Manufacturing Industry* (Paris: OECD, 1983), pp. 41–42. Figures for the United States are based on data from Tech Tran Corp. as cited in Office of Technology Assessment, *Computerized Manufacturing Automation: Employment, Education, and the Workplace* (Washington, D.C.: OTA, 1984), p. 291. Figures for France are based on data in Japan Industrial Robot Association, "Industrial Robot in France," *Robot* (Tokyo), no. 48 (April 1985), p. 29. Figures for Japan are based on data from the Japan Industrial Robot Association.

Sources for total number of robots in manufacturing: The figures given here are for industrial robots that are used within ISIC 3000 (total manufacturing) industries and consequently exclude robots used for R&D or education or used within the energy, agricultural, or mining sectors.

Figures for Austria, the Netherlands, and Sweden are from Economic Commission for Europe, *Production and Use of Industrial Robots,* pp. 57, 65, and annex 6. Figures for Belgium are from *ibid.,* p. 58, and British Robot Association, *Robot Facts 1984.* Figures for France are from Japan Industrial Robot Association, "Industrial Robot in France," pp. 29–30. Figures for Germany and the United Kingdom are from British Robot Association, *Robot Facts 1983; Robot Facts 1982; Robot Facts 1981.* Figures for Japan are from the Japan Industrial Robot Association and exclude industry codes 17 and 18. Figures for the United States are from text Table 7–1.

Sources for share of value added, by industry: Value-added shares for Japan are from Ministry of International Trade and Industry, Research and Statistics Department, *Census of Manufactures, 1983: Report by Industry* (Tokyo: MITI, 1985), pp. 2–13. Value-added shares for other countries are from Organization for Economic Cooperation and Development, *Industrial Structure Statistics, 1983* (Paris: OECD, 1985), pp. 9, 19, 29, 37, 40, 43, 48, 56. Shares for Belgium are also partially from United Nations Industrial Development Organization, *Handbook of Industrial Statistics, 1984* (New York: United Nations, 1985), p. 45.

Source for total value added in manufacturing: Organization for Economic Cooperation and Development, *Historical Statistics, 1960–1983* (Paris: OECD, 1985), pp. 14–16, 44, 59.

within chemicals, only plastic products are considered, France is the leader. In basic metals—the very first commercial application, which the United States pioneered—the United States leads, followed by Germany. In nonelectrical machinery, Sweden is far ahead, trailed by Japan; the United States lags at the end of the list. For electrical machinery, Japan is way out front, followed by Sweden, and much farther back, France and Germany. In transport equipment, Japan again is the clear leader, followed by Sweden; though United States robot use relative to other countries looks much larger when only the automotive subsector is examined.

INDUSTRIAL ROBOT USE IN JAPAN

The essential fact about Japanese robot use is that the vast bulk of robots shipped by Japanese producers have gone to only a handful of industries. At the top of that list are automobiles and electronics. From 1982 on, exports become a significant factor in robot demand and are now the second most important market (trailing only the electrical machinery industry) for Japanese robots. A considerably smaller portion of output is shipped to plastics, metal products, and precision machinery manufacturers. Even smaller slivers of production go to metal processing machinery, nonferrous metals, and chemicals firms.

It is possible to examine the intensity of robot use by application and industry, and interesting results can be obtained. Table 7–8 shows three different measures of intensity of robot use—value of robots per value added, per value of tangible capital assets, and per worker employed—for a variety of applications and industries. The most striking feature is the extent to which particular applications are concentrated in just a few major user industries. The overall pattern is very much one of a few major applications, in just a handful of industries, accounting for most of the robot use.

Table 7–9 summarizes the overall pattern of use by industry and application. Assembly applications in the electrical equipment industry, and welding in autos, along with exports, dominate Japanese robot use. It is important to point out that the very broadest Japanese definition of a robot is used in this analysis.

A more detailed examination of Japanese robot use has been carried out, and it is useful to briefly note the major conclusions of this analysis.[28] In general, the pattern is again one of very specific "islands" of automation scattered in a very disparate fashion across industries.

The earliest use of robots in Japan seems to have been in hazardous, dangerous, and monotonous metal processing and machining tasks, as was also true in the other regions of the world where this technology was first developed. Japan's break from the rest of the pack came in the mid- to late 1970s, when large-scale investments in playback spot welding robots for use in the auto industry were made, well before any other country's auto producers had made such a commitment. This segment of the Japanese robot market

Table 7-8 Relative Intensity of Japanese Robot Use, by Application and Industry

Industry:	Casting	Die Casting	Plastic Forming	Heat Processing	Forging	Pressing	Arc Welding	Spot Welding	Gas Welding	Spraying	Plating	Grinding	Assembly	Inspection	Shipment	Other Uses
Robots/Worker (1000 Yen)	0	0	0	0	0	0	0	0	0	0	0	0	0	663	272	1359
Food Processing	0	0	0	0	0	0	0	0	0	0	0	0	64	176	65	6073
Textiles	0	0	0	0	0	0	0	0	0	1553	0	0	0	76	76	1107
Lumber Products	0	0	0	0	0	0	0	0	0	0	0	0	0	938	177	487
Pulp & Paper	0	0	0	0	0	0	0	0	0	0	0	0	0	0	0	0
Chemicals	0	25	418	59	0	0	385	0	0	32	0	25	427	586	35	5660
Oil & Coal Products	0	0	7429	0	0	0	0	0	0	0	0	0	0	1988	0	8711
Rubber Products	0	0	0	75	0	0	0	0	0	60	0	0	342	1759	313	2893
Ceramic & Stone Products	0	0	0	79	0	33	52	24	0	148	0	82	365	560	201	3907
Steel	878	0	20	24	12510	283	1407	0	0	19	0	3044	255	3611	283	7405
Nonferrous Metals	3845	40726	115	74	699	2668	63	0	0	348	21	816	953	1052	427	3781
Metal Products	152	400	0	2	262	7038	10161	588	0	1565	5289	8075	229	634	33	2496
Electrical Machinery	1	706	3179	9	0	4921	2215	880	0	1445	70	6328	86142	400	922	4265
Precision Machinery	0	588	863	44	1	3077	21	80	0	1801	145	10152	12306	363	1021	9569
Boilers & Motors	3186	0	0	0	0	0	2868	0	0	1562	0	9212	2906	0	0	177
Construction Machinery	194	0	0	0	352	872	96747	1006	85	6529	0	4297	206	99	0	1632
Metal Processing Machinery	6186	767	457	14	66	18451	15816	132	90	493	43	80746	2447	3506	573	2981
Automobiles	1365	5111	4205	785	106	38952	36294	77737	0	10502	277	33484	16311	650	644	7107
Bicycles	1000	3013	0	0	90	23530	108570	2489	0	1351	0	17609	26067	5150	2520	8655
Shipbuilding	0	0	0	0	0	0	6694	0	0	680	0	340	234	0	703	0
Plastics	0	0	129270	0	0	217	15	0	0	10640	0	0	207	272	45	936

Table 7-8 (Continued)

Robots/Capital Stock (1000 yen/million yen tangible fixed assets)

Industry:	Casting	Die Casting	Plastic Forming	Heat Processing	Forging	Pressing	Arc Welding	Spot Welding	Gas Welding	Spraying	Plating	Grinding	Assembly	Inspection	Shipment	Other Uses
Food Processing	0	0	0	0	0	0	0	0	0	0	0	0	0	188	77	386
Textiles	0	0	0	0	0	0	0	0	0	0	0	0	35	96	36	3336
Lumber Products	0	0	0	0	0	0	0	0	0	701	0	0	0	34	34	500
Pulp & Paper	0	0	0	0	0	0	0	0	0	0	0	0	0	135	26	70
Chemicals	0	2	37	5	0	0	34	0	0	3	0	2	37	51	3	495
Oil & Coal Products	0	0	175	0	0	0	0	0	0	0	0	0	0	47	0	205
Rubber Products	0	0	0	23	0	0	0	0	0	18	0	0	104	538	96	884
Ceramic & Stone Products	0	0	0	15	0	6	10	5	0	28	0	16	70	107	38	746
Steel	54	0	1	1	769	17	87	0	0	1	0	187	16	222	17	455
Nonferrous Metals	432	4577	13	8	79	300	7	0	0	39	2	92	107	118	48	425
Metal Products	59	156	0	1	103	2751	3972	230	0	612	2067	3157	90	248	13	976
Electrical Machinery	0	263	1185	3	0	1834	826	328	0	539	26	2359	32111	149	344	1590
Precision Machinery	0	236	346	18	1	1234	8	32	0	723	58	4073	4937	146	410	3839
Boilers & Motors	600	0	0	0	0	0	540	0	0	294	0	1736	548	0	0	33
Construction Machinery	44	0	0	0	80	198	21946	228	19	1481	0	975	47	22	0	370
Metal Processing Machinery	2019	250	149	5	21	6021	5161	43	29	161	14	26349	799	1144	187	973
Automobiles	240	897	738	138	19	6834	6367	13638	0	1843	49	5874	2862	114	113	1247
Bicycles	333	1004	0	0	30	7843	36190	830	0	450	0	5870	8689	1717	840	2885
Shipbuilding	0	0	0	0	0	0	1398	0	0	142	0	71	49	0	147	0
Plastics	0	0	39106	0	0	66	5	0	0	3219	0	0	63	82	14	283

Application:

Table 7-8 (Continued)

Robots/Value Added (yen/million yen value added)

Industry:		Casting	Die Casting	Plastic Forming	Heat Processing	Forging	Pressing	Arc Welding	Spot Welding	Gas Welding	Spraying	Plating	Grinding	Assembly	Inspection	Shipment	Other Uses
																Application:	
Food Processing		0	0	0	0	0	0	0	0	0	0	0	0	0	101	41	207
Textiles		0	0	0	0	0	0	0	0	0	0	0	0	16	43	16	1476
Lumber Products		0	0	0	0	0	0	0	0	0	350	0	0	0	17	17	250
Pulp & Paper		0	0	0	0	0	0	0	0	0	0	0	0	0	130	24	67
Chemicals		0	1	23	3	0	0	22	0	0	2	0	1	24	33	2	318
Oil & Coal Products		0	0	259	0	0	0	0	0	0	0	0	0	0	69	0	304
Rubber Products		0	0	0	11	0	0	0	0	0	9	0	0	52	270	48	444
Ceramic & Stone Products		0	0	0	10	0	4	7	3	0	20	0	11	48	74	27	519
Steel		89	0	2	2	1263	29	142	0	0	2	0	307	26	364	29	747
Nonferrous Metals		445	4711	13	9	81	309	7	0	0	40	2	94	110	122	49	437
Metal Products		26	69	0	0	45	1217	1757	102	0	271	915	1397	40	110	6	432
Electrical Machinery		0	95	430	1	0	666	300	119	0	196	9	856	11653	54	125	577
Precision Machinery		0	100	147	7	0	524	4	14	0	307	25	1731	2098	62	174	1631
Boilers & Motors		280	0	0	0	0	0	252	0	0	137	0	810	256	0	0	16
Construction Machinery		20	0	0	0	37	92	10161	106	9	686	0	451	22	10	0	171
Metal Processing Machinery		868	108	64	2	9	2588	2218	18	13	69	6	11326	343	492	80	418
Automobiles		146	547	450	84	11	4166	3881	8314	0	1123	30	3581	1744	70	69	760
Bicycles		134	404	0	0	12	3158	14573	334	0	181	0	2364	3499	691	338	1162
Shipbuilding		0	0	0	0	0	0	859	0	0	87	0	44	30	0	90	0
Plastics		0	0	19933	0	0	33	2	0	0	1641	0	0	32	42	7	144

Source: Same as Table 7-4.

Table 7-9 Annual Japanese Robot Shipments 1978–84 by Major Application and Industry (Billion Yen)

	1978	1979	1980	1981	1982	1983	1984
electronics assembly	3.51	2.90	22.92	24.88	34.79	48.30	87.98
other electrical machinery	2.80	3.77	4.80	8.23	8.37	10.47	11.36
exports	.47	1.04	.97	1.10	1.42	1.62	3.33
auto spot welding	3.63	4.12	7.29	9.70	11.97	14.48	13.66
auto arc welding	.16	.35	1.58	5.95	9.72	7.81	8.51
other automobiles	6.67	10.17	13.38	15.35	17.62	14.28	16.37
other spot welding	.18	.21	.73	1.44	.51	2.63	9.49
other arc welding	.87	1.52	4.23	8.56	12.52	11.49	13.04
other assembly	.76	2.73	2.05	2.61	12.74	24.12	40.28
all other applications and industries	7.55	11.27	18.97	27.04	34.22	48.59	51.57

approached saturation in the early 1980s, and growth rates for domestic shipments tapered off considerably.

From the late 1970s on, the greatest growth was instead found in a new type of sophisticated robot, generally with NC computer control, often with a sensor system attached. Although it is unremarkable that the steadily plummeting cost of computing power should result in greater programmability and intelligence mated to automated machinery, it is quite important to note that this new genre of sophisticated robot was used in a relatively narrow spectrum of applications, almost exclusively in assembly of electrical and electronics products.

If attention is restricted to the use of sophisticated robots, following the U.S. definition (that is, excluding variable sequence manipulators), use across applications is very concentrated. (See Table 7–10.) Assembly accounted for the bulk of sophisticated robot shipments in 1984, followed by welding (arc and spot). Very much smaller demands for sophisticated robots were generated in spraying, shaving, and inspecting.

The sophisticated robots going to this very narrow set of ap-

Table 7-10 Share of Value of Total Shipments of Sophisticated Robots by Industry and Year

		Percent			Percent			Percent
Food	1978	.00	Steel	1978	.86	Miscellaneous	1978	.43
Processing	1979	.08		1979	9.82	Machinery	1979	1.30
	1980	.08		1980	.36		1980	.62
	1981	.05		1981	.17		1981	1.06
	1982	.13		1982	.48		1982	1.14
	1983	.85		1983	.81		1983	1.67
	1984	.70		1984	.31		1984	2.40
Textiles	1978	.14	Non-ferrous	1978	.13	Automobiles	1978	37.01
	1979	.00	Metals	1979	.38		1979	31.69
	1980	.00		1980	.05		1980	22.54
	1981	.00		1981	.11		1981	28.58
	1982	.11		1982	.15		1982	25.69
	1983	.14		1983	.26		1983	19.18
	1984	.05		1984	.72		1984	14.07
Lumber	1978	.00	Metal	1978	.41	Bicycles	1978	2.35
Products	1979	.35	Products	1979	7.02		1979	1.27
	1980	.00		1980	5.19		1980	.43
	1981	.01		1981	5.05		1981	1.46
	1982	.17		1982	3.20		1982	.54
	1983	.10		1983	3.21		1983	.54
	1984	.22		1984	1.92		1984	.50
Paper	1978	.00	Electrical	1978	45.20	Shipbuilding	1978	.00
Products	1979	.16	Machinery	1979	26.18		1979	.45
	1980	.02		1980	62.11		1980	.19
	1981	.01		1981	46.25		1981	.04
	1982	.14		1982	39.46		1982	.03
	1983	.16		1983	38.73		1983	.28
	1984	.25		1984	45.05		1984	.20
Chemicals	1978	.00	Precision	1978	.73	Plastics	1978	.58
	1979	.09	Machinery	1979	6.67		1979	2.33
	1980	.17		1980	.26		1980	1.34
	1981	.06		1981	.43		1981	2.17
	1982	.24		1982	.92		1982	1.26
	1983	.46		1983	1.37		1983	1.07
	1984	.80		1984	2.21		1984	.54
Oil &	1978	.00	Boilers &	1978	.41	Other	1978	.00
Coal	1979	.00	Motors	1979	2.05	Manufacturing	1979	2.42
Products	1980	.00		1980	.12		1980	.59
	1981	.00		1981	.05		1981	.73
	1982	.03		1982	.08		1982	1.67
	1983	.14		1983	.20		1983	1.37
	1984	.13		1984	.35		1984	.70
Rubber	1978	.14	Construction	1978	2.13	Other	1978	1.33
Products	1979	.00	Machinery	1979	2.39	Industries	1979	.49
	1980	.02		1980	2.09		1980	1.14
	1981	.00		1981	2.19		1981	.38
	1982	.15		1982	2.27		1982	.35
	1983	.31		1983	1.71		1983	.77
	1984	.19		1984	1.56		1984	1.03
Ceramic &	1978	.00	Metal	1978	.49	Export	1978	7.68
Stone	1979	.54	Processing	1979	.30		1979	4.04
Products	1980	.15	Machinery	1980	.20		1980	2.34
	1981	.17		1981	3.52		1981	7.52
	1982	.32		1982	2.44		1982	19.02
	1983	.50		1983	1.54		1983	24.67
	1984	.76		1984	1.10		1984	24.25

plications now account for about three-quarters of the value of Japanese robot shipments (about 80 percent when peripherals are excluded). This represents a major transformation in the market for industrial robots in Japan. As recently as 1978, sophisticated robots were only 30 percent of the market—nonprogrammable manipulators accounted for half of the value of shipments, variable sequence manipulators another fifth of the market. By 1983 unsophisticated robots were down to 25 percent of shipments, almost evenly split between variable sequence manipulators and nonprogrammable machinery.

Table 7–9 shows the exceptional importance of welding applications, in the auto industry, and assembly, in electrical equipment, in driving robot usage in Japan. Exports, as was just remarked, are essentially made up of a mix of the types of robots used in large numbers in these two exceptional industries.

The impact of technological advance in robots, it would seem, has *not* spread out in broad waves, stimulating the widespread marginal application of robotics in small doses in a range of industries. Instead, the pattern has been one of a particular, very specific application—often confined to a single industry—suddenly becoming an economic use of the improved programmable machinery. Often times, the widespread shift to a particular application may have come years after pioneers had experimented with the first such implementations, only to find further use stymied by the limitations of current technology or cost. After this switchover point was reached, though, the new generation of machines and their application then have diffused rapidly, with large jumps in usage over a very short period of time. The pattern has emphatically not been one of slow, gradual growth in use across a variety of industries and applications, though the growth may appear slower and more continuous when all industries are aggregated together.

CONTRASTS WITH THE UNITED STATES

Superficially, at least, the use of robots in the United States seems to greatly resemble the Japanese pattern, albeit with a several-year lag and considerably less pervasiveness. As in Japan, autos and electronics are widely thought to be the leading sectors. And within automobiles, the observed distribution by application greatly resembles the Japanese case.

The apparent similarities mask some important differences. In particular, the applications seen in the Japanese auto industry seem much more inclined to use simpler types of robots (that is, without sensors, using playback control, more dependent on fixturing) in stand-alone workcell configurations. The kinds of new systems U.S. industry leader GM has been installing, on the other hand, seem to be considerably more complex. Descriptions of recent installations at GM stress the use of sensors, particularly machine vision, advanced programming mechanisms, integration into a factory-wide computer network, and close coordination with other plant-level, computer-based information and automation systems.[29] Japanese auto firms seem to place much less emphasis on such state-of-the-art applications and seem to be more inclined to use less advanced, cheaper, less integrated robot systems.[30]

A good example of the difference in approach can be found in the types of installations used for arc welding. As of the end of 1985, both Nissan and Toyota relied exclusively on expensive hardware—fixturing—to hold the workpiece surfaces to be welded together in precisely the correct position.[31] Arc welding applications developed at GM, by way of contrast, often use complex optical sensors linked to a robot's computer controller to position the welding tip correctly and do not require the same degree of accuracy in the orientation of the workpieces. The U.S. approach yields a more general purpose installation (reprogramming is all that is required to shift to a new model), while the Japanese system requires new and expensive fixturing hardware when significant model changes are made. (At Toyota in the fall of 1985, precision fixturing was reported to account for 40 percent of the equipment cost of an arc welding workcell, the robot for 60 percent.[32]) For moderately large production runs, however, the savings on the cost and complexity of the robots required to do the job may more than compensate for the additional investment in special purpose hardware. (Recall that robots used in the Japanese auto industry are overwhelmingly of the simpler playback variety, while the mix in the U.S. auto industry probably leans more toward more advanced CNC robots.)

The deeper question that runs through all comparisons of U.S.

and Japanese investments in robotics technology is an emphatic *why?* Why was it that the United States lagged so far behind for so long, in applying technology that was first developed within its own borders? And why is it that the United States continues to trail behind others? The answers to these questions are neither easy nor obvious, but they are the second major issue to be examined in this chapter.

The Puzzle

Back in the early 1960s, when the potential employment impacts of automation first became a concern for public policy, the United States led in the application of advanced machinery to manufacturing (recall that the first robots were being introduced in the United States at the time). In an assessment prepared for the National Commission on Technology, Automation, and Economic Progress, an analysis of the use of automatic assembly machinery was undertaken.[33] At the time, robots were classified as a "general purpose assembly machine" (the first application that used a manipulator to hold a tool was just being developed in Norway).[34] Industrial robots were then a very rare and exotic item; "special purpose assembly machines" that were relatively inflexible and dedicated to a particular function and application marked the outer limits of the state of the commercial art. Such hard automation had long been used in the mass production of items like light bulbs, radio tubes, tin cans, and razor blades appeared on a small scale in the auto industry in the early 1950s and was spreading again in the early 1960s.[35] Most interestingly, the auto industry accounted for 50 to 70 percent of all automatic assembly installations, the consumer appliance industry for another 5 to 15 percent, and all other industries for the remainder.[36]

Thus, the U.S. auto and electronics industry had led the world in the development and application of hard automation to mass production items. These very same industries were to pioneer the shift to flexible automation and robots a decade later. Yet along the way, the mantle of leadership in applying this new technology was passed along to Japanese manufacturers.

On the face of it, it would seem that manufacturing firms in the United States also had the greatest incentive to pioneer the commercial use of robotics. Almost by definition, robots replace workers in relevant tasks (the total employment impact, of course, also depends on the impact of robot use on the competitive position of a firm, and it is quite possible to find a net positive impact on employment if firm output increases). There is little doubt that wages in Japan have historically been set far below U.S. levels— U.S. Department of Labor comparisons show the total hourly compensation for a production worker in Japanese manufacturing to have been swinging between a peak of two-thirds and a trough of one-half of the U.S. manufacturing labor cost over the 1975 through 1985 period.[37] Judged purely on the basis of labor-saving cost reductions, Japanese manufacturers would seem to have realized one-half to two-thirds of the savings available to U.S. producers.

Even restricting the discussion to labor savings, there are other considerations that may modify this conclusion. The largest Japanese employers have something resembling a "lifetime" employment guarantee for their "permanent" employees, and if one interprets this as making labor a quasi-fixed factor (that is, imposing adjustment costs on reductions in staffing levels), then optimal choice of technology may involve treating the effective cost of labor as above the actual wage paid if there is variability or uncertainty in forecasts of future levels of production. On the other hand, even large Japanese firms employ substantial numbers of "temporary" workers, or subcontract considerable amounts of work to smaller firms with no such employment guarantees (so that subcontractor employment absorbs the effect of unforeseen variation in output). At the margin, therefore, it is not at all clear that the relevant cost for the services of a production worker will necessarily be significantly above the wage actually paid.

The cost of labor, of course, is only half of the story. A robot is a costly capital good, and cross-country variations in the cost of capital may have significant effects on the economic gains from automation. Here, unfortunately, the picture is extremely murky. The user cost of capital services from a particular piece of equipment is computed by multiplying the price of the equipment by a measure of the after-tax cost of funds used to purchase the equipment. Unfortunately, there is little solid evidence on the relative costs of equivalent robots in the United States and Japan.[38] Because there are now considerable robot exports from Japan to the United States, it seems likely that any price differential is small.

Comparisons of the cost of funds in Japan and the United States vary widely in their conclusions. One commonly cited study by Hatsopoulos concludes that the real cost of capital in the United States is better than three times greater than in Japan.[39] In another study, the U.S. Commerce Department interprets available data as showing nominal U.S. capital costs as 25 percent greater than in Japan (though the U.S. Treasury concludes that the same evidence indicates that real costs are about the same in the two countries).[40] A recent study by Auerbach and Ando, on the other hand, finds a very slightly lower return to capital in Japan, a finding that they interpret as indicative of little or no difference in capital cost between the two countries.[41] Still another study by Fuss and Waverman of production costs in the U.S. and Japanese auto industries imputes a slightly *lower* cost of capital to the U.S. industry,[42] while the study by Winston and his collaborators of Japanese-American auto production cost differences generally finds a slightly lower cost of funds in Japan using improvements to the Hatsopoulos data and methodology.[43] Finally, a 1983 survey of management practices in U.S. and Japanese firms found a somewhat longer acceptable payback period from investment projects in Japanese firms compared with U.S. producers (thirty-seven months versus thirty-one months).[44]

Because all these studies use somewhat different methodologies and look at different time periods, comparisons are difficult. On balance, the evidence—though far from conclusive—seems to suggest a somewhat cheaper cost of funds in Japan, but not an enormous difference in recent years.

Further complicating any comparison of the cost of capital is the especially favored treatment accorded to investment expenditure on robots by the Japanese tax system. In 1978 robots were named as a sector eligible for special promotion measures. Special supplemental depreciation allowances can be taken on a long list of qualifying robots. The various schemes apply to "High Performance Industrial Robots Controlled by Computers," "Industrial Robots Used for Operational Safety," and "Promoting Investment to Apply New Technology in Small Enterprises."[45] Coupled with other accelerated depreciation schemes, it was possible to write off over half the cost of a robot in its first year of use.[46] The government has also supplied a limited amount of low-cost capital to the Japan Robot Leasing Corporation (JAROL), which leases robots to business users at reduced rates. The impact of this latter measure is quite limited however—in 1982, for example, it leased 790 units, compared to shipments of almost 10,000 sophisticated robots; in 1986, JAROL expected to add 200 new machines to its stock.[47]

To summarize, then, special depreciation measures coupled to a somewhat lower cost of funds in manufacturing make it likely that the effective cost of capital services from Japanese robots is somewhat lower than in the United States. A fairly extreme position might hold that such services were half as costly in Japan as in the United States. However, because until recently the price of labor services was also about half of the U.S. labor cost, this suggests that the ratio between the price of capital and labor services would then have been about the same in Japan as in the United States, and choice of robot technology about as attractive economically. Clearly, the relative cost of capital and labor services—as imperfectly and imprecisely measured as it seems to be—is not going to explain all of a vastly different rate of adoption and pattern of usage. Let us turn next to a closer scrutiny of the economic factors affecting robot use.

THE ECONOMICS OF ROBOT USE

Choice of technology in manufacturing involves coordinating and organizing a complex set of processes and tasks. The intention of this discussion is to highlight some of the more important considerations when use of robots is contemplated.

Scale of Output.   Perhaps the key factor missing from the discussion thus far is the relation of an economic manufacturing technology to the volume of production in which it is used. An expensive piece of machinery introduces significant economies of scale over some range of output. If used to produce only a few items, the average cost per piece of using that machine is quite high, but as more output is produced, average cost per unit drops. This contrasts with wholly manual methods, where the average cost per unit fabricated may be relatively constant with respect to production levels.

An idealized view of production costs is sketched out in Figure 7-5. Hard automation, which involves a specialized piece of machinery utilized in manufacturing only a single product and often

Figure 7–5. Relation of Optimal Technology to Production Volume.

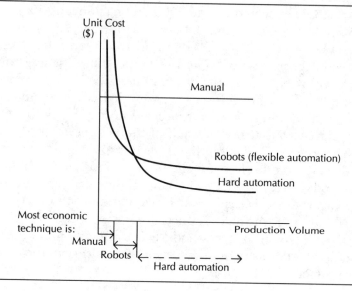

only a single design vintage for that product, tends to be very expensive for small volumes of output but relatively cheap for mass production. For smaller output levels, manual fabrication is most cost effective; for large-scale production, hard automation is economic. Programmable automation essentially extends the economic use of machines inward, toward lower production volumes. The key is the flexibility of the machinery: Because it can be used with more than one type of product or application, the cost of flexible automation can be charged to a variety of products of intermediate volume—so-called batch production.[48] Generality of purpose is traded off for somewhat lower speed and performance on a single application when compared to hard automation, however, and for high-volume mass production, hard automation tends to be most cost effective.

It should be stressed that beyond the mere fact of capacity sharing, the time requirements and relative ease of changeover from one product to another are critical dimensions of "flexibility." What makes robots flexible—and limits less easily programmable machinery—is the rapidity with which they can be reprogrammed. Stopping a production line for four hours—or four days—so it can be switched over to another variety of automobile or microwave oven idles large amounts of very expensive capital equipment. The ease with which robots can be programmed is the key to their flexibility. At the Yamaha Motors plant in Hammamatsu, Japan, for example, which integrates robots extensively with workers on assembly lines producing a mix of snowmobiles, golf carts, motor scooters of various types, and family bikes, it takes approximately one minute to switch a line from producing one type of vehicle to another.[49] On assembly lines at Nissan Motors' Murayama plant, different models are mixed and put together on the same line. An optical scanner reads a product code and instructs welding robots on what program to use in assembling the vehicle.[50]

Flexibility was the crucial, binding constraint on automation in the fabricating industries when the National Commission on Technology, Automation, and Economic Progress examined the issue in the early 1960s. It is advances in computer controllers and programmability that have pushed in the frontiers of automation, beyond mass production, to batch production.

Capacity Utilization. Closely related to the scale and flexibility issues is the question of capacity utilization. As just mentioned, because of shorter changeover and setup times, capacity can be shared among products, utilization of capital equipment is improved, and the scale of production needed for economic use of automated equipment is reduced. But there are further dimensions to improved capacity use that also can be explored with use of programmable automation. If individual work stations are linked together through connections to a communications network, with a central computer controller dynamically scheduling materials use and process flow, production work schedules and allocation of tasks among workcells can be continuously monitored and altered to minimize bottlenecks and idle time on machines. This second class of savings requires computerized control of an entire set of processing tasks and cannot be achieved through the use of robots in isolated, unconnected work stations.

Inventory Reduction. Shorter changeover times clearly result in inventory savings in a multiproduct firm, both because of the reduced idle time that components and work in process must face during product changeover, and because of the reduced need for inventory that results from enabling production to respond quickly to variation in demand. Further sources of savings can also be envisioned. In a typical metal fabrication plant, a considerable amount of work in progress is in transit between work stations, queued for processing, or otherwise placed in temporary storage pending further action. By rationalizing production scheduling to reduce work in progress, and continuously observing the progress of assigned tasks, central computer control linked to computer-based programmable automation offers the possibility of reducing the inventory associated with a particular production rate.

Consistency/Quality Control. On a high-precision, automated machine, tasks are performed with a great deal of consistency. A machine that is functioning correctly will perform the exact same operation over and over again, with little or no variation. The machine never gets tired, or bored, or has its attention wander. When it malfunctions, problems are relatively easy to detect because output will then tend to be consistently bad until the difficulty is corrected. The serial correlation in errors makes them relatively cheap to detect, whereas defects with manual operations tend to be more randomly distributed and hence require greater inspection and testing effort to guarantee a particular level of quality. Highly exacting quality control requirements tend to be cheaper and simpler to meet with automated manufacture. In batch (medium-volume) production where consistency of product is an important consideration, robots can be an economic solution where manual operations might otherwise be cheaper. Where consistency is not critical, the tighter control over manufacturing parameters inherent to the use of robots can reduce scrappage and the cost of reworking defective output.

Constrained Environments. The first commercial use of robots was in hot, dirty, toxic, and radioactive settings. Because of their relative imperviousness to physical conditions that debilitate humans, robots remain the technology of choice in such areas. It may even be argued that this was the primary economic consideration driving their use until the late 1970s. Before they came into widespread use in electronics assembly, the most active applications for robots—welding, painting, die casting, foundry work—were almost exclusively in hazardous environments. And it does not require belief in charitable motives on the part of employers to understand why they were used. There are substantial hidden costs to using labor in occupations where disability claims, lawsuits, and occupational injuries lead to high turnover rates and a continuing stream of job-related training and safety expenditures.

Welding, in particular, required a substantial employer investment in job-specific training for the worker. Yet because of the harsh working conditions, turnover rates have tended to be very high for welders.[51]

A recent trend toward using robots in the clean rooms required for semiconductor fabrication puts a novel twist on the traditional association between robotics and special environments. Perhaps the primary determinant of yields of "good" product—and hence profit—on chip fabrication lines is the amount of contamination found in the manufacturing environment. After appropriate filtering of air and cleaning of equipment, the primary source of particulate contamination in clean room environments is the humans working in them. Replacing human workers on wafer fab lines is economic not because the robots are a cheaper way to perform particular tasks, but because they are cleaner and therefore reduce yields less. As the complexity of integrated circuits increases, control over contamination at all stages of processing will become more and more important, and it seems reasonable to predict that robots will be used on other tasks off the wafer fabrication line (assembly, for example), in order to improve yields. Similar considerations make robotics economic in the manufacture of magnetic fixed-disk drives, which are assembled and hermetically sealed in a clean room environment.

Another area where capabilities not easily replicable with human workers are an important force driving robot use is in precision electronics assembly. In high-precision assembly of very small parts, robots are able to use micromanipulators to fit components into inaccessible spaces where manual assembly would be extremely difficult. New types of denser component packaging, particularly surface mount packages for semiconductors, essentially require the use of robots or hard automation for assembly.

Vertical Integration. The last factor that ought be mentioned is that manufacturing technology is very product and process specific. There is no such thing as an off-the-shelf manufacturing workcell that can be purchased from an independent vendor and plugged into a manufacturing plant like a lathe or an oven. Designing and installing a robot workcell invariably requires heavy development costs, and a considerable knowledge of a manufacturer's proprietary technology must often be integrated into the design of the work station. This has been a considerable barrier to the purchase of robots by smaller firms with a limited potential for replicating an application once the initial development costs have paid for.

Finally, it should be noted that scale of robot use, not just scale of output, can be a significant factor in making robots economically attractive. In 1985 an expert panel estimated that initial planning, development, integration, and installation costs for a robot workcell, on average, amounted to 40 percent of the robot system itself.[52] Much of this cost was a "generic" expenditure, associated with designing and implementing a particular type of application in a particular industrial activity. This one-time cost is relatively fixed in nature—later installations of the same type can be installed for a fraction of this initial cost. This is a source of significant economies of scale in robot use, and a potent explanation for the empirical observation that larger firms seem much more inclined to implement robotics in their manufacturing processes.

Explaining International Variation in Robot Use

Next, consider the puzzle that motivated this discussion: Japan leads, and the United States lags, in applying robotics to manufacturing in the 1970s. There are at least six reasonable hypotheses that might deserve some attention in explaining the "robot gap."

Cheaper Robot Costs.   The first, and most conventional explanation, is that the services of robots are very much cheaper in Japan than in the United States—so cheap, in fact, that despite wage rates set at half of U.S. levels, it is still more economically attractive to use robots in Japan than in the United States. The "cheaper robot" hypothesis requires at least one of two further arguments. One possibility is that capital is generally much cheaper in Japan and the user cost of industrial robot services therefore lower. An alternative explanation, mentioned earlier, is that the price of robotic capital goods is lower to start with in Japan and is further cheapened by special tax breaks. Even with an identical cost of funds, a lower cost for the machines means a lower cost for the services of industrial robots. Both, explanations, in fact might be at work. In addition, to explain very large international differences in robot demand, per unit of output, the demand for robots must necessarily be quite elastic with respect to the cost of robot services.

There are problems in arguing that declines in Japanese robot prices explain variation in robot use across countries. First, measuring some sort of quality-adjusted price for industrial robots is difficult. It is possible to calculate a per unit cost for robots, and this effort[53] does in fact indicate that the cost of Japanese robots, on average, has been lower than robot costs abroad. In part, however, this may reflect the fact that the types of robots sold in the United States are more complex, sophisticated systems compared to the simpler machines widely used in Japan.

Second, it is possible to calculate a unit cost for shipments of industrial robots within Japan, either by industry and application or by type of robot and industry. The apparent trend, however, is of a relatively gradual decline over time, marked by considerable year-to-year volatility, and not one of steep reductions in cost. One might conclude from this exercise that declines in the cost of robotic capabilities—which by descriptive accounts, have been substantial in recent years—have taken the form of improved performance and not sharp cuts in the prices of individual robots, which have held relatively steady.

Estimating some sort of quality-adjusted price index for robots that reflects advances in capability requires more advanced methods and cannot simply be based on observed unit costs. One such effort has been undertaken.[54] With a couple of exceptions, the pattern that emerges for a variety of applications—including welding and assembly, the quantitatively most important ones—is one of relatively stable service prices over the 1975 through 1980 period, followed by declines of varying degrees after 1980. Since there is some evidence that the cost of funds was increasing sharply in Japan during the late 1970s, this would seem to indicate that a declining price for robots was roughly offset by the increasing costs of funds over this period.

Thus, the first big burst of robot use by the Japanese auto industry, after 1975, seems much more likely to have been related to an expansion of auto production than any sharp decline in the cost of robot services. There do seem to have been significant declines in robot service price across a broad variety of applications after about 1980, which played some role in the expansion of demand, especially in assembly.

It is unlikely, then, that international differences in robot cost have played much of a role in explaining cross-country variation in relative intensity of use. Since 1980, though, it appears that significant cost declines for Japanese robot services in a variety of applications have stimulated rapid growth in a number of relatively new areas, and the Japanese lead in these newer applications may well reflect faster access to the latest generation of commercial products.

Demographics.   A second hypothesis might be termed the "long run cheaper robot" theory. In discussions of robotics in Japan, it is quite common to hear references to the "graying" of the Japanese workforce and an expected steep decline in the size of the prime-age labor supply. If Japanese employers are forecasting a dramatic increase in the cost of labor in the medium-term future, and if they believe that by acquiring experience with the capabilities of industrial robots they may develop over time a possible economic substitute for a shrinking human labor supply, then investments in robotics now can be viewed as a longer-run strategy for coping with a labor-short future.

It is often asserted that Japanese firms operate with longer planning horizons than U.S. competitors, but whether this is a significant factor in explaining behavioral differences is difficult to assess. What is certainly true, though, is that different demographic dynamics are at work in Japan and the United States. Figure 7–6 charts projected dependency ratios (young and old dependents per prime age worker) for the United States and Japan through the year 2080.[55] The ratio of retirement age population to prime age workers is expected to double in Japan in the near future—between the years 1995 and 2015. A similar phenomenon will also occur in the United States, but not until the two decades between 2010 and 2030. Although this fifteen-year head start toward an older population age structure may not explain much of the current interest by Japanese companies in robotics, it has played a

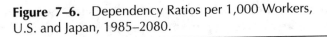

Figure 7–6. Dependency Ratios per 1,000 Workers, U.S. and Japan, 1985–2080.

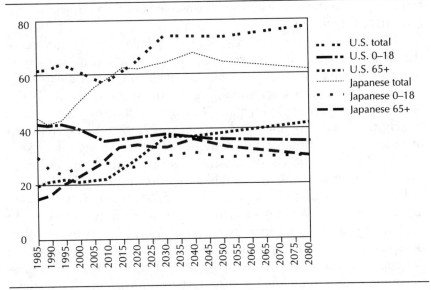

public role in government commitments to support the development and deployment of the technology, and these in turn may have had some impact on the private sector.

The difficulty with this explanation is that over the decade when robot demand exploded in Japan—1975 to 1985—wages remained far below U.S. levels. It is hard to see how expectations of structural change two decades in the future would have prompted private firms—which fundamentally must pay some attention to their bottom-line profit—to actively invest in capital goods that would inevitably be obsolete by the time a demographic crunch arrived, unless it were profitable to do so in the much shorter term for other reasons.

Product Mix Strategy.   It has been argued that one strategy followed by many Japanese manufacturers in the late 1970s to better compete in world markets was to substantially increase product variety.[56] Rather than duplicate the existing product line of a foreign manufacturer, in which the latter may even have enjoyed scale economies, the new tactic was to differentiate, to tailor a product to a particular market niche. Industrial robots, and the flexibility and quick changeover times associated with them, are clearly more attractive in a manufacturing environment oriented toward producing short and medium-size runs of a variety of differentiated products.

This viewpoint has been articulated in Japanese discussions of the advantages of industrial robots. Figure 7–7 is a graphic used by Sony in presenting the rationale for its FX-1 robotic assembly system for tape recorders (including the many varieties of "Walkman" it manufactures). In many Japanese electronics firms, it is clearly the flexibility and versatility of robot assembly systems, in producing a very broad range of models, that is driving their use.[57]

It is also arguable that the same logic pushed the adoption of robot welding systems in the Japanese auto industry in the mid- to late 1970s. Growth in Japanese auto production after the 1973 oil shock was driven by exports; domestic shipments remained roughly constant. This increasing geographical variation in shipments required greater product differentiation—steering columns in export models often had to be placed differently than in domestic models, and mandated safety and pollution control systems varied substantially among export markets. Increasing variety was also used as a marketing tactic—Abegglen and Stalk note that a new product introduction rate of one to 1.5 new vehicles per year in the early 1970s had risen to five vehicles per year by the early 1980s.[58]

Figure 7–7. Trend in Numbers of Models and Related Production Systems in Sony Tape recorders, 1976–1981.

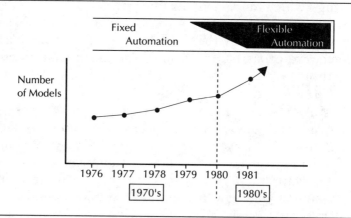

Available data certainly seem to support the claim that there was a considerably different approach to product variety in the Japanese auto industry. In 1975 there were roughly half as many different auto body types per volume of U.S. cars produced as in Japan, and about one-third as many engine types.[59] By 1980, the situation had changed little, and the U.S. industry appeared to have even reduced the diversity of engine types somewhat.[60] On the face of it, different approaches to incorporating product variety into manufacturing might seem a very credible candidate explanation for differences in robot use.

Nonetheless, to argue that Japanese use of robots resulted from a recognition that differentiated, "focused" manufacturing strategies had become economically feasible for the first time, based on a new generation of flexible automation technology, begs the question of why it was the Japanese, and not their competitors, who first realized this. It is possible to make an economic argument as to why it is the outsider—the challenger—who innovates. An argument first formulated by Kenneth Arrow in 1962, popularly known as the "Arrow effect," suggests that an established firm with a position of market power has less incentive to introduce an innovation than a newcomer with no such position.[61] The newcomer, by establishing a position of market power by virtue of innovation, then collects some element of monopoly rent that an established monopolist does not count as net return on its investment in innovation (because profits on its new products come partially at the expense of profit on its now obsolescent older products, a deduction that a new entrant does not make when evaluating returns on the required investment).

In any event, a cheapening of flexible automation does suggest that new manufacturing strategies become economically viable. A continued shift toward differentiated, specialized products may well be the outcome of continued advances in robotics and other forms of programmable automation.

Product Quality Strategy.   Another explanation of early Japanese interest in industrial robots might focus on the quality improvements realized in the manufacturing process. Certainly in the semiconductor industry, superior levels of quality and reliability in Japanese memory chips were used to good effect in their initial penetration of U.S. markets. The increasing use of flexible automation on wafer fabrication lines in the semiconductor industry today is largely based on improvements in cleanliness and therefore yields and product quality, not labor cost.[62]

It is also tempting to attribute some of the initial Japanese interest in using robots in the auto industry to improvements in quality, since superior fit, finish, and overall quality have been widely perceived in the United States to be a distinct selling point for Japanese cars since the mid-1970s. However, as has been noted, use of industrial robots was largely confined to spot welding applications, and weld quality on auto bodies and assemblies is an attribute likely to be neither noted by consumers nor much of a factor in the reliability and maintainability of a finished vehicle.

Labor Relations.   In conversations with visitors, Japanese often attribute some of their lead over U.S. industry in robot use to resistance by powerful U.S. labor unions. There may be a historical germ of truth in this. Certainly GM's bitter experience with a divisive and costly strike at its Lordstown plant in 1972, at the time an experimental outpost of large-scale robot use within that company, may have dulled its appetite for further adventures in flexible automation.

But in general, it is hard to accept this argument. Since the late 1970s, at least, the United Auto Workers have been generally receptive to increased automation as a productivity-enhancing measure. And the U.S. electrical and electronics industry, where unions are virtually nonexistent, lagged the Japanese in introducing assembly robots in much the same way that the auto industry had fallen behind in welding applications.

Vertical Integration.   A final structural difference between U.S. and Japanese industry that might explain the latter's lead in applying robotics to manufacturing involves the degree to which robot users and producers are integrated into the same company. Industrial robots are used in highly customized systems, tailored to the proprietary product and process technology of a particular firm. In certain applications, particularly electronic and precision mechanical assembly, successful robot design has depended critically on close interaction between system designers and system users.[63] A substantial investment in such a custom system makes economic sense only if there is a reasonable probability that a manufacturer will purchase the system. In practice, this requirement for close cooperation between vendor and user has meant that a substantial number of commercial robots can be traced to the internal engineering efforts of particular firms for a specific application, which were later commercialized and marketed externally.

Although the robot industry in both Japan and the United States is quite young, and remains relatively fragmented, with a large number of small producers, that situation is changing rapidly. The largest firms that are now emerging as major forces in the U.S. market—GMF, IBM, GE, Westinghouse—with a few exceptions (principally Cincinnati Milacron) are major users of robots who market systems also used in large numbers on internal applications. The same is generally true in Japan.

This close integration between producer and user may be relevant to Japan's early lead in robot use in the electrical and electronics industry. The SCARA (Selective Compliance Assembly Robot Arm) robot, invented in a Japanese university and developed within a number of Japanese companies, was a major advance in robotics for precision assembly work. It was used internally within these Japanese companies well before it was marketed commercially, and the early Japanese lead in these applications may well reflect a superior technology that was unavailable on the open market for a period. Put most baldly, advances in price performance for industrial robots born in Japan may not have led to increased robot use in the United States simply because they were not for immediate sale. IBM, in fact, entered the commercial robot business in 1982 by marketing a system using a SCARA arm built by Sankyo Seiki and did not announce a SCARA assembly robot of its own design until the spring of 1986.[64]

As of early 1985, all high-precision robots available commercially were manufactured in Japan. The most precise U.S. robot had an accuracy of 25 microns, compared to Japanese robots with a 5 micron accuracy.[65] Certain of the most advanced Japanese systems are simply not sold to external users.[66]

Direct Evidence on the Returns to Robot Use

All of these factors, of course, may have played some role in the rapid progress in applying robotics to manufacturing applications in Japan. There is very limited direct evidence available on the perceived benefits of robot use, to which we next turn.

In the United States, the Society of Manufacturing Engineers (SME), the primary professional organization concerned with manufacturing technology, has had an ongoing project to forecast trends in robot production and use, based on a Delphi (iterative expert panel questionnaire) methodology. The panels of experts were composed of professionals from companies using robots (75 percent), manufacturing robots (17 percent), or engaged in other activities (8 percent, including academics and consultants).[67] One question asked panelists to break down the total savings from robot use in manufacturing, by major category of savings (percentages), by application. The results from this exercise were published in 1985.

In Japan, the Japan Industrial Robot Association (JIRA) sponsored a survey of Japanese robot users that attempted to quantify the effects of robot use.[68] The results, also published in 1985, were based on a survey of 281 robot users and are broken down by application in which the robot was used. The categories of effects are much more detailed than those found in the U.S. study but can be reworked into similar groupings. This has been done for four major types of applications—spot welding, arc welding, assembly, and painting—and the results are shown in Table 7–11.

Some clear differences between U.S. and Japanese perceptions of the economic effects of robot use emerge. One of the most striking discrepancies is the impact on product quality—in both the United States and Japan, quality improvements were thought to provide an important justification for robot use, but (with the exception of arc welding) these gains were judged to be considerably more important in the United States than in Japan. Similar statements can be made about gains from inventory reduction and reduced materials waste, both of which were thought to provide negligible benefits by Japanese users and relatively large savings to U.S. users.

One interpretation of this notable difference that might make sense is that Japanese producers have production management systems in place that raise quality and reduce inventory and materials waste, which are quite separate from use of robotic systems, while U.S. producers use solutions to these problems that rely on the complex factory automation systems in which robots are embedded. Just-in-time (JIT) inventory systems, and quality circles, for example, are oft-emulated Japanese production management techniques that stress direct involvement of the production workers on the shop floor in attacking inventory and quality control problems quite independently from robot use. In the United States, perhaps, greater emphasis is placed on centralized computer control of production (using MRP-type software) and automated inspection and quality control systems that minimize the demands placed on the production worker.

Reductions in direct labor requirements were assigned remarkably similar shares of the overall benefits of robot use in the two countries, despite a labor cost that on average is twice as high in the United States. The exceptions were in arc and spot welding, where direct labor savings were judged to be considerably more significant in the United States than in Japan. It is worth noting that relatively skilled operations—like welding—have largely remained onshore in U.S. industry, and wages paid by firms in these tasks have remained relatively high. On the other hand, less skilled operations like assembly—in electronics and other industries— have increasingly been shifted to low-wage offshore facilities by U.S. firms, so that smaller differences in magnitude of perceived labor-saving benefit in these applications could reflect the lower labor costs faced by U.S. firms when offshore assembly is taken into account.[69]

Most intriguing are the large weights assigned in Japan to factors not even mentioned in the U.S. survey. Improvements in the operating rate (capacity utilization) and reductions in equipment expenditure due to flexibility and improved capacity sharing together accounted for 15 to 20 percent of the economic benefits perceived from Japanese robots.

Simplified management and improved production technology were also assigned a substantially higher weight in Japan than in the United States. And mention of improved working conditions—confined to safety in the United States—was a marginal contributor to the benefits of U.S. robot use, but a major portion (8 to 15 percent) of the impact in Japan. Within that category in Japan, the bulk of the benefit comes from reductions in physically demanding, dirty, and monotonous work, and improvements in workplace morale. Only 5 to 15 percent of the total assigned to working conditions is attributed to improved safety per se.

Table 7-11. U.S. and Japanese Perceptions of the Economic Benefits of Robot Use.

Percent of Benefit Due to:	Application				
	Spot Welding	Arc Welding	Electronic Assembly	Other Assembly	Painting
United States					
Product quality	32%	15%	38%	26%	34%
Direct labor productivity	50	62	38	41	40
Indirect labor productivity	6	8	6	13	6
Energy savings	0	0	0	0	2
Materials waste reduced	6	8	6	5	11
Space savings	0	0	0	0	0
Inventory reduction	6	7	6	6	1
Simplified management	0	0	6	6	0
Safety	0	0	0	3	6
Other	0	0	0	0	0

Japan

Product quality	15	19	14	16
Labor reduction	27	27	35	38
Labor utilization improved	3	3	3	4
Operating ratio improved (capacity utilization)	7	9	13	9
Reduced capital investment required and efficiency of use improved	12	8	8	4
Energy savings	0	0	0	0
Materials waste reduced	0	0	0	1
Space savings	0	0	0	0
Inventory reduction	0	1	0	0
Simplified management and improved manufacturing technology	14	12	13	10
Working conditions improved	15	12	8	14
Development of robot business	0	1	0	0
Safety/Quality image with public	1	1	0	1
All other	3	5	3	2

Source: See text.

Comparison between these two surveys, then, suggests that Japanese firms put considerably less weight on quality improvements, and inventory and materials savings, than U.S. robot users. The Japanese, on the other hand, place much greater emphasis on flexibility and capacity sharing, overall improvement in the quality of the workplace (including safety, but mainly other factors), and simplified management and improvement of production technology.

The impression one gets—coupled with direct observation of Japanese factories—is that automation is used much more to improve the productivity of a trained, motivated, and cooperative workforce than to replace it. In the United States, by way of contrast, robotics seems to be integrated into complex factory automation that is designed to require minimal input from the ordinary production worker, to accomplish tasks (quality control, materials waste control, inventory reduction) that in Japan are often carried out with simpler, more labor-intensive techniques.

The differences between the two surveys should not be read as particularly conclusive. There were large differences between the type, and level of detail, of the questions asked, and the compositions of the samples of respondents. There are also more basic questions that might be asked about the correlation between perceptions of economic benefit and actual benefit, and between the factors identified as important on questionnaires and the considerations actually used when investment decisions are made by profit-seeking firms. Nonetheless, these results cast some doubt on the "product quality strategy" hypothesis and suggest that greater attention be paid to the "product mix strategy" explanation for accelerated adoption of robots in Japan. Also, while differing levels of unionization seem inadequate as an explanation for U.S. and Japanese variation in robot use, the survey results do indicate that labor relations and workplace morale—and more important, cooperation between labor and management—may well play some difficult to quantify role in technological choice in manufacturing.

CONCLUSIONS

This survey has focused on two related issues: how and where industrial robots have been used in manufacturing, and how robot use in the United States compares with manufacturing practice abroad. The facts can be established in a fairly straightforward way; identifying the reasons that underlie them is considerably more difficult.

Robots, it should now be clear, are scattered rather spottily through industry and are used primarily in a number of very specific tasks and industries. Historically, they were first used in hazardous and unpleasant operations associated with metal processing, in relatively small numbers. After 1975, Japanese auto producers began to use them in large numbers for spot welding operations on their assembly lines and late in that decade expanded their field of application to arc welding. Their foreign competitors soon followed suit. Such welding activities greatly expanded the number of robots in use around the world.

Since 1980 a new and much more sophisticated generation of robots has been used on a large scale for assembly, primarily by electrical and electronics producers. Japanese manufacturers again led the way. The vast bulk of robots now in use are found in electronic assembly and automotive welding.

Evidence seems to indicate that once a new application was introduced and found profitable, it was widely and rapidly imitated by competitors. The barrier to wider use of robots is fundamentally economic—there are only a handful of major uses in which they are currently a cost-effective solution for manufacturers. Rather than being part of a pervasive, evolutionary change that is spreading through all industry, industrial robot use has expanded in fits and starts, as small and narrow niches have been conquered by more capable machines.

The fact that such general tasks as "assembly" and "grinding" are performed by robots in just a few industries suggests that the specific content and context of these tasks varies considerably from industry to industry. One would be well advised to be skeptical of technological optimists who, on the basis of broad statistical job classifications for industrial workers, project veritable tidal waves of robots inundating manufacturing in the medium-term future.

The recent experience of Japan is illuminating in this respect. In 1986, after half a decade of growth in the 20 to 30 percent per annum range, largely fueled by electronics assembly systems, industrial robot shipments in Japan grew at very tepid rate.[70] This is almost entirely due to a slump in the electrical and electronics industry. Continued technological advance has not led to the opening elsewhere in the economy of important new markets for robots that have compensated for this decline. The only sector that promises to yield double-digit growth in robot shipments in 1986, in fact, is in that traditional mainstay of robot use, the auto industry, where a demand for new welding robots to be used in the U.S. production facilities of Japanese automakers has pumped up sales. Even in Japan, vanguard of the robot invasion, use of industrial robots remains highly concentrated in a very narrow range of industries and applications.

It is also clear that robot use in the United States has lagged considerably behind that of manufacturers abroad, in Japan and Europe (especially Sweden and Germany). Only in metals processing, the original application for robots first developed in the United States, does U.S. industry continue to lead in its intensity of robot use. In the most important applications and industries—electrical machinery and autos—U.S. manufacturers, overall, continue to lag behind competitors in Japan, Sweden, Germany.

The reasons for this lag can and will be debated. I have argued that the conventional economic arguments—which focus on cross-country variation in the relative prices of capital, labor, and other factors of production—show little promise of explaining very much of these differences. Of the other candidate explanations, one of the most promising—in my view—is a shift to greater product variety (and the required more flexible manufacturing plant) by Japanese firms. Continued technological advance in robotics may yet fundamentally alter international markets for manufactured goods, toward greater product differentiation, in a much broader array of goods extending well beyond autos and electronics.

There is, of course, no reason that there has to be a single explanation for the distinct courses charted by U.S. and Japanese companies. Many small factors may add up to a big difference in behavior.

If one wishes to speculate further on what some of the crucial differences between U.S. and Japanese manufacturing practice are, certain points loom large. Key elements would include the following:

1. Excessive demands being placed on automation as the solution to all problems associated with manufacturing production management in the United States. There is reasonable evidence that inventory reduction and quality control are handled in Japan by production management techniques that have no direct connection to automation, while in the United States, automation is sold as the solution to inventory and quality problems;

2. Possibly, excessive zeal for finding the highest-technology solution to a manufacturing problem. There is a fair amount of evidence that the robots used in Japan tend to be simpler and less complex than those found in the United States. Over the long term, the U.S. approach may in fact make the United States the leader in creating the "factory of the future," but in the near term, the extra development cost makes more complex systems less economically attractive;

3. Insufficient attention in the United States to the economic benefits of greater product variety and flexibility in manufacturing. Flexibility and its economic returns—higher rates of capacity utilizatior, faster responsiveness to market changes, focused products with greater appeal to specific niche markets—may be poorly captured by the simpler economic models used to evaluate capital investments.

In any event, this simple overview of the empirical data on robot use has highlighted the analytical complexity of the microeconomics of choice in manufacturing technology. Distinct manufacturing technologies can offer very different potential economies of scale and scope, radically alter assumptions about capacity utilization, and make economic new strategies to compete on the basis of product variety and quality. Though one is well advised to remain skeptical of overblown claims for advanced technology, it does seem plausible that technical advance in programmable machinery—like robots—is making possible new types of firm strategies. The age of mass production may in fact be yielding—slowly and gradually—to a world of variegated, differentiated products that better address the diverse preferences of consumers, and the flexibility of the industrial robot may be the crucial element that will make this transition economically feasible.

NOTES

The views expressed in this chapter are those of the author and do not necessarily represent those of the trustees, the officers, or other staff members of the Brookings Institution or the National Academy of Sciences.

1. This chapter has benefitted from a research effort supported jointly by the World Bank and the Brookings Institution. A discussion paper outlining the preliminary results of this project was issued as Kenneth Flamm, "International Differences in Industrial Robot Use: Trends, Puzzles, and Possible Implications for Developing Countries," Report No. DRD185 (Washington, D.C.: Development Research Department, the World Bank, July 1986).

2. A variety of different methodologies and data sources produce estimates of quality-corrected computer prices that are remarkably similar in trend. See K. Flamm, *Targeting the Computer: National Policy and International Competition* (Washington, D.C.: Brookings Institution, June 1987), Ch. 2.

3. See Laura Conigliaro, "Robotics and Vision: The Hazards of Covering These Industries," and Peter A. Cohen, "Whatever Happened to the Robot Boom?," in *Robots 9 Conference Proceedings*, Vol. 2 (Dearborn, Mich.: Society of Manufacturing Engineers, 1985); Donald N. Smith and Peter N. Heytler, Jr., *Industrial Robots—Forecasts and Trends* (Dearborn, Mich.: Society of Manufacturing Engineers, 1985), p. 13.

4. This is taken from Robert S. Kaplan, "Financial Justification for the Factory of the Future" (unpublished draft, April 1985). A revised version of this paper was published as "Must CIM be Justified by Faith Alone?", *Harvard Business Review* No. 2 (March/April 1986).

5. My assertion is based on a one-week factory tour of Japanese industrial robot installations sponsored by the Japan Industrial Robot Association in September 1985 and on additional Japanese factory tours taken in April 1984 and March 1986.

    It should be noted that there are some very visible, large-scale flexible manufacturing systems (FMS) installed in Japan that contain large numbers of robots linked together in specially designed, advanced manufacturing plants. These facilities invariably appear on the itinerary of foreign visitors. It is clear from published Japanese statistics on robot use, however, that the vast bulk of Japanese robots are not found in these showcase plants.

6. See for example Richard M. Cyert, "The Plight of Manufacturing: What Can be Done?," *Issues in Science and Technology* vol. 1, no. 4 (Summer 1985).

7. See, for example, National Research Council, *State of the Art Reviews: Advanced Processing of Electronic Materials in the United States and Japan* (Washington, D.C.: National Academy of Science, 1986); *High Technology Ceramics in Japan* (Washington, D.C.: National Academy of Science, 1984); Science Applications International Corporation, *JTECH Panel Report on Computer Science in Japan* (La Jolla: SAIC, 1984); *JTECH Panel Report on Opto- and Microelectronics* (La Jolla: SAIC, 1985); *JTECH Panel Report on Mechatronics in Japan* (La Jolla: SAIC, 1985); *JTECH Panel Report on Telecommunications Technology in Japan* (La Jolla: SAIC, 1986).

8. Programmable machine tools, like robots, have historically come in at least two varieties: the simplest numerically controlled (NC) machine tools essentially "played back" an external program stored on some physical medium using an electromechanical control system; the more complex computer numerically controlled (CNC) machine tools used an electronic digital computer as the controller for the machine tool.

9. A more detailed review of the origins of robotics may be found in Flamm (July 1986), note 1 above.

10. See "Tackling the Prejudices against Robots," *Business Week* (April 26, 1976): 84E.

11. See "Tackling the Prejudice against Robots," p. 84E, note 10 above; Alfred B. Bortz, "Joseph Engelberger: The Father of Industrial Robots Reflects on His Progeny," *Robotics Age*, vol. 7, no. 4 (April 1985): 16; V.D. Hunt, *Industrial Robotics Handbook* (New York: Industrial Press, 1983), p. 11, Fig. 1–5.

12. Robert U. Ayres, Leonard Lynn, and Steven Miller, "Technology Transfer in Robotics Between the U.S. and Japan," in Cecil H. Uyehara, ed., *Technological Exchange: The U.S.-Japanese Experience* (Washington, D.C.: University Press of America, 1982).

13. See M.A. Cusumano, *The Japanese Automobile Industry* (Cambridge, Mass.: Harvard University Press) 1985, p. 228.

14. Eikonix Corporation, "Technology Assessment: The Impact of Robots" (Burlington, Mass.), 30 September 1979, pp. 54–55.

15. According to unpublished tables given to me by the Japan Industrial Robot Association in September 1985.

16. The Frost and Sullivan marketing studies from the mid-1970s, for example, appear to count "pick and place" machines, and therefore cannot be compared to later RIA counts. See Frost and Sullivan, *U.S. Industrial Robot Market* (New York, N.Y.) 1974; *The Industrial*

Robot Market in Europe (New York, N.Y.) 1975; Eikonix (1979), note above, pp. 163–64. The Frost and Sullivan numbers appear to be the basis for figures referring to the mid-1970s found in OECD, *Industrial Robots* (Paris: OECD, 1983), Table III.9.

17. For example, the RIA estimates make a conscious point of noting that variable sequence ("pick and place") manipulators are included in French statistics, contrary to U.S. practice but do not mention that the very same class of machines are included in the statistics cited for the Japanese robot population. See the Robotics Industry Association, *Worldwide Robotics Survey and Directory,* various years.

18. See Kenneth Flamm, "International Differences in Industrial Robot Use" (July 1986).

19. The manufacturing workforce estimates come from OECD, *Labor Force Statistics, 1963–1983* (Paris: OECD) 1985. The robot population comes from Table 1.

20. Brian Rooks, "The Cocktail Party That Gave Birth to the Robot," *Decade of Robots,* Special 10th Anniversary Issue of *Industrial Robot* (1983):8. see Joseph F. Engelberger, "The Ultimate Worker," in Marvin Minsky, ed., *Robotics* (Garden City, N.J.: Omni Press, 1985).

21. In spot welding, two pieces of metal are aligned, clamped between two electrodes, and an electric current passed through the electrodes to heat and weld the work pieces at the weld point. With arc welding, the edges of two metal pieces are aligned, and then heated by an electric arc at their seam. Filler material is then applied as the arc follows the seam between the work pieces. Because it requires the application of filler material and continuous following of a seam (versus applying the arc to two fixed points), arc welding is considerably more difficult for a "blind" robot that is not equipped with sophisticated and complex sensor systems.

22. Rooks (1983), p. 9, note 20 above.

23. See "The European Market: Still Looking Good," in *Decade of Robotics, Industrial Robot* (December 1983): 123; Rooks (1983), p. 10, note 20 above.

24. Note that the statistics shown for Japan in Table 7–2 include variable sequence manipulators, which are heavily concentrated in the plastics industry, where they are used in removing items from plastic injection molding machines. Making these statistics more comparable with the U.S. definition of a robot would increase the share of assembly robots considerably.

25. The only estimates of robot use by industry based on a statistical sampling technique are those found in *American Machinist,* which publishes a survey-based estimate of the stock of various types of metalworking machinery in U.S. manufacturing establishments, at roughly five-year intervals. Unfortunately, robots are scattered

among a variety of classifications—automatic assembly machinery, programmable robots, and apparently, automatic electric arc-welding machines.

(I make this last inference because *American Machinist* reported only twenty-one welding robots in use in the entire U.S. motor vehicle industry in 1983—out of a total of 124 robots of all kinds for that industry. On the other hand, there were 1,989 automatic arc welding machines reported to be in use, which leads me to believe that playback robots—which are not "programmed" by means of a formal, written computer program—must be included in the latter category.)

Unfortunately, lumping robots with other categories of equipment makes these numbers unsuitable for analysis of robot use. See "The 13th American Machinist Inventory of Metalworking Equipment 1983," *American Machinist* (November 1983).

26.   See John Holusha, "Sales Rise for Robots," *New York Times* (April 22, 1986): D19; James Just, et al., of DHR, Inc., "Robotics Technology: An Assessment and Forecast," in Robert U. Ayres, et al., *Robotics and Flexible Manufacturing Technologies* (Park Ridge, N.J.: Noyes Publications, 1985), p. 306.

A recent study by W. Leontief and F. Duchin—*The Future Impact of Automation on Workers* (New York: Oxford University Press, 1985)—vividly illustrates the problems created by the poor quality of U.S. data on robot use. Leontief and Duchin use the 1979 Smith and Wilson Delphi forecast estimates of *numbers* of robots *shipped in 1979*—shown in Table 9–5—along with supplementary data from a 1979 Frost and Sullivan marketing survey to allocate the *value* of the *stock* of robots among industrial sectors in 1980. In their synthetic estimate, for example, Leontief and Duchin show the auto industry with about 20 percent of the robot stock in 1980, holding that share of the robot stock steadily through 1990 and 2000. By 1985, on the other hand, Smith and Wilson's revised Delphi survey of the "experts" had the *value* of shipments to the auto industry climbing to a 50 percent share, while light manufacturing dropped from a 37 percent share to 7 percent. Leontief and Duchin, by way of contrast, generally show the share of light manufacturing industries (excluding electronics and metal products) increasing after 1980. The lesson to be drawn, perhaps, is that forecasting the future depends critically on the adequacy of data on the present. In this case, the data are clearly not up to the job. See also David Howell, "The Future Employment Impacts of Industrial Robots—An Input-Output Approach," *Technological Forecasting and Social Change* 28 (1985): pp. 297–310.

27.   The U.S. International Trade Commission estimated in 1983 that

the U.S. auto industry purchased almost 2,800 industrial robots from U.S. producers over the 1979–83 period. Of these robots, 60 percent were for spot welding. See U.S. International Trade Commission (USITC), *Competitive Position of U.S. Producers of Robotics in Domestic and World Markets*, USITC Publication 1475 (Washington, D.C.: USITC, December 1983). and author's calculations.

28. See Flamm (July 1986), note 1 above.

29. See for example Frank A. DiPietro, "Robotics State of the Art Applications," in Society of Manufacturing Engineers, *Robots 9 Conference Proceedings* (Dearborn, (Dearborn, Mich.: SME, 1985).

30. This statement is based on the tabulation of installations data presented above, the author's tours of Nissan and Toyota plants, and conversations with knowledgable manufacturing engineers.

31. I toured the Toyota Motomachi frame welding line for Crown model automobiles in September 1985 and found 95 percent of the frame arc welding done by robots. The arc welding relied totally on precision fixturing of the parts to be welded. My tour group at Toyota was told that little use of so-called intelligent robots was made, though the possible application of such complex systems to inspection was under study. Similarly, on a tour of Nissan's Murayama plant in spring 1986, I observed fixturing used to permit the limited amount of robotic arc welding done in that assembly plant. The situation is in marked contrast to the United States, where GM has widely publicized its use of intelligent, optical-sensor based, seam tracking robots for arc welding.

32. Briefing at Toyota Motors factory tour, September 1985.

33. See Eugene S. Schwartz and Theodore O. Prenting, "Automation in the Fabricating Industries," in National Commission on Technology, Automation, and Economic Progress, *Technology and the American Economy, Appendix Volume 1, The Outlook for Technological Change and Employment* (Washington, D.C.: U.S.G.P.O., February 1966).

34. Schwartz and Prenting (1966), pp. I-342–43, note 33 above.

35. *Ibid.*, p. I-298.

36. *Ibid.*, p. I-350.

37. The Bureau of Labor Statistics has calculated the following index for hourly compensation for a production worker in Japanese manufacturing (with the U.S. = 100):

	1975	1976	1977	1978	1979	1980	1981	1982	1983	1984	1985
Index of hourly compensation:	48	48	53	67	61	57	57	49	51	50	51

Unpublished data supplied by the Office of Productivity and Technology, Bureau of Labor Statistics, U.S. Department of Labor (January 1986).

38. Some fragmentary evidence is presented in Flamm (July 1986), note 1 above.

39. See George N. Hatsopoulos, *High Cost of Capital: Handicap of American Industry* (Waltham, Mass.: Thermo Electron Corporation, 1983).

40. See U.S. Department of Commerce, International Trade Administration, *A Historical Comparison of the Cost of Financial Capital in France, the Federal Republic of Germany, Japan, and the United States* (Washington, D.C.: U.S.G.P.O., April 1983).

41. Albert Ando and Alan Auerbach, "The Corporate Cost of Capital in Japan and the U.S.: A Comparison," NBER Working Paper No. 1762 (Cambridge, Mass.: NBER, November 1985).

42. See Melvyn Fuss and Leonard Waverman, "The Extent and Sources of Cost and Efficiency Differences between U.S. and Japanese Automobile Producers," NBER Working Paper No. 1849 (Cambridge, Mass.: NBER, March 1986). Fuss and Waverman calculate the residual return to capital for auto producers in the two countries and select a year (1976) in which the average returns over the 1969–80 period were approximately equal to the annual return per unit capital. They then calibrate capital service cost indices in manufacturing for the two countries based on the relative return across countries in that year, and find the cost of capital to work in favor of the U.S. industry.

43. See Clifford Winston and Associates, *Blind Intersection? Policy and the Automobile Industry* (Washington, D.C.: Brookings Institution, 1987).

44. See Seiichi Fujita and L.J. Turvaville, "A Comparison of Engineering Economy Practices in the U.S. and Japan with Productivity in Mind," *Proceedings of the 1984 Annual International Industrial Engineering Conference* (Institute of Industrial Engineers, 1984). The study also found a big difference in minimum "hurdle" rates of return for investment projects—a 10 percent minimum attractive rate of return in Japan versus 21 percent in the United States. The study made no attempt to correct for differences in industrial distribution between the U.S. and Japanese samples; the samples themselves were also not randomly selected (firms' employees participating in engineering economy professional activities were polled).

45. Lists of qualifying robots for each of these schemes are published in the Japan Industrial Robot Association's bimonthly magazine, *Robot*. The extra depreciation accorded a "High Performance In-

SECTORAL PATTERNS OF TECHNOLOGY ADOPTION

dustrial Robot Controlled by Computers," for example, amounted to an additional 13 percent of the purchase price deductible immediately, in 1980, and has been set at 10 percent since 1982. See also U.S. General Accounting Office, *Industrial Policy: Japan's Flexible Approach* (June 1982), p. 62.

46. Testimony of Paul Aron before Subcommittee on Investigations, Committee on Science and Technology, House of Representatives, 97th Cong., 2d sess., *Robotics,* June 2 and 23, 1982 (Washington, D.C.: U.S.G.P.O., 1983), p. 27.

47. For 1983, see USITC (1983), p. 23, note 27 above; total shipments are from official JIRA statistics. For 1986, see Richard Brandt with James B. Treece, "Retool or Die: Job Shops Get a Fix on the Future," *Business Week* (June 16, 1986): 108.

48. For a more detailed general discussion of economies of scale in fabrication industries, with a particular emphasis on capacity sharing, see Thanong Lamyai, Yung W. Rhee, and Larry E. Westphal, "Economies of Specialization as a Source of Technological Change in the Mechanical Engineering Industries" (Washington, D.C.: Development Policy Staff, The World Bank, unpublished, no date).

49. Factory tour, September 20, 1985.

50. Factory visit to Nissan Motors, Murayama, March 17, 1986.

51. See Vincent J. Pavone and Edward Phillips, "Arc Welding Robots—An Overview," in *Robots 9 Conference Proceedings* (1985), pp. 5-120–21.

52. See Donald N. Smith and Peter Heytler, Jr. *Industrial Robots— Forecasts and Trends* (Dearborn, Mich.: Society of Manufacturing Engineers and the University of Michigan, 1985), pp. 93–95. Note also that the robot itself accounted for only half of the cost of the robot system, on average, with peripherals, controllers, and materials handling equipment accounting for the remainder of system cost.

53. See Flamm (July 1986), note 1 above.

54. Flamm (July 1986), note 1 above, describes the methodology and results for one application, assembly.

55. These population projections are taken from Gregory Spencer, *Projections of the Population of the United States, by Age, Sex, and Race: 1983 to 2080,* Current Population Reports Series P-25, Bureau of the Census, (Washington: U.S. G.P.O., May 1984), p. 6; Japan Statistics Bureau, Management and Coordination Agency, *Japan Statistical Yearbook 1985,* (Tokyo: Japan Statistical Association), p. 25.

56. James C. Abegglen and George Stalk, Jr., two experienced observers of Japanese industry, write:

The rapid increase in the product variety of the kaisha is not limited to the automobile. Product variety has literally exploded for the Japanese manufacturers of trucks, audio equipment, air conditioning equipment, home appliances, diesel engines, calculators, and more. In almost every case, their Western competitor counterparts are holding their product variety constant or are trying to reduce variety.

See Abegglen and Stalk, *Kaisha, the Japanese Corporation* (New York: Basic Books, 1985), p. 89.

57. See the discussion in D. Brandin, et al., *JTech Panel Report on Computer Science in Japan*, prepared for U.S. Department of Commerce (La Jolla, Calif.: Science Applications International Corporation, December 1984), pp. 3–5 to 3–14.
58. Abegglen and Stalk (1985), p. 89, note 56 above.
59. See David Friedman, "Beyond the Age of Ford: The Strategic Basis of the Japanese Success in Automobiles," in John Zysman and Laura Tyson, ed., *American Industry in International Competition* (Ithaca, N.Y.: Cornell University Press, 1983).
60. Friedman gives the following statistics for 1975 and 1980, describing distinct component types per 1,000 autos produced by the major companies:

1975	*Body Types*	*Engine Types*
Nissan	1:15	1:82
Toyota	1:12	1:170
GM	1:25	1:370
Ford	1:29	1:225
1980		
Nissan	1:8	1:13
Toyota	1:9	1:14
GM	1:33	1:215
Ford	1:27	1:185

See Friedman (1983), Table 1, note 59 above. An error in Friedman's table for Toyota, 1980, has been corrected here.
61. See Kenneth J. Arrow, "Economic Welfare and the Allocation of Resources for Invention," in R. Nelson, ed., *The Rate and Direction of Inventive Activity* (Princeton, N.J.: Princeton University Press and NBER, 1962).
62. See the discussion in J. Nevins, et al., *JTech Panel Report on Mechatronics in Japan*, prepared for U.S. Department of Commerce (La Jolla,

Calif.: Science Applications International Corporation, March 1985), pp. 3–58 to 3–60.

63. To quote the JTECH Panel report on Mechatronics, "the excellent performance of Japanese high precision robots is . . . primarily attributable . . . to painstaking and original mechanical design and construction. The robots are usually constructed by a group which is closely connected with the needs of the precision application which has a direct and beneficial bearing on the design process." See Nevins et al. (1985) p. 3–53, note 61 above.

64. See the description of the IBM 7576 assembly robot given in *Electronics*, vol. 59, no. 18 (May 5, 1986): 59.

65. Nevins et al. (1985) p. 3–52, note 61 above.

66. The JTECH Mechatronics Panel gave the example of a high performance automated wire bonding machine for semiconductor assembly within Hitachi that is simply not for sale. See Nevins et al. (1985) p. 3–60, note 61 above.

67. See Smith and Heytler (1985), p. 5.

68. See Japan Industrial Robot Industry Association (JIRA), *Sangyōyō Robotto Dōnyū ni Tomonau Keizai Kōka Bunseki: Chōsa Kenkyū Hōkokushō (The Introduction of Industrial Robots with an Analysis of their Economic Effect: An Investigative Research Report)*, (Tokyo: JIRA and Japan Machine Industry Association) 1984. An English summary of the study was published as Kanji Yonemoto, Yukio Hasegawa, Kenji Shiino, and Toru Hatano, "Method of Estimating Economic Effects of Robot Introduction—ROBEQ," in *Proceedings of the 15th International Symposium on Industrial Robots* (Tokyo: Japan Industrial Robot Association, September 1985).

69. For an extensive description and analysis of the development of offshore manufacturing by U.S. firms, see Joseph Grunwald and Kenneth Flamm, *The Global Factory: Foreign Assembly in International Trade* (Washington, D.C.: Brookings Institution, 1985).

70. See "Industrial Robot Market Slowing Down after Sharp Growth," *Japan Economic Journal* (March 1, 1986): 12.

8 THE INDIRECT EFFECT OF TECHNOLOGY ON RETAIL TRADE

Walter Y. Oi

Goods do not costly move from producers to consumers. *Distribution* can be defined as a set of activities that facilitate this move. These may include the movement of goods in space and time, the provision of product information, matching buyers and sellers, and last, but not least, reducing the transaction costs of consummating mutually beneficial exchanges. The distributive trades, which accounted for only 6.1 percent of total employment in 1880, provided fully 20.6 percent of all jobs in 1980. Jobs are not the same, and a typical job in wholesale/retail trade meant a substantially shorter work week of only 32.3 hours versus 40.9 hours in producing goods in 1980. At the turn of the century, H. Barger (1955) reported that men in the trade sector received weekly earnings that were higher than the earnings of men in manufacturing. The relative returns to all employees measured by a ratio of average hourly earnings in trade versus manufacturing was less favorable; the wage ratio was .875 in 1950 and fell to .753 in 1980. The comparison is even worse if the hourly earnings in retail trade (rather than wholesale/retail trade) are put in the numerator.

The functions required for the distribution of goods can be performed by producers, middlemen, and consumers. Direct sales by farmers and manufacturers were important in the early nineteenth century, and some consumers still drive out to the country

329

to get goat's milk. Retailers are specialists whose services are demanded by both consumers and producers. They rarely quote explicit "prices" for their services but receive their returns by introducing a wedge between retail and wholesale prices. The quantity and quality of retail services have varied over time. An obvious example is that department stores offer fewer delivery and credit services today than they did thirty-five years ago. Less apparent is the declining importance of the department store buyer who at one time was responsible for finding good bargains for the store's clientel. More and more manufacturers make direct appeals to consumers via national advertising campaigns. The economic analyses of distribution by Henry Smith (1948), W. Arthur Lewis (1970), and Bob R. Holdren (1960) have mainly dealt with the retail store. In their studies, location differentiates stores and results in a market equilibrium characterized by monopolistic competition.

Economists, with the notable exception of Charles Ingene (1984), have paid less attention to the role of the consumer. Shopping is costly both in time and money. A household balances lower shopping costs, meaning fewer shopping trips per month, against higher home inventory costs. A rational consumer will patronize the store that offers the lowest *full price*, which is the sum of the retail price plus the implicit unit costs of shopping at that store and holding the resulting inventory at home.

When we study the effects of technology on industry performance, we ordinarily look at the direct effects. The mechanical cotton picker and the cranberry paddler reduced the marginal cost of picking crops and allowed for a substitution of capital for unskilled labor. The development of cellophane benefited some but harmed others who produced waxed paper. Technology has profoundly affected the distributive trades, but it has exercised its impact in an indirect fashion. Technological advances in the production of goods during the nineteenth century led to the establishment of large-scale enterprises where workers specialized in producing certain goods, while others specialized in transporting and distributing them. The extent of this specialization was reinforced by cost-reducing innovations in transportation and communications. Households acquired more cars, larger refrigerators, and bigger houses, due, in large measure, to rising real incomes and falling relative prices for these consumer durable goods. These acquisitions altered consumer shopping patterns. Average transaction sizes got larger, which supported the construction of large super stores. Lower advertising rates encouraged more national advertising of standardized brands that reduced a shopper's search costs. The shopper no longer had to rely on the reputation of the Broadway to stock good underwear; he could simply buy Jockey. Some of the retail functions that were previously supplied by middlemen have been shifted forward to the customer or backward to the manufacturer. The remaining retail services can evidently be more economically produced by less skilled employees. The increasing number of two-car families, and the greater use of national advertising are, in my opinion, largely responsible for the retail revolution of the postwar years.

GROWTH OF THE DISTRIBUTIVE TRADES

The distributive trades are a consequence of specialization and the division of labor. In the early nineteenth century, most Americans combined production and consumption activities on family farms. Technological advances in transportation, manufacturing, and agriculture were responsible for greater specialization and large-scale enterprises. People moved out of agriculture when they found that they could command higher real incomes by separating the activities of making and spending money. Specialization created the need for more markets. Retail firms were established to facilitate the process of exchange as goods moved from producers to ultimate consumers. The transactions involved in transferring goods are costly and have to be "produced" like other economic goods and services.

The resources allocated to the distributive trades have grown in relation to the resources employed in producing goods. This fact was persuasively documented by Harold Barger (1955), who assembled employment data by sectors for the period 1869 through 1950. In panel A of Table 8–1, I reproduce Barger's data to which I append the postwar employment statistics.[1] Some 70.1 percent of the 17.4 million persons in the 1880 labor force were engaged in the production of goods (agriculture, fisheries, forestry, mining, and manufacturing), while only 6.1 percent were in the wholesale/retail trade sector. Of the 97.6 million employed persons in 1980, 26.5 percent were in goods producing industries, and 20.5 percent

Table 8–1. Employment, Hours, and Earnings by Major Industry, 1880–1980.

	1880	1900	1920	1940	1950	1960	1970	1980
A. Employees (thousands)								
A.0 All industries	17,390	29,070	41,610	53,300	55,813	64,647	78,678	97,639
A.1 Goods producing	12,185	18,020	23,510	22,190	22,368	22,528	23,454	25,857
A.1a Manufacturing	3,170	6,340	10,880	11,940	14,469	17,530	19,369	21,915
A.2 Transport/utilities	860	2,100	4,190	4,150	4,346	4,459	4,504	7,087
A.3 Trade	1,057	2,391	4,060	7,180	10,385	11,798	14,922	19,934
A.4 Other[a]	3,288	6,559	9,850	19,780	18,714	25,862	35,798	44,761
A.5 Ratio, trade/goods[b]	8.67	13.27	17.27	32.26	46.43	52.37	63.62	77.09
B. Average weekly hours								
B.1 Goods producing	52	51	48	43	43.8	41.5	41.2	40.9
B.1a Manufacturing	55	52	46	38	40.5	39.7	39.8	39.7
B.2 Trade	66	65	56	48	40.5	38.6	35.4	32.3
B.2a Wholesale	n.a.	n.a.	n.a.	41.3	40.7	40.5	40.0	?
B.2b Retail	n.a.	n.a.	n.a.	43.2	40.4	38.0	33.8	30.2
B.3 Ratio, trade/goods	127	127	117	112	92	93	86	79
Workhours per week (millions)								
B.4 Goods producing	634	919	1,128	954	980	935	966	1,058
B.5 Trade	70	155	227	345	421	455	528	644
B.6 Ratio, trade/goods[c]	11	17	20	36	43	49	55	61

C. Average hourly earnings

C.1 Goods producing	$0.102	$0.128	$0.384	$0.454	$1.39	$2.17	$3.26	$7.16
C.1a Manufacturing	0.142	0.175	0.477	0.633	1.44	2.26	3.35	7.27
C.1b Mining	0.164	0.198	0.759	0.886	1.82	2.7	3.85	9.17
C.1c Agriculture	0.084	0.087	0.275	0.253	0.75	0.97	1.8	3.9
C.2 Trade	0.144	0.175	0.48	0.536	1.26	1.93	2.71	5.34
C.2a Wholesale	n.a.	n.a.	n.a.	n.a.	1.48	2.3	3.44	6.96
C.2b Retail	n.a.	n.a.	n.a.	n.a.	0.98	1.52	2.44	4.88
C.3 Ratio, goods/trade	70.8	73.1	80	84.7	110.3	112.4	120.3	134.1
C.4 Ratio, manufacturing/ trade	98.6	100	99.4	118.1	114.3	117.1	123.6	136.1

D. Real Wages

D.0 Price index (1967 = 100)	$29	$25	$60	$42.0	$72.1	$88.7	$116.3	$246.8
D.1 Goods producing	0.49	0.70	0.80	1.51	2.00	2.55	2.88	2.95
D.1a Manufacturing	0.50	0.70	0.80	1.28	1.75	2.18	2.33	2.16
D.2 Trade	n.a.	n.a.	n.a.	n.a.				
D.2a Wholesale	n.a.	n.a.	n.a.	n.a.				
D.2b Retail	n.a.	n.a.	n.a.	n.a.	1.36	1.71	2.1	1.98

a. Other industries include services, construction, government, and so forth.

b. Ratio is defined as employment in trade divided by employment in goods producing industries.

c. Ratios are multiplied by 100.

Sources: Employees, hours, and earnings 1880–1940 are taken from Harold Barger (1955): Tables 1 and 5, and A–1. Data for 1950 to 1970 obtained from Historical Statistics and Statistical Abstract of the U.S. (selected issues). Data for 1980 obtained from the Population Census and Statistical Abstract.

were distributing them. This dramatic shift in the industrial distribution of the labor force took place in a growing economy where the employment to population ratio remained stable from 1880 to 1940, but it rose in the last forty years.[2] Line A.5 of Table 8–1 presents another way of describing this shift in the allocation of labor. For every 100 workers who produced goods in 1880, 8.6 persons held jobs in trade; this figure rose to 46.4 in 1950 and to 77.1 in 1980.

Employment counts make no allowance for variations in the length of the workweek and hence provide an imperfect measure of the input of labor services. The workweek has been shortened in all industries, but this decline has been especially steep in trade. Over the century, 1880 to 1980, the workweek shrunk from 52 to 40.9 hours in the goods-producing industries compared to a drop from 66 to 32.3 hours in the trade sector.[3] Estimates of workhours per week (employment times average weekly hours, $M = NH$), are presented in line B.5. The ratio of manhours in trade to that in goods was $(M_T/M_G) = 0.11$ in 1880, but it climbed to 0.43 in 1950 and 0.59 in 1980.[4] The relative importance of distribution to production has clearly increased measured either by a ratio of employment counts or by a ratio of manhours of labor input.

The composition of the workforce in the distributive trades differs from that in manufacturing. It contains larger proportions of females, part-time employees, and a smaller proportion of individuals in the prime working ages, twenty-five to forty-four years of age. Female labor force participation rates rose steadily throughout the postwar period. In the late 1960s, the labor market began to absorb the entry of those born during the baby boom.[5] These two trends have been analyzed in the literature. Less attention has been paid to the rising importance of part-time employees. In 1950, 13.0 percent of all employees held part-time jobs, but by 1980 they accounted for 20.9 percent. In retail trade, the share of part-timers has nearly doubled, which can be seen in the data of Table 8–2. A rising ratio of part-time to full-time employees (and by implication, of unskilled to skilled workers) is observed even when one controls for the type of retail establishment; confer the lower panel of Table 8–2. B. Bluestone (1981) attributed this "de-skilling" of the retail labor force to the national marketing of standardized, branded goods—product information is conveyed by the media rather than by trained sales personnel. When "retail

Table 8–2. Part-Time, Prime-Age, and Female Employees, 1950, 1980 (percentages of all employees).

Industry	Female 1950	Female 1980	Part-Time 1950	Part-Time 1980	Prime Age 1950	Prime Age 1980
A. Major industries						
0. All industries	30.02%	41.78%	13.03%	20.85%	65.83%	63.67%
1. Agriculture/forestry/fisheries	8.31	16.50	19.47	23.48	55.90	54.73
2. Mining	2.35	11.68	20.08	6.77	74.11	68.81
3. Construction	2.81	7.55	15.83	7.04	69.05	65.98
4. Manufacturing	24.98	31.91	8.52	9.65	69.18	66.93
5. Transportation/Communications/utilities	15.59	24.35	6.13	10.25	68.72	72.25
6. Wholesale trade	21.30	26.60	6.39	11.75	70.14	65.16
7. Retail trade	40.14	50.28	12.46	24.52	64.98	49.36
8. Finance/insurance	44.61	57.78	7.70	15.48	58.98	64.60
9. Business repair	14.96	33.21	9.05	21.06	72.44	63.63
10. Personal services	70.08	69.23	27.07	40.03	62.43	55.78
11. Entertainment/recreation	25.99	39.81	29.53	41.23	57.50	51.53
12. Professional services	62.00	66.02	18.21	32.03	65.68	67.89
13. Public administration	24.52	40.47	5.59	10.64	72.99	61.00
B. Retail Trade						
7.1 Hardware	13.72	25.42	6.32	20.95	68.39	54.30
7.2 General merchandise	70.16	69.97	11.37	39.10	60.08	49.52
7.3 Food and dairy	26.60	45.29	14.13	39.45	62.03	48.28
7.4 Auto dealers/supplies	12.86	16.88	3.82	11.58	74.65	61.30
7.5 Gas stations	2.72	15.72	8.89	26.45	64.39	47.14
7.6 Apparel stores	59.30	69.43	13.85	41.27	64.03	45.18
7.7 Furniture/appliances	21.33	32.80	8.11	22.76	70.15	59.55
7.8 Eating/drinking places	54.63	59.09	15.06	43.07	68.16	40.71
7.9 Drugstores	48.52	59.90	18.70	39.99	56.40	48.87
7.10 Other retail trade	33.78	53.50	12.63	34.84	64.29	57.09

services" are homogenized, they can be mass-produced by individuals with little firm-specific training.

The changing skill mix of retail workers is reflected in the historical trend of relative wages. In 1880 production workers in manufacturing earned average hourly earnings that were *below* the hourly wages of wholesale/retail trade employees. However, by 1950 the wage ratio (manufacturing over trade) had climbed to 1.143 as shown in panel C. Relative wages continued to climb, so that by 1980 it reached 1.361. At least three factors contributed to the relative decline in retail wages: (1) The industrial composition of employment within the trade sector had shifted toward the low-wage industries within trade; (2) within each three-digit industry, firms increased the ratio of unskilled to skilled workers; and (3) retail trade unions lost some of their economic power.[6] I try, in this chapter, to untangle the extent to which these three factors are the results of changing technologies and preferences.

THE "PRODUCT" OF A RETAIL FIRM

Some firms are mainly producing goods—fermenting wines, growing catfish, mining coal, or fabricating metals. These "goods" may undergo further processing before they are ultimately consumed. Part of the additional "processing" involves the services and functions that accompany the *distribution* of goods to final consumers. Direct sales by farmers and manufacturers were important at one time in our history and are still important in some economies and for particular firms. A firm that distributes its own products is vertically integrated into distribution. As specialization spread, a class of middlemen emerged. They specialized in facilitating transactions and in providing "services" that were useful to consumers and producers. These middlemen do not, in general, charge explicit fees for their services. They earn a return by introducing a gross margin—a spread between retail and wholesale prices. The size of the gross margin will depend on sales volume, the services supplied by the retailer, input prices, and competitive market forces.

A retail establishment or store can properly be regarded as a multiproduct firm, or should we call it a multiservice firm? Stores differ in the services that they supply, and the service mix of a particular store can vary over time. Retail services can include some or all of the following: (1) a product line—an array of goods that is assembled and made available for customers, (2) point of sale services—providing information, waiting on customers, and so forth, (3) convenience—store location, store hours, price and product information, (4) delivery, credit, return privileges, and so forth, and (5) packaging, processing, and producing.[7] The demands for these retail services are derived from the consumer's willingness to pay for final goods (at the times, places, and conditions offered by the store) and the supply prices at which wholesalers and producers will make goods available to the store. If a retailer disseminates product information in its local ads or sets aside shelf space for displays, it may be able to get price concessions from the favored manufacturer. Self-service stores, especially cafeterias, have lower gross margins because these stores supply less point of sales services. Variations in the quantity and quality of services could thus generate a dispersion of retail prices across stores and over time.

The production function for the distribution of goods shares many of the properties of the production functions in transportation and education. The production of a person-trip calls for the input of the trip-taker's time in combination with the inputs needed to supply the transportation mode. One had to combine the input of the Duke of Buccleigh and one moral philosopher in order to produce one student who could matriculate at Oxford. In an analogous manner, the customer supplies an essential input in the distribution of goods. Every consumer allocates time and money to shopping, which includes the activities involved in searching for information about prices and qualities, locating the appropriate store at which to make purchases, and arranging for the delivery of those goods to the place and time of final consumption.[8] The kinds of services that a store offers are likely to affect its sales volume. Advertising can, for example, reduce the customer's costs of searching for information. Shopping time is related to a store's location, and layout, the number of checkout lanes, and the size of inventories, which affects the probability of stockouts. A theory of the distributive trades ought to explain just what kinds of services will be supplied in distributing goods as well as the division in providing those services among producers, middlemen, and consumers.

In the early stages of growth, stores added delivery, credit, and free home trial privileges to a basic retail service package of a product line, transaction, and customer service. According to Barger, these added services were responsible for the increase in gross margins of department stores in the early twentieth century. Some services like the home delivery of milk were turned over to specialists who charged for this service in higher prices. However, department stores did not explicitly charge higher prices to customers who bought on credit. Credit costs were spread across all customer classes.[9] These services were evidently valued by producers and consumers who expanded the demand for retailers. The percentage of consumable goods passing through retail outlets rose from 72 percent in 1869 to 87 percent in 1929 (see Barger 1955: 58).

In the postwar period, retailers have altered the service mix in response to technological changes that occurred outside of the trade sector. The automobile enables a customer to provide his own delivery of milk and other consumables. Technological advances in communications and computers have revolutionized the credit market. Banks and financial institutions introduced credit cards, which now supply much of the consumer credit that might otherwise have been supplied by department, appliance, and furniture stores. The national advertising of branded goods is, in some ways, a substitute for the information that could have been conveyed by an experienced clerk or by the product line assembled by a reputable store. The service mix that constitutes the "product" of a retail firm is endogenously determined by technology, preferences, and input prices.

The background model implicit in the preceding paragraphs is one in which technical change is responsible for the growth of the distributive trades. Reductions in transport costs combined with economies of scale led to greater specialization. There was a sharp increase in the volume of transactions that could be more economically consummated by an expansion in traders who specialized in the distribution of goods. The model provides a plausible explanation for growth in a period where one observes a rising fraction of goods passing through retail outlets. But when this fraction reaches saturation, how do we explain the continued growth of distribution? This latter phenomenon can be explained by a model in which sectors differ in rates of technical progress. Given fixed proportions, balanced growth means that more resources have to be allocated to sectors with lower rates of technical progress. To test this explanation, we need to define and measure "output." Hall, Knapp, and Winston (1961) measured the output of retail trade by constant dollar sales. This procedure assumes that retail services are proportional to sales implying that changes in gross margins are due to changes in input prices.[10] An alternative procedure assumes that retail "output" is proportional to value-added, which is approximately equal to sales times the gross margin.[11] Both measures of output were used by Ratchford and Brown (1985), who estimated total factor productivity for retail food stores. Over the two decades, 1959–79, their output index based on sales increased by more than 60 percent, $(S_{79}/S_{59}) = 1.606$. Using constant dollar margins, the output index was $(V_{79}/V_{59}) = 1.909$.[12] Based on the sales measure, total factor productivity in retail food stores grew at an annual rate of 0.47 percent compared to 2.07 percent in manufacturing.[13] This finding led Ratchford and Brown to conclude that ever increasing shares of resources will have to be allocated to distribution. They go on to argue that unless new methods of distribution are developed, the rising costs of distribution might inhibit the growth of the economy as a whole.

SHOPPING AND HOUSEHOLD PRODUCTION

The household production model of G.S. Becker (1965) provides a useful framework to analyze the shopping activities that are undertaken to reduce the full costs of consumption. Consumers must select the supplier or suppliers with whom to make transactions. They must decide on the frequency of patronizing each supplier and on the amounts and kinds of goods that will be purchased on each trip.[14] Although these decisions are simultaneously made, it is analytically convenient to assume that they are made sequentially.

Turn first to the frequency of shopping trips. Goods are not acquired in continuous flows but are purchased in discrete lots or batches. Suppose that there is only one good and one store. In the inventory model of T.N. Whitin (1952), the total cost of consuming Q units per unit time period is the sum of three components: (1) outlays at the store, PQ, (2) costs of shopping trips, $C_S = ST$, where S is the cost of a trip, and $T = (Q/q)$ is the shopping trip frequency when the purchase or lot size is q, and (3) costs of holding home inventories, $C_H = b(q/2)$, where b is the unit holding cost for the average home inventory of $(q/2)$ units.

$$C = PQ + C_S + C_H = PQ + \frac{SQ}{q} + \frac{bq}{2} \qquad (8.1)$$

Buying in larger lots and making fewer trips will reduce the total cost if $(C_S' + C_H') < 0$. When S and b are constants, the optimum lot or purchase size that minimizes C is given by the familiar square root formula.

$$q^* = [\frac{2QS}{b}]^{\frac{1}{2}} \qquad (8.2)$$

For a given consumption level Q, the cost-minimizing number of shopping trips per period is seen to be,

$$T^* = \frac{Q}{q^*} = [\frac{Qb}{2S}]^{\frac{1}{2}} \qquad (8.3)$$

If equation (8.2) is substituted into (8.1), we get the minimum total cost.

$$C = PQ + [2QSb]^{\frac{1}{2}}$$

Dividing by Q, we obtain the *full price* when the purchase size q is optimally chosen. It is the sum of the retail price plus the unit cost of shopping and holding home inventories denoted by c.

$$P^* = \frac{C^*}{Q} = P + c \qquad c = [\frac{2Sb}{Q}]^{\frac{1}{2}} \qquad (8.4)$$

Full prices thus vary across consumers being lower for those who incur lower shopping and holding costs or who demand larger total quantities Q per period. Spoilage, storage, and interest costs affect the unit holding cost b. Refrigerators, freezers, and more cupboard space tend to reduce b, which, in turn, leads to fewer shopping trips and a larger transaction size per trip, q^*. There are reasons to suppose that S and b will be increasing functions of q. These diseconomies will tend to reduce the optimum purchase size q^* meaning more shopping trips. More precisely, if E_S and E_H denote the elasticities of the average shopping and holding costs, the cost-minimizing purchase size is given by a modified square root formula.[15]

$$q^* = [(\frac{2QS}{b})(\frac{1-E_S}{1+E_H})]^{\frac{1}{2}} \tag{8.2'}$$

The customers who patronize a particular store are likely to exhibit a dispersion of optimum purchase sizes, q^*. Rightward shifts in the distribution of q^* will affect both the costs and pricing policies of retail stores.

Turn next to the choice of a store. Retain the assumption of a single good, but allow for the presence of two stores. The i-th consumer will patronize store A (meaning $Q_{bi} = 0$), if A offers a lower full price.

$$Q_{ai} > 0, \text{ if } P_a^* = (P_a + c_{ai}) < (P_b + c_{bi}) = P_{bi}^* \tag{8.5}$$

All consumers confront the same retail prices (P_a, P_b), but they may face different full prices at the two stores, which are differentiated by location or retail services.[16]

Consider a model like that in Hotelling (1929) or Smithies (1941) in which a consumer buys only one unit and hence makes only one shopping trip per period. If D_{ji} is the distance to store j, and k_i is the travel cost per mile, the unit shopping cost is, $c_{ji} = k_i D_{ji}$.[17] Store A will be chosen by the lower full price rule if

$$P_a + k_i D_{ai} < P_b + k_i D_{bi} \tag{8.6a}$$

If retail prices are the same, $P_a = P_b$, customers and stores are matched by proximity; customer i goes to store A if $D_{ai} < D_{bi}$ irrespective of his travel cost parameter k_i. Store A can offset a locational disadvantage, $D_{ai} > D_{bi}$, by offering a price concession. Customers with low travel costs k are more likely to make a longer trip to enjoy a lower retail price. Equation (8.6a) can be rearranged to show this result, namely, k_i has to be less than the price differential per additional mile of travel.

$$P_{ai}^* < P_{bi}^*, \text{ if } k_i < (\frac{P_b - P_a}{D_{ai} - D_{bi}}) \tag{8.6b}$$

Differences in k_i could result in an equilibrium with overlapping markets. Suppose, for example, that there are two types of custom-

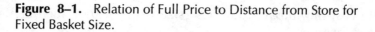

Figure 8–1. Relation of Full Price to Distance from Store for Fixed Basket Size.

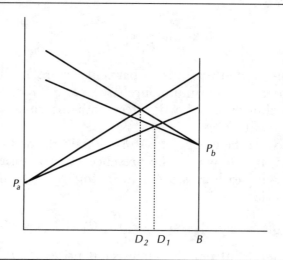

ers with $k_1 < k_2$. They are located along a straight line living across the street from one another. In Figure 8–1, we show the full prices facing the two types of customers who reside between two stores. Type 1 customers to the left of D_1 attain a lower full price by choosing store A; those to the right of D_1 go to store B because for them, $P_{a1}^* > P_{b1}^*$. The watershed for type 2 customers who face a higher travel cost occurs at D_2.[18] The overlap is observed in the interval, $D_2 D_1$ where type 2 customers go right to store B, while type 1 customers choose to go left to enjoy store A's lower retail price. High-k customers can thus support the survival of high-priced stores like B, which will be interspersed between low-priced stores like A. Decreases in travel costs k enable store A to expand its market area as D_2 and D_1 both shift to the right. This simple model thus shows how a dispersion in travel costs can sustain an equilibrium in which stores charge different prices for the *same* good Q.

The preceding model can be extended in at least three directions: (1) The number of shopping trips and hence the purchase size are variable, (2) the cost of a shopping trip depends on variables other than distance, and (3) total demand Q and inventory holding cost b are variable. For analytic ease, I compare the full prices at two spatially differentiated stores for a given customer;

this allows me to suppress the i subscript. For the first extension, the cost of *one* shopping trip is proportional to the distance to store j, namely $S_j = kD_j$. When purchase is optimally chosen, the unit cost of shopping at store j and holding the resulting home inventory is given by

$$c_j = [\frac{2kD_j b}{Q}]^{1/2} = \alpha\sqrt{kD_j}, \qquad \alpha = [\frac{2b}{Q}]^{1/2} \tag{8.7}$$

The interesting case arises when one store, say A, quotes a lower price. The full price, $P_a^* = P_a + \alpha\sqrt{kD}$, rises at a decreasing rate as D, the distance between customer and store, is increased. Let $D_b = (B - D)$ denote the distance to store B. Plotting in the reverse direction, $P_b^* = P_b + \alpha\sqrt{k(B - D)}$ rises as $(B - D)$ is increased, and at D_1, $P_a^* = P_b^*$. Those to the left of D_1 go to store A. However, persons located closer to the store incur a lower trip cost, $S_a = kD$; those near A make more trips and buy less on each trip. In this model, the market area captured by store A will expand as P_a, α, or k are reduced.[19]

The second extension acknowledges that shopping may entail a set-up cost S_O as well as a component that varies with distance D_j and purchase size q_j. If the cost per trip is approximately linear, $S_j = S_O + kD_j + k'q_j$, we can solve for the optimum lot size q_j^*, trip frequency $T_j^* = Q/q_j^*$, and full price.[20]

$$P_j^* = P_j + [\frac{2S_j b}{Q}]^{1/2} = P_j + k' + \alpha\sqrt{S_O + kD_j} \tag{8.7'}$$

where again, $\alpha = \sqrt{2b/Q}$. The addition of two cost parameters, (k', S_O), results in an upward shift in the full price curve and flattens its slope with respect to distance.[21] The set-up cost S_O reduces the frequency of shopping trips and increases the optimum purchase size q_j^*.

The last extension is straightforward. A fall in b or an increase in Q reduces α, which flattens the full price curves in Figure 8–2. As α is reduced, the market area of the low-price store A expands. An increase in real income is ordinarily accompanied by a larger total demand Q. Further, the acquisition of a car is likely to reduce k, while refrigerators and larger houses tend to reduce b. On balance, these forces lead to larger market areas for stores that can offer lower retail prices. Moreover, these stores will experience rightward shifts in the frequency distribution of purchases $\phi(q)$.

Figure 8–2.   Relation of Full Price to Distance from Store for Variable Basket Size.

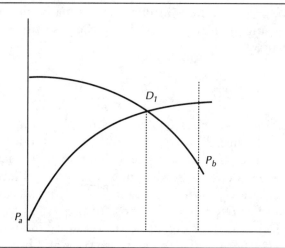

Consumer shopping behavior is influenced by more than transport and inventory costs. We demand many goods, and the store that offers a wider product line can attract more customers. By consolidating purchases at one store, the transport cost S can be spread thereby spreading the full total cost for a fixed-consumption vector, $(Q_1, Q_2, \ldots Q_G)$. In the special case of additive inventory holding costs, the consumer chooses his shopping trip frequency T to minimize total full cost.[22]

$$T^* = [\frac{b_1 Q_1 + b_2 Q_2 \cdots + b_G Q_G}{2S}]^{\frac{1}{2}} \qquad (8.3')$$

There are obvious economies of "one-stop shopping" that are likely to prevail even if we relax the assumptions of a fixed transport cost per trip S and inventory costs. If the demand vector $(Q_1, \ldots Q_G)$ is exogenous, stores are chosen to minimize the full total cost. If, however, demands are endogenous, and retail price vectors vary across stores, the customer's choice of a store is determined by utility maximization.[23]

We often shop to find a particular good or to search for prices and products that can elevate us to higher utility levels. Failure to find a preferred brand means that we must either substitute another brand or incur the added cost of visiting another store. Stores that maintain wider and deeper inventories can reduce these "regret" costs of stockouts. Information about prices and the availability of products can be obtained via direct customer search or by advertising. Firms and manufacturers who advertise provide consumers with a valuable service whose relative importance depends on the relative costs of advertising versus the implicit costs of customer search. Individuals differ in the values that they attach to customer service by trained clerks. If I want to buy only a few items, I am more likely to delegate the task of selecting and gathering those items to an "agent." However, as the variety and volume of purchases increase, I may assume the tasks of selecting and gathering at a self-service store.[24] In short, the demands for various retail services depend not only on the implicit "prices" for those services, but also on factors such as store location, ownership of a car and refrigerator, the opportunity cost of time, and so forth, which affect the position of these service demand functions.

COSTS AND THE SUPPLY OF RETAIL SERVICES

Distribution resembles transportation in at least two respects: Demands are stochastic, and costs are largely fixed or quasi-fixed. Holdren (1960) claimed that fully 90 percent of a supermarket's total costs (other than the costs of goods sold) are fixed costs or discretionary fixed costs.[25] The costs of assembling and displaying goods are similar to the costs of supplying available seat-miles. Statistical studies often relate costs to a store's sales volume X and the number of customer transactions N.

$$C = C(X,N) \tag{8.8}$$

If Holdren is right, $C(X,N)$ should exhibit short-run scale economies.

A store in G. Heall's (1980) model brings goods Q to a site closer to consumers who are located on the circumference of a circle. The cost per mile of transporting *one* unit of the good exhibits increasing returns, $c = c(Q)$ with $c'(Q) < 0$. A retailer exploits these economies by consolidating the demands of customers in its market area. His transport costs are covered by the gap between retail and wholesale prices, which is just the gross margin.[26] Heall examined the properties of equilibrium for alternative market organizations—a public enterprise that minimizes total social costs, independent competitive firms, and a monopoly chain. The equilibrium number of stores and the size of a representative store measured by its sales, $Q = r\theta q$, are determined by the quantity q demanded by each consumer and the structure of the transport cost function. As demands by individuals rise, the market can support more stores. Each store will be larger even though it spans a smaller geographic market area. A shift in the cost structure $c(Q)$ that increases the relative importance of overhead costs (either via a rise in the fixed set-up cost α or a fall in total variable costs, βQ) leads to fewer but larger stores.[27] All stores attain the same size Q, which depends on q and $c(Q)$.

Retail costs depend not only on sales volume X and the number of customer transactions N, but also on the mix of services and input prices. B. Nooteboom (1983) derived a cost function for a class of stores that were similar with respect to the width and breadth of the product line, types of departments, and annual store hours. Based on data for a cross-sectional sample of Dutch supermarkets, Nooteboom found that the annual labor input measured in manhours M_i was a linear function of annual sales X_i.

$$M_i = \beta_O + \beta_1 X_i + e_i \tag{8.9a}$$

The fixed labor input β_O depends in the long run on annual store hours and types of departments, while β_1 is the marginal labor input needed to service an additional dollar of sales. Part of the variable labor input is required to wait on customers, while part is employed in handling goods, $\beta_1 = \tau_1 + \tau_2$. Given a stochastic arrival rate of customers, Nooteboom imposes a decision rule wherein the ratio of queueing to serving times is kept within narrow limits. This rule thus determines the demand for labor. Thus, τ_1 is the ratio of the mean time to wait on a customer σ to the mean transaction size q, which is a proxy for service time, $\tau_1 = \sigma/q$. Sorting, stocking, and sweeping entail a labor input that is proportional to sales.[28] Substituting for β_1 yields a manhours labor requirement equation that is linear in the number of customer transactions N_i and annual sales X_i.

$$M_i = \beta_0 + \sigma N_i + \tau_2 X_i + e_i \tag{8.9b}$$

The scale economies with respect to the number of transactions N are smaller than those with respect to purchase size q. Schwartzman (1968) recognized this fact and argued that the upward trend in average transaction size q could explain much of the secular growth in labor productivity. Changes in the level and mix of retail services, the product line, or store hours will affect the parameters of a labor requirements equation like (8.9b), and hence labor costs.[29] Other inputs (floor space, parking area, equipment, utilities, advertising) will be functions of transactions N as well as the services and amenities A supplied by the store. The cost function of equation (8.8) can be expanded to,

$$C = C(X,N,A,W), \; C_X = \frac{dC}{dX}, \; C_N = \frac{dC}{dN}, \; C_A = \frac{dC}{dA} \tag{8.8'}$$

where W is a vector of input prices. This retail cost function exhibits increasing returns meaning that a doubling of X and N, which holds constant the average purchase size, $q = X/N$, is accompanied by less than a twofold increase in operating costs.

Suppose that competing stores handle a single homogeneous good and the demand facing a single store is given by

$$X = X(P,A,Z), \; E_P = -(\frac{P}{X})(\frac{dX}{dP}), \; E_A = (\frac{A}{X})(\frac{dX}{dA}), \tag{8.10}$$

where P is the retail price, and A is an index of services offered by the store. The Z vector includes variables describing the incomes and characteristics of potential customers, the prices and service levels of competing stores, and the store's location in relation to its competitors. Changes in P and A can affect X through their impact on either the number of customer transactions N (the traffic flow through the store), and/or on the quantity demanded per shopping trip q. Each store faces both a demand for customer visits N and a demand for goods per visit, q.

$$N = N(P,A,Z_1) \; \mu_p = -(\frac{P}{N})(\frac{dN}{dP}) \; \mu_A = (\frac{A}{N})(\frac{dN}{dA}), \tag{8.10a}$$

$$q = q(P,A,Z_2) \; \epsilon_P = -(\frac{P}{q})(\frac{dq}{dP}) \; \epsilon_A = (\frac{A}{q})(\frac{dq}{dA}) \tag{8.10b}$$

The exogenous variables Z_1 in the visits demand equation are likely to differ from those in the per trip demand equation Z_2. Total profits for a single firm will be a function of two decision variables, P and A.

$$\pi = (P - P_w)X - C(X,N,A,W) \tag{8.11}$$

where P_w is the wholesale price of the good. I shall use subscripts to denote partial derivatives, such as $X_P = (dX/dP)$. The first-order conditions for profit maximization are given by

$$\pi_P = \frac{d\pi}{dP} = X + (P - P_w)X_P - C_X X_P - C_N N_P = 0 \tag{8.11a}$$

$$\pi_A = \frac{d\pi}{dA} = (P - P_w)X_A - C_X X_A - C_N N_A - C_A = 0 \tag{8.11b}$$

If elasticities are substituted for partial derivatives, we get[30]

$$P[1 - \frac{1}{E_P}] = P_w + C_X + (\frac{\mu_P}{E_P})(\frac{C_N}{q}) \tag{8.11a'}$$

$$[P - P_w - C_X - (\frac{\mu_A}{E_A})(\frac{C_N}{q})]X_A = C_A \tag{8.11b'}$$

In equilibrium, the marginal revenue (the left side of (8.11a')) will be equated to the marginal full cost, which is the sum of three components: the wholesale cost P_w, the marginal selling cost C_X, and a portion of the marginal customer cost deflated by the purchase size. The marginal cost of more services C_A will be equated to its marginal value product, which is the product of the net margin per unit sold (given by the bracketed term in (8.11b')) times the increment in sales X_A. In both (8.11a') and (8.11b'), the marginal cost of serving another customer, C_N, is divided by the mean purchase size q and weighted by the portion of sales generated by more shopping trips relative to the total increment due to more N and more q per visit. Let c^*_{XP} and c^*_{XA} denote the full marginal costs when sales variations are induced by price and service level changes. If $g = (P_w/c^*_{XP})$ is the share of the full MC attributable to the cost of goods sold, then the store's gross margin is given by[31]

$$m = \frac{P - P_w}{P} = 1 - g(1 - \frac{1}{E_P}) \tag{8.12}$$

Gross margins are in the neighborhood of 0.20 for supermarkets and 0.34 for convenience stores. If $g = .95$, these margins imply own price elasticities E_P of 6.33 and 3.28.

Each store in this model views Z as exogenous. If the choice of P and A results in losses, the store withdraws from the industry, which affects the Z vector for other stores. Positive profits are likely to invite entry, which adversely affects the N and q demand functions facing the profitable store. In Heall's model, consumers are alike, and stores confront the same cost function. As a consequence, identical firms quote the *same* retail price and attain the same size. However, in an economy where consumers differ in their locations, preferences, and shopping/inventory costs, equilibrium in a retail market could easily be characterized by a dispersion of retail prices as well as the survival of stores of varying sizes. It was shown above that when transport costs k differ, customers may patronize stores that charge different prices; see Figure 8–1 and equation (8.6). The one product store could be placed in a linear world to study the properties of long-run equilibrium. Such models can be found in the literature, and I shall not try to extend them here.[32]

Turn next to the case in which each differentiated store handles many goods and has some monopoly power in pricing. C. Bliss (1985) assumed that the store is a multiproduct monopoly. Its ability to set prices is, however, limited by competition that imposes on it a zero profit constraint. Optimal departures from wholesale prices (the margins, $p_j - p_{wj}$) follow the Ramsey rule; namely, mark-ups are chosen to achieve equi-proportional reductions in demand. Margins are higher for goods with less elastic demands, but given identical consumers, there are no loss leaders.[33] The demand elasticities that determine the Ramsey prices presumably reflect preferences. However, "one-stop shopping" does not give a store monopoly power over *all* of the items on the price list. Customers often frequent several shops. The pertinent demand elasticities for goods that are carried by several shops (such as cigarettes and shampoo) are obtained from the residual demand curve facing the store in question.

Bob R. Holdren (1960) developed a pricing model that incorporates his analysis of the actual price margins of some eight to ten supermarkets in a New England city. Setting prices for thousands of items in a product line poses a formidable problem, which managers allegedly solve by mimicking the mark-ups of a price leader, such as A&P. This allegation is roundly rejected by the observed dispersion of prices across stores. Holdren argued that for pricing purposes, goods can be placed into four classes: (1) externally fixed prices due to consignment selling or resale price maintenance, (2) noncompetitive goods whose prices are un-noticed, (3) goods with wide pricing latitude due to ignorance, small budget shares, or diversity in quality, and (4) highly competitive goods that are in his "k class." This last group includes those items that are important in choosing a store.[34]

Consider a store whose product line contains G goods, $(x_1, x_2, \ldots ,x_G) = X'$. Total profits are

$$\pi = \sum_j (p_j - p_{wj})x_j - C(X,N,A) \tag{8.13}$$

where N is the number of customer transactions called shoppers for short. The first-order conditions are

$$\frac{d\pi}{dp_j} = x_j + \sum_h (p_h - p_{wh} - c_{xh})(\frac{dx_h}{dp_j}) - C_N(\frac{dN}{dp_j}) = 0,$$

$$(j = 1, \ldots ,G) \tag{8.13a}$$

$$\frac{d\pi}{da} = \sum_h (p_h - p_{wh} - c_{xh})(\frac{dx_h}{dA}) - C_N(\frac{dN}{dA}) - C_A = 0 \tag{8.13b}$$

where $c_{xh} = (dC/dx_h)$ is the marginal selling cost for item h; this could include the store's inventory carrying cost. Let $e_{hj} = (\epsilon_{hj} + \mu_j)$ denote the elasticity of demand for x_h in response to a relative change in p_j.[35] The adjusted price margins m_j, which maximize profits, satisfy a system of G equations.

$$\sum_h k_h(\epsilon_{hj} + \mu_j)M_h = -k_j[1 + (\frac{C_N}{q_j})\mu_j]. \quad (j = 1, 2, \ldots ,G) \tag{8.14a}$$

where $k_h = (p_h x_h /R)$ is the share of outlays at the store expended on x_h, and M_h is an adjusted gross margin that allows for marginal selling costs.

$$M_h = p_h - p_{wh} - c_{xh} \tag{8.14b}$$

These equilibrium conditions contain a term μ_j that describes the additional number of shoppers attracted by a relative decline in p_j. The traffic generated by a price reduction will be greater if x_j

accounts for a larger share of outlays. Low margins appear to be set for goods with high transfer effects that extend to goods other than x_j.[36] Finally, the size of μ_j will depend on transport costs and consumer densities. The availability of an auto allows a consumer to switch to a store that offers lower prices even though that store is considerably further than his current store.

RETAIL FOOD STORES: CONCENTRATION AND COMPETITION

A theory of the distributive trades ought to explain several features of the postwar retail revolution including (1) concentration of sales in larger stores, (2) diminution in point of sales services, delivery, and credit, and (3) increasing reliance on part-time, unskilled workers. My empirical observations are largely drawn from grocery stores, which are now fewer in number and larger in size. In 1940, there was one food store for every seventy-eight households, but by 1980 the ratio of households to stores had climbed to 481. Over the forty years spanned by the data of Table 8–3, food purchases per household in constant 1967 dollars increased by 38 percent, and cars per household nearly doubled. The share of food sales captured by chain stores rose from 35.2 to 46.7 percent. The average sales size of a chain store doubled from 1940 to 1950. It doubled again in the decade of the 1950s but remained virtually stable in the 1960s (see line 3.2). The upward trend in store size resumed in the next decade. R. Parker (1986) found that the simple average of the four-firm sales concentration ratios for 196 SMSAs rose from 45.3 percent in 1954 to 55.8 percent in 1977.[37] Larger stores and increasing market concentration can, I believe, be explained by rising real incomes and declining relative prices for autos, household appliances, and communications.

The census data reveal wide cross-sectional differences across states in the size distribution of food stores. Two measures of average store size were constructed from the 1958 *Census of Business* and the 1977 *Census of Retail Trade;* the log of sales per establishment X^* and the log of employees per establishment E^*. They were related to three explanatory variables, (1) Met = the log of the fraction of a state's population residing in SMSAs, (2) Inc = the log of per capita income, and (3) 2 cars = the log of the percentage of households in the state owning two or more cars.[38]

Table 8-3. Food Stores: Sales and Related Variables, 1940–80.

	1940	1950	1960	1970	1975	1980
1. Number of stores (in thousands)						
1.1 Independent	405.0	375.0	240.0	174.1	142.73	112.6
1.2 Chain	41.35	25.70	20.05	34.20	23.08	18.70
1.3 Convenience					25.0	35.0
1.4 Total	446.35	400.70	260.05	209.30	191.80	167.10
2. Sales (in millions of constant 1967 dollars)						
2.1 Independent	16,563	22,752	36,534	40,331	40,080	41,353
2.2 Chain	9,034	13,611	22,316	36,619	38,056	40,501
2.3 Convenience					3,124	4,870
2.4 Total	25,597	36,362	58,750	76,950	81,260	86,724
3. Sales per store (in thousands of constant 1967 dollars)						
3.1 Independent	41	61	152	232	281	367
3.2 Chain	218	530	1108	1071	1649	2166
3.3 Convenience					125	136
3.4 Total	57	91	226	369	424	519
4. Number of households (000)	34,949	42,867	53,021	63,450	71,920	80,390
5. Number of registered autos						
5.1 Privately owned (000)	27,372	40,191	61,420	88,775	106,077	120,866
5.2 Per household	0.78	0.94	1.16	1.40	1.47	1.50
6. Disposable Income (in constant 1967 dollars)						
6.1 In billions of 1967 dollars	183.6	273.9	377.4	577.0	668.2	765.9
6.2 Per household	5,253	6,390	7,118	9,094	9,291	9,527

Sources: Panels 1 and 2, *Progressive Grocer* (April 1983: 48, 66); panel 5, MVMA, *Facts and Figures*; Panel 6, *Economic Report of the President*

Table 8–4. Relation of Store Size to Car Ownership, Income and Urbanization.[a]

	1958 Sample		1977 Sample	
	Coefficient	T-Ratio	Coefficient	T-Ratio
A. X = LN sales per establishment				
Mean	5.5972		6.9938	
Standard deviation	.2769		.2637	
Intercept	.162		−6.369	
Two-car	.416	6.02	.228	1.01
INC	.810	8.06	1.493	7.39
MET	.158	4.31	.018	.3
R^2	.8507		.6508	
MSE	.1106		.1611	
B. E = LN employees per establishment				
Mean	1.8969		2.5722	
Standard Deviation	.2268		.2492	
Intercept	−1.643		−9.094	
Two-car	.299	4.48	.058	.24
INC	.54	5.56	1.29	5.97
MET	.184	5.19	.043	.67
R^2	.792		.5528	
MSE	.1069		.1722	

a. OLS estimates based on cross-sectional samples for forty-eight states. All variables in natural logarithms.

The OLS regression results for a log-linear model are reported in Table 8–4. For the 1958 sample, the model explained 85.1 percent of the variance in X_i^*, the sales size of stores, and 79.2 percent in the variance of employment size E_i^*. The elasticity of X^* with respect to income was .810 and .416 for two-car ownership. These elasticities were smaller for the employment size measure, .540 and .299. I suspected that in states with high female labor force participation rates, households would try to economize on shopping time by frequenting larger stores. This implies that the coefficient of FLFPR ought to be positive, but when this variable was included in the model, the coefficient was insignificant and sometimes had the wrong sign. The results were poorer for the 1977 sample; the R^2 fell to .6508 for the sales size regression and .5528 for employment size. These state regressions add some additional, albeit weak, support for the hypothesis that stores are larger in localities with higher incomes and car ownership rates.

In the 1960s, a large supermarket had a selling area of 10,000 to 15,000 square feet compared to today's super stores of 40,000 square feet. The sales volume needed to support such large super stores could not have been generated by the kinds of consumers who were around in 1960. Technological advances reduced the relative prices of cars, refrigerators, and storage space. The consequence was a rapid growth in the number of two-car families that could hold large home inventories of food. Consumer shopping patterns changed. Innovative entrepreneurs advertised low prices that attracted customers from a far wider geographic area. The scale economies in supplying retail services allowed the large supermarket to cut margins and still retain competitive profit rates.[39]

Size differentiates food stores, not only in the breadth of the product line but also in the organization of production. Labor productivity measured by sales per full-time employee, shown in line 6.4 of Table 8–5, is higher in bigger stores.[40] Part of the higher productivity is due to a higher capital/labor ratio, but we cannot measure this latter ratio from the data of Table 8–5. Capital is more than floor space or buildings. It must also include equipment, warehouses, inventories, and parking space. Larger stores use more and newer equipment as evidenced by the fraction of stores that use scanners, line 13. The super stores also have larger delicatessens, bakeries, and meat departments, which tend to be more capital intensive.[41]

Aside from inventories, capital costs are largely overhead costs. A high capital/labor ratio gives larger stores an incentive to operate longer hours. A colorable case can be made that when a supermarket expands store hours, it will simultaneously reduce price-cost margins. Imagine a city in which blue laws prohibit Sunday sales. Closing on Sunday could have been optimal (absent blue laws) if the marginal cost of supplying retail services on Sundays, c_2, exceeded the threshold or choke price at which consumers demand *no* Sunday retail services. The repeal of the blue laws is analytically equivalent to a fall in the marginal cost c_2 below the threshold price. I show in Appendix 8B that under plausible conditions, a decrease in c_2 will lead to a fall in the "price" for weekday sales. There is, however, an important difference between a supermarket and the usual multiproduct monopoly. Retailers do not

Table 8–5. Selected Statistics for Independent Supermarkets by Store Size, 1982.

Item	Annual Sales in Millions of Dollars						
	2–4	4–6	6–8	8–10	10–12	12+	Average
1. Weekly sales (000)	$55,833	$95,289	$135,342	$170,021	$206,154	$351,127	$91,184
2. Transactions (000)	$6,023	$8,366	$9,626	$11,026	$14,652	$18,212	$7,734
3. Average transaction size ($)	$9.27	$11.39	$14.06	$15.42	$14.07	$19.28	$11.79
4. Inventory turns	17.1	20.8	21.9	24.0	27.9	30.6	20.7
5. Percent net profit	1.8%	1.6%	1.6%	2.0%	2.1%	2.0%	1.8%
6. Employees	21.0	32.3	42.6	50.6	63.1	96.0	30.4
6.1 Full-time	14.2	20.7	27.1	32.1	43.2	59.5	19.9
6.2 Part-time	13.6	23.1	30.9	37.0	39.8	73.0	21.0
6.3 Part-time/full-time	0.96	1.12	1.14	1.15	0.92	1.23	1.06
6.4 Sales per employees ($)	$2,659	$2,950	$3,177	$3,360	$3,267	$3,658	$2,999
7. Number of checkouts	4.4	5.9	7.1	8.4	9.0	11.1	5.5
8. Selling area (square feet)	10,276	13,798	17,635	19,838	22,750	28,679	13,122
8.1 Per employee	489	427	414	392	361	299	432
9. Total area (square feet)	13,578	18,297	24,067	27,192	32,311	42,141	17,715
10. Coupons total	1,463	2,754	3,758	5,972	5,671	11,572	2,445
10.1 Per 1,000 transaction	243	329	390	542	387	635	316
11. Store hours	85	93	93	89	100	107	89
12. Percent open twenty-four hours	3%	8%	8%	7%	7%	19%	6%
13. Percent scanning	7.8%	23.5%	35.0%	34.2%	47.4%	59.5%	17.8%

Source: *Progressive Grocer* (April 1983: 25–26).

charge explicit "prices" for their services but receive their returns via discriminatory mark-ups. Different mark-ups could, in principle, be set for sales that are differentiated in time. Restaurants often charge different prices for lunches and dinners. However, most vendors of consumables behave as if the costs of varying mark-ups over time are prohibitively high. If stores do indeed adhere to the constraint of stable mark-ups by time and day, it has to reduce the potential gains from longer store hours.[42] The largest independents who participated in the *Progressive Grocer* survey were open some twenty-two hours per week longer than the littlest supers, as shown in line 11 of Table 8–5. The lower retail prices for the giant supers may reflect both the pricing of related "services" by a multiservice retailer, and lower selling costs arising from possible economies of scope.

At a Mom and Pop store in the 1950s, one could be served by a clerk, but self-service is now the rule, even at convenience stores. Cashiers and stock clerks are specialists and provide little help to customers who have to punch computers to find the dog food. National advertising has been substituted for the services of trained clerks who could inform us about the qualities of various products. Store-specific human capital might have been a valuable asset when customers actively sought information and valued familiar clerks who could be trusted to supply useful information and to honor implicit warranties in the event of faulty goods. Impersonal transactions can be standardized, and the costs of consumating an exchange no longer appear to depend on durable ongoing relations between customers and clerks who know one another. Stockers load shelves, and cashiers scan packages and make change. The human capital needed to perform these standardized tasks is quite general and easily transferred from one store to another. Retail trade, which once provided stable, long-term jobs, has become an industry where employees get low wages and experience high turnover rates.

The analyses of retail labor markets by Belton Fleischer (1981) and Walter Wessels (1980) have ignored the way in which stochastic shopper demands affect wages and employment. A supermarket's demand is analogous to the demand for repairmen in queueing models. Goods are assembled and clerks are hired to await customer arrivals and to make exchanges. The demand for labor depends on, among other things, the mean arrival rate of shoppers

λ, average transaction size q, and the quality of service that can be measured by the utilization rate of capital and labor, or equivalently, the mean length of customer queues. Clerks wait for shoppers, or shoppers wait for clerks to serve them; someone is almost always idle. Let μ be the mean of an exponential distribution of serving times. If the ratio of customer arrivals to clerks, (λ/C) is small relative to (μ/λ), clerks will be idle and their measured productivity low. Conversely, a high value of (λ/C) results in longer customer queues and possibly balking. Mulligan (1983) and De-Vaney (1976) showed that a system with stochastic demands and queues, is characterized by economies of massed reserves. More precisely, if the mean arrival rate λ and the number of clerks C are doubled, the store will realize more than a twofold increase in the number of transactions, and hence, the sum of the mean waiting times of clerks and customers will fall; that is, the utilization rate of a clerk's time as well as of a shopper's time rises with the scale of operation.[43] The principle of massed reserves reinforces the economies of large purchase size q emphasized by Nooteboom and Schwartzman. The average transaction in a large super entailed an outlay of $19.28, which was 2.080 times larger than the mean purchase size, $q = \$9.27$, in a small super; see line 3 of Table 8–5. The mean service time μ is likely to be longer for larger purchases, and in the queueing model of Mulligan, the optimal ratio of shoppers to clerks (λ/C) ought to be smaller in bigger stores. Transaction per full-time equivalent employee (line 2 divided by 6) was 287 per week in the small supers and 190 in the largest.

As food stores and chains of food stores grew larger, employers made smaller investments in store-specific human capital. The relative demands for part-time employees climbed as evidenced by the data of Table 8–2. Hourly wages of part-time and full-time clerks were positively related to store size. A surprising feature of the cross-sectional data is that the biggest supermarkets demanded proportionally more part-time employees even though they had to pay them relatively higher wages. Across the six size categories, the ratio of part-time to full-time employees varied from 0.96 to 1.23. The greater reliance on part-timers may be due to a higher capital/labor ratio and longer store hours.[44] The hourly wages paid by the largest supers are some 35 to 40 percent above the wages in small stores. In spite of this wage differential, the largest supermarkets have grown and have done so by charging lower retail prices and offering better services. Massed reserves imply that workers in the largest stores with high shopper arrival rates are more intensively utilized and hence are more productive. Part, if not all, of the wage differential between large and small stores may simply be a compensating difference for greater work effort.

Technological advances that raised real incomes and lowered relative prices for cars, refrigerators, and storage, have indirectly altered consumer shopping patterns. An upward trend in average purchase size q led not only to larger stores but also to changes in the kinds of retail services supplied by the typical store. Stores no longer offered free delivery whose value fell when nearly all shoppers owned cars. Cash and carry accompanied the shift to self service and accounted for the disappearance of grocery store credit.[45] Advertising strategies shifted away from weekend sales, trading stamps, and tournament games to "deals" meaning coupons and double coupons. Eppen and Liebermann (1984) argued that stores introduced dealing to shift inventory holding costs from the store to the customer. This model implies that stores incurring higher holding costs ought to offer more deals. Inventory turnover rates (line 4) are positively related to store size, but this may be due to the economies of massed reverses and not to holding costs.[46] Coupons per thousand transactions (line 10.1) rise with store size. It is surely in a store's self-interest to have its customers hold larger home inventories, not for the reason of differential inventory holding costs but rather (1) larger home inventories are accumulated by buying in larger lots, and (2) consumption rates and waste are likely to be higher when commodities are readily available in the refrigerator or on the shelf.[47] Deals are designed to attract customers who will buy in bulk and generate the retail complementarities emphasized by Holdren. Their effectiveness obviously depends on the availability of a car and a place to store the goods.[48]

CONCLUDING REMARKS

In *Distribution's Place in the American Economy*, Harold Barger reminded us that we had been ignoring an important and rapidly growing sector of the economy. Measured by employment, wholesale/retail trade has enjoyed a spectacular growth for more than a century. The trade and service sectors have created many of the jobs for the influx of teenagers and women to the labor force in the decades of the 1960s and 1970s. The earnings that go along with the jobs in retail trade have not kept pace with earnings in manufacturing jobs for at least two reasons. First, average weekly hours have declined sharply in retail trade. Second, relative hourly earnings have declined as retail employers shifted their labor demands toward more use of unskilled and part-time employees. Stores, especially supermarkets, are considerably larger now than they were some thirty years ago.

As real incomes rose, consumers demanded larger total quantities of food. A simple inventory model implies that as total demand Q rises, the average transaction size q also increases thereby supporting the establishment of larger stores. The availability of a car and lower storage costs reinforce the incentives to make fewer shopping trips and to hold larger home inventories.[49] The growth in the number of two-car families greatly expanded the potential market area that could be served by a super store.

The rise in the labor force participation rate of married women has been proposed as an alternative explanation for the trend toward larger purchase sizes and fewer shopping trips. The labor force participation rate of married women and car ownership rates are highly collinear in the time series data, which makes it difficult to disentangle the partial effects of each. The cross-sectional regression analysis reported in Table 8–4 lends some weak support to the car ownership explanation. Additionally, price competition via deals relies on the ability of a store to generate shopper traffic from a wider geographic area.

The specialization of labor in the large supermarkets has been accompanied by a decrease in point of sale services. Delivery and credit have largely been abandoned. Stores apparently have little incentive to establish repeat business with particular named customers; it is only the aggregate flow of shoppers that matter. Transactions are impersonal, and retail employees possess general human capital that can be easily transferred across employers. Stores have consequently increased their use of part-time employees who can be hired at lower wages and who receive few, if any, fringe benefits. Retail stores have thus responded to higher wages in the economy as a whole by substituting unskilled for skilled, trained employees and by shifting more of the distributive functions forward to customers and backward to manufacturers. As communication and transportation costs continue to fall, we can conjecture about future changes in the structure of the distributive trades.

APPENDIX 8A. ON THE NUMBER OF STORES IN THE HEALL MODEL

In Heall's model, a retailer moves goods from a central supplier (wholesaler or manufacturer) to a store on the circumference of a circle that is closer to the customer's fixed location. The cost of moving one unit over one unit of distance, $C(q)$ is characterized by scale economies, $C' < 0$. If the circle has radius r, the retailer's total freight cost is

$$RC = QC(Q)r \qquad (8A.1a)$$

The store serves a market defined by the arc $r\theta$ where θ is the angle in radians to the circumference served by one store. Each customer is assumed to demand q units so that $Q = r\theta q$. The implicit moving cost incurred by a customer located $D = rx$ units from a store is $C(q)qrx$. The aggregate costs for all of one store's customers, the circumference moving costs (CC) is

$$CC = 2\int_0^{\theta/2} C(q)qrxdx = C(q)qr(\theta^2/4) \qquad (8A.1b)$$

Since customers are uniformly distributed on the circle, there are $N = 2\pi/\theta$ stores. Hence the total transport costs are

$$TC = (2\pi/\theta)[RC + CC] = 2\pi[C(Q)r^2q + C(q)qr(\theta/2)]. \qquad (8A.2)$$

where $Q = r\theta q$ is the sales volume of one store. As the arc served by each store $r\theta$ is increased, the unit cost incurred by a retailer falls, but the moving costs incurred by customers given by (8A.1b) rises. The total social cost of moving goods to consumers is a minimum when

$$-C'(Q) = C(q)/(4r^2q) \qquad (8A.3)$$

Heall obtains closed form solutions for three cost functions. His second example is consistent with the linear retail cost function derived by B. Nooteboom (1983) and I shall adopt it in this appendix. The cost function will be given by

$$C(Q) = (\alpha/Q) + \beta \qquad (8A.4)$$

The length of the arc $r\theta$ served by one store can be chosen to minimize total cost. The angle θ^* and hence the optimum number of stores $N = 2\pi/\theta^*$, which minimizes TC is obtained from (8A.3) and (8A.4). Specifically, we have

$$\theta^* = [4\alpha/(\alpha + \beta q)]^{\frac{1}{2}} \qquad (8A.5)$$

An increase in the fixed cost α or a decrease in the variable cost βq will expand the optimal arc $r\theta^*$ spanned by each optimal sized store. In fact, it can be shown that

$$(\alpha/\theta)(d\theta/d\alpha) = -(\beta/\theta)(d\theta/d\beta) = -(q/\theta)(d\theta/dq) =$$
$$(1/8)[\theta^2(4-\theta^2)] \qquad (8A.6)$$

If we measure store size by sales volume, $Q = r\theta q$, then an increase in purchase volume by each consumer q will lead to an increase in the optimum number of stores meaning a fall in θ^*. However, $Q = r\theta^*q$ will expand because the elasticity of θ^* with respect to q is numerically less than unity.

Turn next to the case of one atomistic store that can set price P for the availability of a unit of the good at the edge of the circle. This store's total profits are given by

$$\pi = \pi(\theta) = [P - C(Q)r]Q \qquad (8A.7a)$$

The sales volume Q and the price that the store can set are constrained by the fact that each consumer can move the good from wholesaler to the customer's site at a cost of $C(q)rq$. Hence, the profit equation is constrained by

$$P + C(q)rx < C(q)r, \qquad Q = r\theta'q \qquad (8A.7b)$$

where θ' is the maximum angle (which fixes the length of the arc $r\theta'$) served by the store. In the limit, the marginal customer will patronize the store if $x \leq r\theta/2$. If we substitute these constraints into (8A.7a) and maximize π with respect to θ, we find that the profit maximizing arc is defined by

$$\theta' = \alpha /(\alpha + \beta q) \qquad (8A.8)$$

Again, we see that θ' will be larger (implying fewer stores, $N = 2\pi/\theta'$) when α is larger or when βq is smaller.

If a store sets price to cover average costs $P = C(Q)r$, its market area is determined by the condition that price plus the customer's moving cost from the store must be less than direct purchase from the center.

$$C(Q)r = C(q)r\theta'' \le C(q)r \qquad (8A.9)$$

Since $Q = r\theta''q$, equation (8A.9) turns out to be a quadratic equation in θ given the cost function of (8A.3).

$$\theta^2 + (2\pi/(\alpha + \beta q))\theta + (2\pi/(r(\alpha + \beta q))) = 0 \qquad (8A.10a)$$

The roots to this equation are

$$\theta'' = (\alpha/(\alpha + \beta q))\{1 \pm [(2(\alpha + \beta q)/r\alpha)]^{1/2}\} \qquad (8A.10b)$$

Heall shows that the smaller root, meaning more stores, will attain if q is large.

In these models, a higher fixed cost α or a lower variable cost βq will result in an expansion in the market area $r\theta$ spanned by a single store. Although an increase in quantity demanded q by each customer leads to a decrease in θ, the sales volume of each store $Q = r\theta q$ expands. There are, however, two disturbing features. The model assumes, among other things, that (1) customer demands q are inelastic and do not depend on retail mark-ups, and (2) the cost function $c(q)$ is the same for both the retailer and consumers as well as for movements from the center to the edge and for moves along the circumference. It is surely reasonable to suppose that consumers face a different cost function, and it would seem promising to explore this extension.

APPENDIX 8B. STORE HOURS AND MONOPOLY PRICING OF RELATED GOODS

Suppose that a firm has a monopoly over two related goods whose demand curves (and their linear approximations) can be written.

$$X_1 = D_1(P_1, P_2) = A_1 - D_{11}P_1 + D_{12}P_2 \qquad (8B.1a)$$
$$X_2 = D_2(P_1, P_2) = A_2 + D_{21}P_1 - D_{22}P_2 \qquad (8B.1b)$$

The two goods are assumed to be substitutes. Marginal costs are assumed to be constants so that the firm's total profits are given by

$$\pi = (P_1 - c_1)X_1 + (P_2 - c_2)X_2 \qquad (8B.2)$$

Using the linear demand equations, the first-order conditions for profit maximization are

$$A_1 - 2D_{11}P_1 + (D_{21}+D_{12})P_2 = -D_{11}c_1 + D_{21}c_2 \quad \text{(8B.2a)}$$
$$A_2 + (D_{12}+D_{21})P_1 - 2D_{22}P_2 = D_{12}c_1 - D_{22}c_2 \quad \text{(8B.2b)}$$

Let $B_1 = A_1 + D_{11}c_1 - D_{21}c_2$, and $B_2 = A_2 - D_{12}c_1 + D_{22}c_2$. Equations (8B.2a) and (8B.2b) can be solved to yield,

$$P_1 = \frac{1}{[D]}(2D_{22}B_1 + (D_{12}+D_{21})B_2) \quad \text{(8B.3a)}$$

$$P_2 = \frac{1}{[D]}((D_{12}+D_{21})B_1 + 2D_{11}B_1) \quad \text{(8B.3b)}$$

From (8B.1b), we can solve for a threshold price $\hat{P}_2$ at which the consumer demands *none* of X_2. This is

$$\hat{P}_2 = (\frac{1}{D_{22}})(A_2 + D_{12}P_1)$$

If c_2 exceeds $\hat{P}_2$, the firm will withdraw from that market, and P_2 no longer appears in equation (8B.1a); the intercept changes to A_1^* $= A_1 + D_{21}\hat{P}_2$. The slope may also change, but the profit maximizing price for X_1 is now given by

$$P_1^* = \frac{1}{2}[(A_1^*/D_{11}) + c_1] \quad \text{(8B.5)}$$

If c_2 falls, the firm can increase profits by introducing X_2. The own price elasticity of X_1 will get numerically larger as shown by E. Rothbart (1941). As c_2 is successively reduced, the price of the original product will be reduced if

$$\frac{dP_1}{dx_2} > 0, \text{ if } D_{12} > D_{21} \quad \text{(8B.6)}$$

If a tennis shoe monopoly enters the market for sandals due to a fall in c_2, the price of the first good P_1 will be reduced if (dX_1/dP_2) $> (dX_2/dP_1)$. The decrease in P_1 will be reinforced if there are economies of scope in producing the two kinds of shoes. The analogy is seen by relabeling variables; let X_1 stand for food sales on weekdays, and X_2 for Sunday sales. The decision to open on Sundays is isomorphic to a multiproduct monopoly extending the width of its product line. Longer store hours will generate additional sales for at least three reasons: (1) some existing customers may substitute Sunday for weekday shopping, (2) new shoppers may be attracted from other stores, or (3) the introduction of Sunday shopping has to reduce the full price P^*, which induces more food consumption. Finally, the profitability of longer store hours will be higher for larger stores, which have higher capital/labor ratios. Whinston and McCoy (1974) showed that capital is more intensively utilized via shift work, the higher is the instantaneous capital/labor ratio.

NOTES

1. Barger's data pertain to the labor force, while the Census data for the postwar years refer to employed persons. If we are mainly interested in the percentage distribution of employees across industries, use of data on the labor force does not involve much bias if industrial differences in unemployment rates were small in the period, 1880 to 1940.

2. Employment, as a percentage of the adult population age fourteen and older, rose from 49.6 percent in 1950 to 57.7 percent in 1980. The employment/population ratio fell for men, but this decline was more than offset by the rapid increase in the female ratio.

3. Reference to the *Economic Report of the President, 1986*, Table B–41, revealed that average weekly hours in retail trade fell from 40.4 hours in 1950 to 30.2 in 1980; it went down to 30.0 in 1984. The sharpest reductions in weekly hours were in eating/drinking places. Increases in the use of part-time employees account for part of this decline, but we also find reductions in the workweek of full-time workers.

4. The data describe hours paid and not hours actually worked. According to Kent Kunze (1985), the ratio of hours worked to hours paid in 1985 was 0.914 in manufacturing and 0.960 in retail trade. If (H_w/H_p) was around unity in both industries in 1880, the use of the hours paid series imparts an upward bias in estimating manhours actually worked. Moreover, the bias is larger in manufacturing where workers received more paid leisure. With the hours paid series $(M_T/M_G) = 0.59$ in 1980—the ratio of labor inputs in trade versus goods production—the ratio of manhours actually worked would have been higher, $(M_{aT}/M_{aG}) > 0.59$.

5. The women who entered the labor force in the decade of the 1950s

came from the middle and the left of the educational distribution. The more highly educated women exhibited the sharpest increases in labor force participation rates in the decade of the 1970s. This pattern has been analyzed in the literature. The number of teenagers entering the labor force began to climb in the mid-1960s and reached a peak in the early 1970s.

6. Within retail trade, eating/drinking places that paid the lowest wages experienced the fastest employment growth. The extent of coverage by retail trade unions fell due to a changing industrial mix (less unionism in eating places) and to expansion of the independent chains. In addition, the wage ratio of union to nonunion employees in the same industry also declined.

7. These last two "services" constitute vertical integration by retail firms. A store that offers delivery and credit has engaged in downstream integration into transportation and finance. The manufacturing of baked goods, prepared salads, and canned goods under private labels represents upstream vertical integration. Barger (1955) pointed to other examples of ancillary services such as nurseries for children, lunch counters at variety stores, and so forth.

8. The "product" of higher education might be defined by a standard college graduate. If the "raw materials" incorporated some processing, an argument can be used to measure "output" by the value-added by higher education. The objective of the *entire* distribution process is to locate goods at a time and place where they will be consumed. Delivery may involve carrying the goods home or arranging to be at home when the goods are delivered. Customers usually carry their fresh fish home but ask to have their sand and firewood delivered.

9. Department stores practiced cross-subsidization, which makes sense if cash customers had less elastic demand curves. Customer credit was historically supplied by sellers. H. Barger (1955: Ch. 2, fn. 35) claimed that one of the first examples of credit being extended by a third party occurred in financing automobile sales in 1915. That sellers had to supply credit for other consumable goods such as carriages may have been due to the fact that the seller was better informed about product quality.

10. An untenable implication of this assumption is that the derived demand for inputs in the trade sector are inelastic with respect to input price changes.

11. The margins measure of retail output was adopted by Barger (1955) and Schwartzman (1968). Value-added will be less than the product of sales times the gross margin rate when the store buys inputs such as accounting or janitorial services. This latter measure assumes

that changes in the margin are due to changes in the value of retail services supplied by the store; input prices are presumed to move in direct proportion to retail goods prices.

12.   Ratchford and Brown disaggregated the data and examined separate time series for sales and margins in three departments—groceries, meats, and produce. The year-to-year rates of change in each department were combined into a weighted average rate of change to obtain a sales output index and a margin output index. The series are reported in their Table 1.

13.   If they use a margins measure of output, the growth rate of total factor productivity over the 1959–79 period climbs to 1.34 percent. Table 6 in Ratchford and Brown reveals the uneven trend in productivity growth. Total factor productivity in retail food stores rose at 1.20 percent a year from 1959–66, which slowed to an annual rate of 0.91 percent. In the last seven years, 1972–79, total factor productivity actually declined; they estimated a negative growth rate of −0.62 percent. Recall that the growth rate for the entire period based on a sales measure of output was +0.47 percent.

 Productivity growth measured either by labor productivity or by total factor productivity, slowed down in nearly all industries in the period 1972–79.

14.   The supplier is usually a retail outlet, but it could be a producer, a mail-order firm, or direct sales by phone.

15.   The relation between the cost per trip S and purchase size q is denoted by $S(q)$, while C'_S = the change in total shopping trip costs

$$C'_S = (\frac{Q}{q^2})[qS'(q) - S] = (\frac{SQ}{q^2})[E_S - 1]$$

Likewise if $b = b(q)$, total inventory costs are, $C_H = qb(q)/2$, and we have

$$C'_H = \frac{1}{2}[b + qb'(q)] = \frac{b}{2}[1 + E_H]$$

Equation (8.2′) is obtained by substituting the above equations into the cost minimization condition, $(C'_S + C'_H) = 0$. The square root formula of (8.2) is the special case in which $E_S = E_H = 0$. Over some range of inventory levels, $b'(q)$ may be zero (constant unit holding costs), but eventually, limits on storage space will result in diminishing returns with $E_H > 0$. Differences in storage costs and interest rates will result in a distribution of b across consumers. Since more shopping time is required to make larger purchases, $S'(q) > 0$.

16.   The unit costs of a shopping trip (S_{ai}, S_{bi}) depend on the store's location, services, and the implicit cost of the shopper's time. In equation (8.4), c_i is the sum of implicit shopping trip and inventory holding costs spread across the Q units consumed per unit time period. If $S_{ai} < S_{bi}$, then $P^*_{ai} < P^*_{bi}$ when $P_a = P_b$. Notice that full prices can vary across stores and consumers.

17.   The implicit cost per mile is the sum of the implicit time cost of the shopper V_i (which depends on the travel mode) plus m_i, the cost for gas or fares. If a customer walks, V_i is his time cost given his walking speed. Ignoring shoe leather wear, one might set $m_i = 0$, but it ought to include the implicit effort costs of carrying groceries, and so forth. I assume that k_i is the same for trips to either stores A or B.

18.   In Figure 8–1, store A is located at the origin, and store B is located B miles to the right. If D is the distance to A, then $(B - D)$ is the distance to store B for a customer residing between the two. The full price including the unit shopping cost is the same for a customer located at D_1 where

$$P_a + k_1 D_1 = P_b + k_1(B - D_1), \text{ or } D_1 = \left(\frac{B}{2}\right) + \left(\frac{P_b - P_a}{2k_1}\right)$$

Type 2 customers who face a higher travel cost, $k_2 > k_1$, will be indifferent when they are located at D_2.

19.   The location D_1 at which $P^*_a = P^*_b$ is attained when

$$\alpha[\sqrt{kD_1} - \sqrt{k(B - D_1)}] = P_b - P_a$$

It is apparent that D_1 is smaller, the larger is α, k, or P_a. If $\alpha\sqrt{kB} < (P_b - P_a)$, no one patronizes store B. If $S = kD$, customers who live very close to a store confront a very small shopping trip cost. Hence, they make many trips and buy little on each trip. Indeed, as D approaches zero, $T = Q/q$ approaches infinity.

20.   The cost per period of shopping trips is given by

$$C_S = (S_0 + kD_j + k'q_j)\left(\frac{Q}{q_j}\right) = (S_0 + kD_j)\left(\frac{Q}{q_j}\right) + k'Q$$

where k' is the marginal cost of buying an additional unit on a given shopping trip. Notice that k' leads to a parallel upward shift in P^*_j and acts like an excise tax. Further, near D_1, customers to the right (who patronize B) make more shopping trips than those located just to the left. This follows from equation (8.3) above.

Finally, one could include other variables such as store layout and size that might affect the cost per shopping trip. Indeed, one would expect to find that low-priced stores will attract more distant customers on weekends when store congestion raises S_O.

21. Let P^*_{aO} = the full price when $S_O = k' = 0$; that is, when S_j is proportional to distance. Further, let P^*_{a1} = the full price defined in equation (8.7′). Differentiate with respect to distance to obtain

$$dP^*_{aO}/dD = \alpha k[kD]^{-1/2} \qquad dP^*_{a1}/dD = \alpha k[S_0 + kD]^{-1/2}$$

22. In equation (8.1), I chose the lot size q as the decision variable. It is convenient now to select the trip frequency T as the decision variable so that the purchase size of the g-th good is $q_g = Q_g/T$. Total cost is then given by $C = \Sigma P_g Q_g + ST + (\frac{1}{2T})\Sigma b_g Q_g$, where b_g is presumed to be constant—that is, there are no diminishing returns in holding inventories.

$$\frac{dC}{dT} = S - (\frac{1}{2T^2})\Sigma b_g Q_g$$

Let $X = \Sigma b_g Q_g$ so that $X/2T$ is the average inventory holding cost. The minimum full total cost of purchasing and holding the fixed Q vector is

$$C^* = EP_g Q_g + [2SX]^{1/2}$$

It is evident that consolidating purchases in one trip reduces C because

$$\Sigma(\sqrt{b_g Q_g}) > \sqrt{\Sigma b_g Q_g}$$

23. The marginal full cost of Q_g is obtained by differentiating the full total cost defined in note 22:

$$MC_g = dC^*/dQ_g = P_g + Ab_g. \quad A = \frac{1}{T}(\frac{C_S}{C_S + C_H})$$

In equilibrium, the consumption vector Q will be adjusted so that the marginal rate of substitution between any two goods will be equated to the ratio of their marginal full costs. This will be true even in the case of a single good where changes in S, b, or Q that lead to changes in full price will affect consumption. The indirect utility of shopping at a particular store will depend on C^*.

24. The shopping time at the store depends on the volume and variety of purchases. It seems reasonable to suppose that there are scale economies in selecting and gathering goods. Hence, individuals

who make larger purchases confront lower "costs" of choosing and collecting groceries and dry goods.

25. A license fee is, for example, a discretionary or avoidable fixed cost that is positive if and only if the output is positive. Opening on Sunday entails outlays for labor, lights, and so forth, which are not directly related to sales on Sunday.

26. Heall assumed that the wholesale price was zero. Hence, the gross margin is equal to the retail price. Suppose that an individual is located D miles from the store and r miles from the supplier located at the center of the circle, $D < r$. The maximum demand price for this individual is the savings in transport costs $(r - D)c(q)$, where q is the amount demanded of the good.

27. It is assumed that consumers are uniformly distributed on the circumference of a circle, and individuals demands q are inelastic. The effects of purchase size q and $c(Q)$ on the equilibrium store size Q are derived in Appendix 8A.

28. Given stochastic arrivals, clerks will sometimes be idle. Nooteboom assumed that at such times, clerks could be employed in "pre and post purchase activities" (stocking, sweeping, and so forth). The ratio of clerks to shoppers ought to depend on relative costs—clerk's wages and the opportunity cost of the shopper's time. This consideration is ignored by Nooteboom, but it is incorporated into a theory of the optimal utilization of capital by Arthur S. DeVaney (1976).

The number of transactions per year N_i is the product of the mean customer arrival rate per hour λ times annual store hours H_i, $N_i = \lambda_i H_i$. Annual sales is simply, $X = Nq$ where q is the mean transaction or purchase size.

29. The elasticities of manhours M with respect to N and q are

$$(\frac{N}{M})(\frac{M}{dN}) = \frac{\sigma + \tau_1 q}{M} = 1 - \frac{\beta_0}{M} < 1, \quad (\frac{q}{M})(\frac{dM}{dq}) = \frac{\tau_1 N}{M}$$

$$= 1 - (\frac{\beta_0 + \sigma N}{M}) < 1$$

Ratchford and Brown (1985) disaggregated sales by department because sales per workhour varied across departments being lower in dry groceries and higher in meat and produce. The relationship between sales per workhour, X/M, and store size X is confounded by a covariation in the share of sales generated in each of the major departments.

30. The relationship between slope X_P and elasticity E_P is seen from equation (8.10), namely, $X_P = -XE_P/P$. Since $X = Nq$, $X_P = (qN_P$

370 SECTORAL PATTERNS OF TECHNOLOGY ADOPTION

+ Nq_P), and $E_P = (\mu_P + \epsilon_P)$. The price and service elasticities clearly depend on Z_1 and Z_2 because variations in Z will alter the composition of a store's customers.

31. When sales X is increased by reducing price, the full marginal cost of selling an additional unit of X is

$$c_{XP}^* = P_w + C_x + a)C_N/q)$$

where $a = (\mu_P/E_P)$. If a lower price attracts no more shoppers, $\mu_P = 0$, and C_N adds nothing to the full marginal cost. A higher service level could take the form of more parking, less congestion at checkouts, and so forth. If this attracts more customers meaning a large value for μ_A, it will raise c_{XA}^*, the increment being larger, the smaller is the mean purchase size q.

In deriving equation (8.12), I rewrite equation (8.11a') noting that $c_{XP}^* = P_w/g$. This yields

$$1 - (1/E_P) = c_{XP}^*/P = P_w/gP$$

Multiply by g and subtract from one to obtain equation (8.12).

32. The models of Heall (1980), Hotelling (1929), and Stahl (1986) assumed fixed residential sites and stable demands. Each firm chooses a location and sets a price. Each customer patronizes only one store and makes one trip per period. Some useful insights have been gained from these models, but I am mainly interested in the reasons for the changing structure of retailing.

33. Bliss assumes that each store maximizes profits subject to the constraint that the customer gets "good value," meaning that the utility from buying at the store equals a given utility level U_O. Mark-ups are the same because each store faces an identical, representative consumer with an indirect utility function, $V(P, I - T)$. If x_{Oj} is the quantity demanded when the mark-up is zero, mark-ups are such that $(x_{Oj} - x_j^*)/x_{Oj}$ is the same for all j where x_j^* is the utility maximizing demand with mark-up m_j.

34. The consumer maximizes $U(X,Y)$ subject to the budget constraint, $\Sigma p_j^* x_j + p_y Y = I$, where Y is an outside good, p_j^* is the full price of x_j, and I is full income. Let P_α^* denote the vector of full prices offered at store α, and P_β^* the full prices at store β. A shopper will choose α if $V(P_\alpha^*, p_y, I) > V(P_\beta^*, p_y, I)$, where V is the indirect utility function. A rise in p_j leads to a utility loss

$$\frac{dU}{dp_j} = -x_j(dp_j^*/dp_j)$$

An item with a high transfer effect is one for which $-(dU/dp_j)$ is large. A good with a high transfer effect has the following traits: (1)

The buyer knows the price, (2) price differentials across stores are perceptible, (3) x_j has a large budget share, (4) demand is predictable, (5) demand is inelastic, and (6) a price differential will not be confused with a quality differential. Holdren claims that as incomes rise and transport costs fall, more items will move into the k-class of high transfer effects.

35. Contrary to equation (8.10), I now define elasticities that are *not* converted to positive values. Thus, $e_{jj} < 0$ and $\mu_j = (p_j dN/N dp_j)$ < 0. If x_h and x_j are substitutes for a shopper $\epsilon_{hj} > 0$ but $e_{hj} = (\epsilon_{hj} + \mu_j)$, and since $\mu_j < 0$, the induced loss in the number of shoppers could result in a gross cross-price elasticity e_{hj} that is negative. In this event, we have what Holdren calls, retail complementarities.

36. This point is nicely illustrated by Holdren (1960: 80) who writes

> The low margin on corn meal was a surprise to this writer, but according to supermarket operators, people who buy corn meal buy it relatively frequently and tend to be "careful shoppers" and "big eaters." Thus to attract and hold their trade, lowering the price of corn meal is relatively efficacious.

37. The weighted averages for the sample of 196 cities were 44.2 and 53.8 percent as shown in Table 3–7 of Parker. When he limited the sample to eighty-two cities with unchanged SMSA definitions, the weighted average of the four-firm concentration ratio climbed from 43.0 percent in 1954 to 53.5 percent in 1977.

The establishment data from the *Census of Retail Trade* confirm this trend. Big food stores with fifty or more employees captured 11.7 percent of sales in 1954 and 48.1 percent in 1982. Their share of industrywide paid employees climbed from 14.7 percent in 1954 to 45.6 percent in 1982.

38. The population censuses for 1960 and 1980 provided the data for Met and Inc. Car ownership data were taken from the 1960 *Census of Housing and Motor Vehicle Facts and Figures* (1986: 42).

39. W. Blozan (1986: 16) traced the origins of the supermarket to M.J. Cullen who built an auto-oriented market on the outskirts of New York City in 1929. He reports that in 1932, an imitator in Elizabeth, New Jersey, relied on newspapers to advertise prices that were some 8 to 15 percent below the prices of competing grocery stores.

Profits as a percentage of sales for independent supers are shown in line 5 of Table 8–5. They rise with store size, but the dispersion is small, from 1.6 to 2.1 percent. Gross margins data that appear in *Progressive Grocer* reveal smaller margins in bigger stores.

40. Employees on line 6 are the sum of full-time plus one-half of the

part-time employees. Holdren (1960) found that sales per employee followed a bell-shaped curve, but he counted part-time and full-time employees without weighting them. The *Census of Retail Trade* data also exhibit the bell-shaped curve for sales to all paid employees.

41. Selling area per full-time equivalent employee (line 8.1 of Table 8–5) declines with size, but store hours are positively related to size. However, after adjusting for the hours effect, I still find that the biggest supers have less floor space per worker. The value of buildings that is related to the price per square foot is higher for larger stores. Big supers are newer with an average age in 1981 of 9.8 years versus 16.6 years for the little supers with sales of $2–4 million; the average age of all independent supers was 15.5 years (See *Progressive Grocer* April 1982: 23).

    The "output mix" varies; giant supers offer video rentals, delicatessens, and bakeries that require less floor space per employee but more capital. If all these adjustments could be made, I am convinced that the capital/labor ratio is positively related to store size.

42. Restaurants may change the quality of the meals and service depending on the time of day. Supermarkets often close the fresh fish, delicatessen, and bakery departments at night. In this way, retail services can be differentiated not only by hour of the day but also between weekdays and Sundays.

43. Douglas and Miller (1972) found that an airline's load factor on a given route, $L_i = (X_i/K_i)$ was positively correlated with the firm's share of capacity supplied on the route. DeVaney (1976) reported that if a hospital supplies a majority of the bed capacity, and if arrivals are proportional to bed capacity, then the largest hospital that supplies, say 60 percent of available beds will capture far more than 60 percent of the occupied beds. This principle might also explain why gasoline stations tend to cluster.

44. *Progressive Grocer* (April 1983) reported that the hourly wages of part-time clerks in small supers was $3.99 and in the largest independent supers, $5.69. The hourly wages of full-time clerks were respectively $4.79 and $6.48. The wage ratio of part-time to full-time clerks was .833 in small and .878 in large supers. If production functions were homothetic, the big supers should have exhibited a lower ratio of part to full-time employees, but the ratio there was 1.23 versus 0.96 in the smallest independent supers. The MRS of part to full-time workers (MP_{PT}/MP_{FT}) evidently rises as one moves to larger stores.

45. The consumer credit, which used to be supplied by department stores, has been taken over by banks and financial institutions, due, I suspect to the sharp drop in telephone and communication rates.

In the 1950s, the costs of free delivery and credit were *not* allocated to those who used these services but were instead spread across all customers. Those who paid cash thus subsidized those who bought on credit. This practice may have been a way to exercise implicit price discrimination.

46. A can of tuna like Ted Edward the Bear waits to be bought. It will have a shorter shelf life in bigger stores that have higher customer arrival rates and larger purchase sizes. The principle of massed reserves implies that a given stockout probability can be attained with a smaller inventory/sales ratio in larger supermarkets.

47. I find that the chips and apples are eaten faster when we keep an inventory of them. In urban transportation, studies have been conducted on persons who shift from private autos to taxis. An individual calculates that the total costs of his usual car trips per month could be reduced by using taxis. However, if he does sell his car, he nearly always finds that he takes fewer trips per month. An inventory of potential car trips at a low marginal cost leads to "overconsumption" of trips. I suspect that this same result would be observed for soap, beer, and facial tissues.

48. *Progressive Grocer* (1958) reported that 78 percent of a supermarket's customers lived within a one-mile radius of the store. Deals could have attracted more distant customers on weekends when the one family car was available. The traffic generating elasticity μ_j was probably small until the number of two-car families became large.

Blozan (1986) rejected the inventory-shifting hypothesis and argued that dealing is a competitive mechanism to affect the consumer's choice of a retail store. Blozan also pointed out that deals involve a random element in the sense that the same products are couponed at irregular time intervals. Randomized behavior would make sense if the firm is trying to attract and retain customers who might be induced to buy in bulk.

49. According to *Progressive Grocer*, shoppers made an average of 2.31 trips per week to supermarkets, which was down from the peak of 3.04 trips per week in 1973. The writer attributed this decline to higher gasoline prices. I question this explanation because gasoline is only a small part of the total cost of a shopping trip.

REFERENCES

Barger, Harold. 1955. *Distribution's Place in the American Economy Since 1869.* Princeton, N.J.: Princeton University Press.

Becker, G.S. 1965. "A Theory of the Allocation of Time." *Economic Journal* vol. 75 (September): 493–517.

Bliss, Christopher. 1985. "The Economic Theory of Retailing." Paper presented at the World Congress of the Econometric Society, Cambridge, Mass., August 1985.

Blozan, William Jr. 1986. "Retail Dealing as Competition for Customer Store Choice: Theory and Evidence." Ph.D. dissertation, University of Rochester.

Bluestone, Barry. 1981. *The Retail Revolution.* Boston: Auburn House.

DeVaney, Arthur. 1976. "Uncertainty, Waiting Time and Capacity Utilization: A Stochastic Theory of Product Quality." *Journal of Political Economy* vol. 84 (June): 523–42.

Douglas, G.W., and J.C. Miller III. 1974. "Quality Competition, Industry Equilibrium, and Efficiency in the Price. Constrained Airline Market." *American Economic Review,* vol. 64: 657–669.

Eppen, G.D., and Y. Lieberman. 1984. "Why Do Retailers Deal? An Inventory Explanation." *Journal of Business,* vol. 57: 519–530.

Facts and Figures, 1985. Detroit, Mich.: Motor Vehicle Manufacturers Association.

Fleischer, Belton W. 1981. *Minimum Wage Regulation in Retail Trade.* Washington, D.C. American Enterprise Institute.

Hall, H., J. Knapp, and C. Winsten. 1961. *Distribution in Great Britain and North America.* London: Oxford University Press.

Heall, G. 1980. "Spatial Structure in the Retail Trade: A Study in Product Differentiation with Increasing Returns." *Bell Journal of Economics,* vol. 11, no. 2: 565–83.

Holdren, B.R. 1960. *The Structure of a Retail Market and the Market Behavior of Retail Units.* Englewood Cliffs, N.J.: Prentice-Hall.

Hotelling, Harold. 1929. "Stability in Competition," *Economic Journal,* vol. 39: 41–57.

Houthakker, H.S., and Lester Taylor. 1970. *Consumer Demand in the U.S.: Analyses and Projections.* Cambridge, Mass.: Harvard University Press.

Ingene, Charles A. 1984. "Productivity and Functional Shifting in Spatial Retailing: Private and Social Perspectives." *Journal of Retailing,* vol. 60, no. 3: 15–36.

Kunze, Kent. 1985. "Hours at Work Increase Relative to Hours Paid." *Monthly Labor Review* (June): 44–46.

Lewis, W. Arthur. 1970. "Competition in Retail Trade." Chapter V in *Economics of Overhead Costs.* New York: Augustus M. Kelly.

Mulligan, J.G. 1983. "The Economies of Massed Reserves." *American Economic Review,* vol. 73: 725–34.

Nooteboom, B. 1983. "Productivity Growth in the Grocery Trade." *Applied Economics,* vol. 15: 649–64.

Oi, Walter Y. 1983. "Heterogenous Firms and the Organization of Production." *Economic Inquiry*, vol. 21, no. 2: 147–71.

Parker, R. 1986. *Concentration, Integration and Diversification in the Grocery Retailing Industry*. Washington, D.C.: Bureau of Economics, Federal Trade Commission, March.

Progressive Grocer. 1982. (April), 25–26.

Ratchford, Brian T., and James R. Brown. 1985. "A Study of Productivity Changes in Food Retailing." *Marketing Science*, vol. 4, no. 4: 292–311.

Rothbart, E. 1941. "The Measurement of Changes in Real Income Under Conditions of Rationing." *Review of Economic Studies*, vol. 8: 100–107.

Schwartzman, D. 1968. "The Growth of Sales per Man-Hour in Retail Trade, 1929–63," pp. 201–29, in *Production and Productivity in the Service Industries*, edited by Victor Fuchs. New York: National Bureau of Economic Research.

Smith, Henry. 1948. *Retail Distribution*, 2nd ed. Oxford: Oxford University Press.

Smithies, Arthur. 1941. "Optimum Location in Spatial Competition." *Journal of Political Economy*, vol. 41: 423–39.

Stahl, Konrad. 1986. "Theories of Urban Business Location." Mimeo, Universitat Dortmund.

Steiner, P.O. 1957. "Peak Loads and Efficiency Pricing." *Quarterly Journal of Economics*, vol. 71: 585–610.

Stone, Richard. 1954. *The Measurement of Consumers' Expenditure and Behavior in the United Kingdom, 1920–1938*. Cambridge: Cambridge University Press.

Syrquin, Moises. 1972. "Returns to Scale and Substitutability in the Repairman's Problem," *Econometrica*, vol. 40: 937–41.

Wessels, Walter J. 1980. "Minimum Wages, Fringe Benefits, and Working Conditions," American Enterprise Institute, Washington, D.C.

Whinston, G.C., and T.O. McCoy. 1974. "Investment and the Optimal Idleness of Capital." *Review of Economic Studies*, vol. 41: 419–28.

Whitin, T.M. 1952. "Inventory Control in Theory and Practice." *Quarterly Journal of Economics*, vol. 66: 505–21.

9 COMPUTERS AND JOBS
Services and the New Mode of Production

Larry Hirschhorn

Technology does not shape jobs in a vacuum. The conventional assumption that productivity-increasing investments reduce employment is certainly too simple. But it is equally insufficient to argue that growth in aggregate demand can simply compensate for the employment-reducing impacts of technological change. Rather, we must understand the economic and technical forces that shape the pace and scope of aggregate demand itself. This chapter argues that the current phase shift to a postindustrial economy provides the overall context for economic decisions. Companies, organizations, managers, and workers are erecting a new mode of production that changes the organization of manufacturing and services and the relationship between the two. A new prototypical enterprise is emerging that secures its markets and profits in new ways.

The chapter is divided into five sections. The first introduces the theme of a new "mode of production" linking it to services growth and computer technology. The second, drawing on data from Comtec, a market research company, examines the growth in computer systems and personal computers for the economy as a whole and for different sectors from 1980 to 1985. The third presents two cases studies of organizational change that highlight the emerging new combinations of capital, labor, and organization in a postindustrial economy. Drawing on the case studies the fourth section outlines a two phase model of the diffusion of computers within the economy. Finally, the fifth section summarizes the results of the prior sections arguing that the macro and micro evidence suggests that on balance computer technology is currently creating jobs.

Nonetheless, examining the new enterprise's strategic use of labor, the section concludes with the the argument that the new mode of production, while not destroying jobs, may be undermining the institutional system that once created and sustained an orderly sequence of jobs. People are developing a more contingent and less predictable relationship to the job market.

A NEW MODE OF PRODUCTION

The growth in services, the emergence of information work, the relative employment decline in the older industrial sectors of auto, steel, and rubber, suggest that we are entering a new phase of economic development in which services and high-technology industries dominate. This conventional picture is oversimplified. Although we are entering a new period of economic development, it is characterized not as much by the decline of industry and the rise of services as by the changing relationship between the two. The manufacturing base has not disappeared nor have services replaced goods production. In 1950 manufacturing contributed about 25 percent of total value added to GNP and in 1983 manufacturing contributed 23 percent of total value-added (Hicks 1985: 44). Similarly, manufacturing's share of total employment fell five percentage points over this twenty-three-year period, hardly a rapid or debilitating shift, masking the more important shift of workers *within* manufacturing from blue-collar to white-collar jobs. By 1982 almost half the workers in manufacturing worked in sales professional, management, clerical, or service jobs (Hunt and Hunt 1986: Table 3.2).

The shift to a postindustrial economy means that we are developing a new *mode of production,* a new combination of labor, capital, and organization that restructures work in every sector, while rearranging the relationships between them. As Hicks brilliantly

argues, changes *within* sectors, plants, and offices are just as important as shifts in the flow of resources *between* them.

Thus, for example, the service sector, which has grown significantly over the last thirty years, can no longer be described as a
sector composed of low-wage or low-skill jobs in department stores
and fast-food shops. Instead, the fastest-growing sector within the
service category is composed of business, professional, and financial services, such as banking, insurance, legal, accounting, advertising, training, engineering, and management consulting. Increasingly, firms within this sector are providing high value added
and customized services using both skilled labor and much computer power. As Hicks notes, "This sector is growing at a faster
rate than any other in the economy, and it currently accounts for
25% of total GNP" (Hicks 1985: 51), just about the same as manufacturing's contribution to value added. Indeed, from 1970 to 1980,
manufacturing created 9.9 percent of the net new jobs, finance and
insurance created 10.2 percent, while "other services" (which includes the bulk of the business and professional services) contributed 38.5 percent of all new jobs.

This growth of the service sector suggests that services do not
replace manufacturing but rather contribute in new ways to the
production of value added *within* manufacturing. In turn, manufacturing is more dependent on such advanced services because its
mode of production is being transformed. As manufacturing companies automate their machinery, train their line workers to cope
with more complex and risky industrial systems, develop telecommunication systems for control and communication, organize
their marketing and research efforts to produce more customized
products, and enter into joint ventures with suppliers and buyers,
they are reversing some of the classical principles of manufacturing management. To produce high profits they must frequently
customize production rather than produce mass-marketed goods,
delegate greater control and authority to line workers who maintain continuous process systems, decentralize their operations to
improve marketing, and recalibrate staff-line relationships to free
line managers from bureaucratic controls (Hirschhorn 1984).

This new role for services had led to a significant capitalization
of the service worker. In contrast to conventional wisdom, service
work is no longer labor intensive but is instead organized through
a new mode of production as well. From 1960 to 1980, for example,

real investment per employee increased five times in finance and insurance but increased only 50 percent in the private nonfarm economy as a whole. Similarly, "in a study ranking 145 industries by capital intensity, it was discovered that nearly half of the top fifth were service industries" (Hicks 1985: 31).

Finally, the growth in the measured service sector masks the fact that goods production increasingly resembles a service. Companies selling durable and nondurable goods find that they add value to their product most when they differentiate between markets, match products to customers, produce to high standards of quality, respond flexibly to changing market demands, and service their customers effectively. As they transcend the practices they inherited from the period of mass production for mass markets, they must discover and invent principles of effective service in order to meet customer demands for specificity, quality, timeliness, and continuity of service. The service and goods sector are converging. While services are industrialized—service companies are capitalizing their workers while applying cost accounting and productivity measures inherited from the manufacturing world—manufacturing systems become increasingly like services. The convergence of the two and the new relationship between them are the best indicators of the emerging postindustrial economy.

These developments upgrade the workforce. Looked at broadly, their has been a secular shift from the "operator" as the paradigmatic worker to the professional and technical as the worker of a postindustrial economy. From 1960 to 1982 the number of professional and technical workers increased from 11 to 17 percent of the workforce, while the number of operatives fell from 18 to 13 percent. Looked at in greater detail, in 1982 professional and technical workers constituted 37 percent of the workers in producers services, and 15 percent in durable goods manufacturing, while operators constituted 33 percent of durable goods workers and only 3 percent of producer services (Hunt and Hunt 1986: Table 3.3). The slow but steady shift to services, and the slow but steady decline of blue-collar workers in manufacturing, upgrade the labor force.

The focus on customized production, the used of skilled labor, and the emphasis on flexibility in the provision of goods and services suggest that a new mode of production is replacing the standard combination of labor, capital, and organization that has characterized our industrial economy for almost the last century. As

Table 9-1. A Stylized Comparison of the Industrial and Postindustrial Modes of Production.

Industrial Firm	Postindustrial Firm
Reduces costs by investing in large-scale integrated systems	Reduces exposure to unpredicted market shifts by reducing reliance on large-scale integrated systems
Controls costs	Controls quality
Produces a standard good; turns goods into commodities	Produces a variety of goods; customizes output
Vertically integrates production	Subcontracts: vertically disintegrates
Labor contributes effort and creates expense	Labor contributes intelligence and enables adaptability
Sees labor as variable cost	Sees labor as capital

Table 9–1 suggests, a new prototypical enterprise different from its industrial counterpart is emerging to take advantage of and further develop this new mode of production.

Thus the new firm considers labor as an asset rather than an expense, it focuses on quality rather than costs, it sustains its flexibility by reducing its exposure, and it makes profits by controlling its quality rather than its costs. As this table suggests, technology is a necessary but not sufficient condition for shaping and creating the new mode of production. Although computers play a critical role in helping organizations create flexible production and delivery systems, managers must also create new forms of organizations and shape new labor relations and human resource policies if the new technologies are to be used effectively. The new mode of production is based on a new *configuration* of capital, labor, and organization, not on any single factor alone.

THE GROWTH IN COMPUTERS

Computer technology plays a critical role in reorganizing the services and manufacturing sectors and in shaping the new prototypical enterprise. Manufacturing plants can become more like a service by serving many markets and producing a much wider array of goods because the computer helps engineers, managers, and workers schedule the flow of materials, control quality, and change production lines as the demand for different products rises and falls. The factory becomes flexible insofar as it becomes computer dependent. Similarly, business, professional, and financial services can expand to meet the growing need for value-adding services by using computers to increase the flow of work while maintaining and increasing its quality.

Comtec, a market research firm that provides proprietary data on the acquisition and use of computers throughout the economy, has provided the author with data on the growth of personal computers and computers systems from 1980 to 1985 for the national economy, for different industrial sectors, and for establishments of different sizes. Comtec surveys 8,000 businesses selected from twelve industry groups and sixty vertical markets and gets information from over 35,000 business executives. Weighting the survey data by an estimate of how many like establishments a particular firm represents, it extrapolates its results to fit a national total of 6.63 million businesses. Comtec reports its basic sectoral data in codes that roughly correspond to the two-digit SIC codes, though, for example, they break out services, separate durable from nondurable manufacturing, and combine agriculture, mining, and construction. (Further information on the Comtec sample and the relationship between its coding and the SIC codes is shown in Appendix Table 9A–1.)

As Table 9–2 shows, there has been extraordinary growth in both computer systems and personal computers from 1980 to 1985, with the latter growing ten times as fast than the former over the same period. Because overall employment grew by about 6 percent in this same period, it is clear that industry is capitalizing its employees with computers at a very rapid rate. The "information power" available to employees has grown exponentially in a very short period of time, suggesting that increasing numbers of workers may be employed in high value-adding production systems.

More important, Tables 9–2 and 9–3 highlight the emerging links between computers and the new mode of production. Excluding educational services—that is, schools, colleges, and univer-

Table 9–2. Growth in Personal Computers and Computer Systems, 1980–1985.

	Total PCs		Percentage of Total		Percentage of Change
	1980	1985	1980	1985	1980–85
Universe	165,131	5,350,034	100.0%	100.0%	3,140%
AgMinCon	8,634	230,712	5.2	4.3	2,572
MnfDur	9,482	398,529	5.7	7.4	4,103
MfrND	3,834	205,178	2.3	3.8	5,251
TransUtil	4,130	295,740	2.5	5.5	7,061
Retail	24,470	348,671	14.8	6.5	1,325
Whlsle	6,049	279,194	3.7	5.2	4,516
Finance	11,322	612,916	6.9	11.5	5,313
BusProf	32,177	660,352	19.5	12.3	1,952
MiscServ	6,292	305,968	3.8	5.7	4,763
Health	3,394	143,720	2.1	2.7	4,134
EdServ	51,035	1,677,014	30.9	31.3	3,186
Govt	4,311	192,040	2.6	3.6	4,354

	Total CSs ·		Percentage of Total		Percentage of Change
	1980	1985	1980	1985	1980–85
Universe	209,851	610,480	100.0%	100.0%	191%
AgMinCon	13,782	35,785	6.6	5.9	160
MnfDur	20,239	53,490	9.6	8.8	164
MfrND	15,269	39,120	7.3	6.4	156
TransUtil	15,946	44,407	7.6	7.3	178
Retail	17,495	93,757	8.3	15.4	436
Wholesale	22,452	78,127	10.7	12.8	248
Finance	25,883	61,447	12.3	10.1	137
BusProf	32,489	78,474	15.5	12.9	142
MiscServ	8,549	34,514	4.1	5.7	304
Health	8,458	28,162	4.0	4.6	233
EdServ	17,036	28,103	8.1	4.6	65
Gov't	12,254	35,094	5.8	5.7	186

Table 9–3. Personal Computers and Computer Systems Per Worker 1985.

Sector	Per Worker
PCs per employee in 1985	
Total Sample	0.052702
AgMinCon	0.036032
MnfDur	0.034378
MfrND	0.026207
TransUtil	0.048643
Retail	0.019109
Whlsle	0.047826
Finance	0.092697
BusProf	0.089975
MiscServ	0.035832
Health	0.018259
EdServ	0.187647
Govt	0.030830
CSs per employee in 1985	
Total Sample	0.006013
AgMinCon	0.005588
MnfDur	0.004614
MfrND	0.004996
TransUtil	0.007304
Retail	0.005138
Whlsle	0.013383
Finance	0.009293
BusProf	0.010692
MiscServ	0.004041
Health	0.003577
EdServ	0.003144
Govt	0.005634

sities—business and professional services are the highest users of personal computers and the second highest users of computers systems. The dominance of this sector cannot be explained by the large number of workers it employs because as Table 9–3 shows it had the second highest per worker use of personal computers and the second highest per worker use of computer systems in

1985. Moreover, since as Table 9–2 shows, its dominance as a computer user fell from 1980 to 1985, the business and professional services sector was an *early adopter* of the new technology. Because in the the last decade it was among the fastest-growing sectors in the economy, the data suggest that computer usage has been inextricably linked to its growth and profitability. Finally, because customized, and high value-added services are particularly important and profitable in this sector, these two tables together suggest that the recent and explosive growth in computer usage has been critical to the emergence of a new mode of production in the economy as a whole. (See Appendix 9B for graphical displays of the growth of computer usage in a sample of sectors over the years 1980, 1983, and 1985.)[1]

Figure 9–1 also highlights the link between computers and the new mode of production. As was argued in the prior section, scale economies become less significant as organizations try to minimize risk by limiting their commitments to fixed assets. As this Figure shows, the diffusion of computers through the system of firms creates a distinctively *U*-shaped curve. Small establishments are generally more capitalized than larger ones, while the largest firms have higher capitalization rates than intermediate ones.[2] This suggests that smaller firms, possessing flexibility and unable to mass-produce goods and services, may be best able to create and sustain the new mode of production, while larger firms facing organizational inflexibilities, lacking great resources, and perhaps unable to shift to a more high-value added production process are less able and willing to capitalize their employees with computing power. Overall, the tables and data suggest that smaller firms in the business and professional services sector represent the "leading edge" in the development and elaboration of the new mode of production.

FORECASTING EMPLOYMENT: TWO CASE STUDIES

What effect will computer technology have on employment? Frequently we imagine that computers displace people since the computer does common tasks faster and automatically. Yet it is

Figure 9–1. Computer Systems per Employee, 1985 (all industry groups).

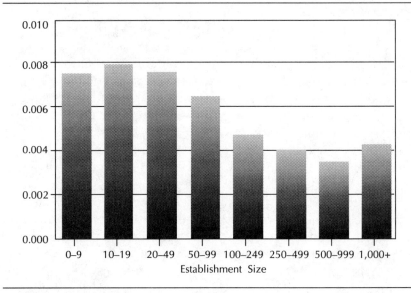

■ CSs per employee

Figure 9–2. Personal Computers per Employee, 1985 (all industry groups).

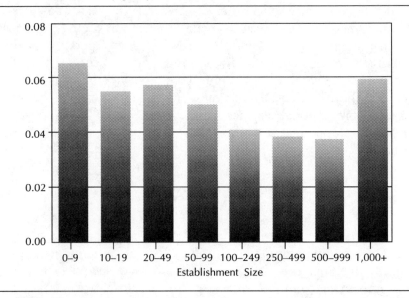

■ PCs per employee

also true that computers enable firms and organizations to increase the quantity or quality of a service or good produced, so that on balance employment may remain either the same or in fact grow. Because in the recent past computing power has grown very rapidly, employment has expanded, and the unemployment rate has fallen, it is difficult to argue on its face that computers on balance destroy jobs. Moreover, we know that business and professional services have been among the fastest-growing sectors of the economy and that small firms have produced a greater number of jobs in the recent past than large firms. Yet as we have also seen, business and professional services have the highest rate of computing power per employee while small firms are more heavily capitalized than larger ones. Thus the data suggest that computer power actually creates rather than destroys jobs. Indeed, Feldberg and Glen show that prior to 1980, despite the automation of many clerical functions, the total number of clerks doubled from 1960 to 1980, the number of secretaries almost tripled, the number of bookkeepers and typists both doubled, and the number of programmers increased sixtyfold. (Feldberg and Glen 1983).

A forecasting study by researchers at the Bureau of Labor Statistics researchers is suggestive here as well. Using econometric techniques, they projected a vector of final demand for goods and services in 1995 and then derived the intermediate demand for the output of different sectors by applying input-output coefficients to the final demand vector. They then assessed the impact of technological change on employment and occupations by predicting the future value of the productivity coefficients that link the intermediate output of any sector to its demand for labor (Hunt and Hunt 1986: 19).

Hunt and Hunt assessed the resulting BLS projections by decomposing the overall change for occupational classes projected by BLS into changes due to overall economic growth during 1982–1995, changes in staffing ratios (changes in the occupational structure of individual industries), and changes resulting from differences in industry growth rates. Table 9–4 shows that changes in industry structure and changes in the occupational structure of industries play a major role in the employment outlook. The BLS projected that growth in clerical employment will result largely from growth in the overall economy and relatively rapid growth

Table 9–4.   Decomposition of Employment Change, 1982–1995.

	Projected Percent Change, 1982–1995	Percent Change, Holding 1982 Industry and Occupation Mix Constant	Percent Change Due to Industry Growth Differences	Percent Change Due to Staffing Ratio Changes
Professional	33.9	28.1	−0.7	6.5
Manager	38.5	28.1	2.1	8.3
Sales	30.4	28.1	2.4	0.0
Clerical	26.5	28.1	1.6	−3.1
Craft	30.5	28.1	0.4	2.1
Operatives	19.1	28.1	−4.5	−4.4
Laborers	21.9	28.1	−3.6	−2.5
Service	28.8	28.1	3.8	−3.1

Source: Hunt and Hunt (1986: Table 4.2).

in the industries that are heavy employers of clerical personnel. These two factors more than offset the changes in occupational structure (many of which reflect technological change) that are less favorable for clerical employment growth. The two highest skill occupational categories, professionals and managers, are projected by BLS to grow the fastest *primarily* because of changes in the occupational structure of industry—measured by changes in staffing ratios—rather than because these occupations are concentrated in rapidly growing industries. This finding suggests that technology's primary impact is on the structure—the mode of production—rather than on the amount of jobs.

This suggests that to deepen our understanding of technology's impact on the economy we need to understand its impact on the structure of jobs, on the *configuration* of labor, capital and organization. The following two case studies, one of Macy's department store (Noyelle 1987) and one of a regional bank together highlight three key points. First, computers are introduced increasingly as part of a company's strategic thrust to help mediate strategy and structure. Computers are *not* introduced to reduce costs, head counts, and rationalize production. Second, computers can increase employment and upgrade the labor force as the company sells more profitable services because employees can make better decisions. Profits rise not because the company has reduced its costs but because employees now add greater value to the final service. Third, computers are introduced as part of a broader process of organizational change through which basic roles, relationships, and careers are reconfigured. Again, computers help create a new mode of production. Let us briefly review each case.

Macy's

Macy's department store, with $4.1 billion in sales, four regional divisions, and ninety-five stores in a fifteen-state market, was the twelfth-largest merchandiser in 1984. But it has succeeded only because it changed its structure and strategy as discount chains challenged department chains in the early 1960s. Initially, Macy's tried but failed to compete by lowering its prices, and profits fell. In 1965 it changed its strategy by emphasizing high quality and high fashion and developed a new divisional structure and new technology to match this strategy.

A historical perspective is helpful here. After World War II chief buyers managed different product lines for the Macy system as a whole and were responsible for purchasing and selling as well as managing the sales effort. They were the "merchant princes" of the system. But as the number of branches grew buyers found it increasingly difficult to pay attention to the tactics of merchandising itself, to decide which products to sell, when, and at what price. Media competition became more important, but buyers found it difficult to coordinate the media campaigns of the different stores while retaining overall coherence. Finally, the buyers were burdened by excessive administrative responsibilities of managing personnel and administering salaries and could not attend fully to their marketing tasks. Consequently, they were unable to meet the new discount store competition.

In response, the Bamberger division created a separate merchandising department in 1963, separating it from store management, such as facilities management, sales management, procurement, and sales. In addition, they created separate functional lines to manage personnel and audit finances. While offering economies of scale in management this functional arrangement posed new problems of coordination because store management and merchandising were ultimately integrated on the sales floor. To integrate these two lines of management and prevent conflict between them Bamberger created a new career ladder so that executives moved back and fourth between them. Managers who advance through this new career track are loyal to Macy's because they develop an integrated view of its operations. The new career ladder reduced management turnover, even though buyers typically have a high turnover rate.

As the old merchant princes lost control the new store managers rationalized the selling effort. They increased the proportion of part-time workers in the stores to control labor costs and reduced fringe benefits, while keeping stores open past 5:00 P.M. Young and minority workers substituted for white women over age thirty. In addition, store managers invested in computer systems to rationalize back office operations such as inventory control, payroll, and timesheet accounting, and they gave sales clerks electronic cash registers to facilitate credit checks. Most important, they integrated inventory, purchase, and sales data to create a timely record of sales. Buyers and their merchandising managers could reduce the risks of stockouts or excess inventory by assessing if merchandise was moving too fast or too slow. This same data enabled merchandisers to pursue a high-quality strategy. Buyers, selling expensive and specialty items, could make focused judgments about consumer habits by assessing timely information about sales and purchases. By separating merchandising from store management, rationalizing the sales effort, and creating a database that helped buyers and merchandising manager make decisions in real time, Macy's was able to meet the challenge of the discounters by carving out a higher-priced, higher value-added market.

These changes have created new career lines. In the past, a person on the sales floor could move into a junior management position and then rise to become a store manager. But as selling was rationalized and new management career ladders were devel-

Table 9–5. Macy's Changing Occupational Structure, 1966–1982.

	1966	1982	Rate of Change	1966	1982
Managers	3,037	6,085	100 %	8.1	12.5
Professionals	758	1,130	49.1	2.0	2.3
Technicians	106	97	−8.5	.3	.2
Sales	19,388	25,568	31.9	51.6	52.7
Clerical	6,869	7060	2.8	18.3	14.5
Craft	309	528	70.9	.8	1.1
Operatives	343	217	−36.7	.9	.4
Laborers	561	661	−17.8	16.5	14.8
Total	37,565	48,550	29.3	100.0	100.0

oped, floor jobs no longer led to the top. The bridge jobs between the nonexempt and exempt ranks disappeared. Instead, to rise to the top a person needed a college degree and had to start as a management trainee. A new dualism based on education was introduced.

Table 9–5 highlights the employment impact of these structural changes. The table shows that as the selling, managing, and merchandising efforts became more sophisticated, Macy's relied increasingly on managers and professionals and less on clerks and sales personnel. Technology played a mediating role. It helped store managers reduce selling costs, while it helped merchandisers carve out a high-quality market niche in the face of competition from the discounters. Technology mediated strategy and structure.

In introducing computers, Macy's also developed new career ladders, upgraded the skills and productivity of the buyers, and changed roles and relationships between store managers and buyers. A new system for retailing emerged, in which costs were controlled by back office automation, while revenues rose because of front office automation. The computer's impact cannot be differentiated from the total reorganization of the selling system. Finally, because Macy's was successful, total employment actually rose, despite higher levels of automation. Macy's had developed a new mode of production.

Metrobank

Metrobank experienced significant financial problems in the late 1970s and early 1980s. Its loan officers made bad international loans, it was locked into longer-term certificates at a time of rising interest rates, and it was squeezed for cash. The board replaced a new president who promptly sold off a large number of unprofitable branches, reduced the bank's assets and liabilities by half, and hired a new executive team.

Metrobank's dilemmas were shaped by a changing economic context for banking. In the late 1970s and early 1980s deregulation, lower inflation, lower interest rates, increased capital spending, technological changes, and the advent of a multitude of new products and services had decisively reshaped banking practice during the bank's crisis period. The result was increased competition on all fronts. Federal legislation removed interest rate ceilings; allowed money market accounts; authorized mutual savings banks to make commercial, corporate, and business loans, and savings and loans to make commercial loans; permitted payment of interest on demand deposits; and permitted interstate and intrastate merges of financially troubled institutions. The protected markets of the past were dismantled as companies in the financial services industry offered similar products and competed directly for customers.

These changes threatened bank solvency. In 1984 the Federal Deposit Insurance Company (FDIC) reported a large number of problem banks, with seventy-nine bank failures, more than in any other year in its history. Changes in the banking laws have increased merger activity and industry concentration. In anticipation of total deregulation, banks are increasing the scope of their operations by acquiring or merging with other banking institutions. Regional institutions can increase their size and scope within their area to become a more formidable regional competitor when deregulation finally occurs. Finally, this new competition reduced the "spread" between the lending rate and the cost of funds, as bank customers began to "shop around" for banking services as they would for any other products. The demand for banking and financial services continues to grow but cost pressures, new delivery technologies, and bank mergers mean much slower job growth.

To cope with the new competition in a deregulated market place, the executive team at Metrobank decided to focus its marketing strength on the consumer side of the bank in its region and no longer compete with the national banks. To retain market share it decided to offer customers a wide range of services and loans rather than low prices. It would be a high-quality bank providing responsive and customized services.

To implement this strategy the bank both rationalized its operations as well as augmented and developed its selling effort. In 1982, it began reducing its teller force by computerizing the teller stations and installing automatic teller machines in all its branches. As older tellers retired, it hired part-time tellers, drawing on more educated women who wanted to work part-time for a decent hourly wage. While reducing teller costs, however, it increased its marketing, product development, and selling costs. It created a product-development division and designed fourteen bank products that might fit its customers' wide range of needs, such as Mastercard, money-saver checking, CDs, and car-leasing. To sell these products the executive team decided to train platform-workers, the clerks who once only filled out loan applications and ordered checks for customers. They purchased a selling-skills program from a local training company vendor, designed an information system to track workers' sales productivity, and began a program of placing automated terminals on every worker's desk. The terminals, now placed in 12 percent of the branches, enable platform workers to call up customer files, to correct customer demographic information, to perform "what if" calculations for customers, and to explain basic product features to customers. The terminal does not replace the platform worker but rather augments his or her capacity to serve and sell the customer.

This program of rationalizing teller service while augmenting the sales effort, was matched by an organization redesign. The senior managers realized that the staff-heavy and downtown-oriented corporate system obstructed the bank's responsiveness to its customers. Branch managers and their supervisors, the area mangers, who once functioned as operations managers funneling money and paper to headquarters, had to be empowered to think more broadly, develop marketing plans for their trade area and work closely with the product managers. Thus beginning in 1985 and continuing into the present the bank is slowly placing staff functions such as training, product development, and commercial lending under the direction of the branch system chief.

To work and manage in this new setting, both platform workers and branch managers must develop new skills and work habits. Predictably this has not always been easy. For example, platform workers do not find it easy to take a sales roles. Some feel uncomfortable pressuring customers to buy new products, while others feel that they are so busy serving customers they have little time to sell.

Analysis suggests that platform workers must develop two new skills—planning their time and selling to customers in a complex interpersonal game—if they are to sell successfully. For example, before the bank changed its strategy, customer requests for service shaped the platform workers' cycle of activities. Never initiating activity, they responded in succession to each customer waiting on the service line. When they began to sell, they had to make more complex judgments: "How much time should they spend helping customer X?" "After satisfying customer Y's request for information, should they show her other bank services and products?" "If customer Z is unresponsive at this point, should they call him back in a month's time?" For the first time the worker had to think tactically, had to allocate limited time resources across a set of competing activities, and had to think ahead.

Similarly, in learning how to sell, the platform worker had to develop new emotional relationships to the customer and the bank. Because platform workers simply served the customer before the bank changed its strategy, they took no risks when talking to the customer. When selling, however, they took the risk of turning a customer off and of feeling rejected. "I'm afraid," said one platform worker, "that a customer will say, 'Forget this person and forget this bank.' " As we have seen, all the workers took a course on selling skills, learning how to explain bank products and deal with customer objections. But a bank-sponsored study showed that workers on the average did not know how to "close a sale." Lacking the aggression to push the sales process forward they simply explained a product's features and benefits to a customer and then stopped selling. Clearly, the bank still faces difficult training and career development issues. The senior managers may conclude that they have to attract a new kind of platform employee, perhaps one with more education and more ambition, if they are to transform the work at the branch level.

Table 9–6 highlights the effect these changes have had on employment at the bank. The table is for the bank as a whole including both commercial lending and trusts. Therefore it reflects some developments not discussed here. Nonetheless several trends are apparent. First, despite the reduction in total employment, due primarily to the bank's cutbacks in the early 1980s, the number of managers rose, while the number of office and clerical workers fell. Similarly, both the rise in the number of technicians and operatives reflects the continued growth in back office automation as bank operations are computerized (the operators in this case run the mainframes as well as the standard check clearing machines). The decline in the number of professionals reflects the cut back in central office functioning as the new executive team rationalized bank operations. If we add the professional and managerial workforce to the workforce associated with computerization, we find that this subgroup rose from 39 to 43 percent of employment over this eight-year period while clerical jobs fell by 2 percent (the decline in the teller workforce is

Table 9–6. Changes in the Occupational Structure of Metrobank, 1977–1985.

	1977	1985	Percentage Change 1977	1985
Managers	1,108	1,101	26 %	28 %
Professionals	363	319	8.5	8.0
Technicians	52	96	1.2	2.5
Sales	7	10	.16	.2
Office and clerical	2,486	2,181	58.0	56.0
Craft	—			
Operatives	103	197	2.4	5.0
Laborers	—			
Service	124	67	2.9	1.7
Total	4,243	3,881	100.0	100.0

hidden in this number; platform workers as well are counted as clerks not sales). The front office workforce is being slowly professionalized, while back office computerization expands the number of semiskilled jobs available.

Once again technology mediates structure and strategy. The bank pursued a customized production strategy, and to do this, it used technology to reduce teller costs, augment its sales effort, and rationalize its clerical operations. Platform worker skills were upgraded, while line managers on the retail side gained more authority and staff resources. The bank is developing a new mode of production stressing customization, field operations, and front office computers. The workforce is being slowly professionalized.

THE IMPACT OF COMPUTERS

This review of the data on the growth and spread of computing power, study of the two cases, and assessment of the general history of computing, point to the following two-phase model of the utilization and impact of computing. In the first phase, the computer comes to the back office of banks, insurance, companies, brokerage houses, and the administrative offices of large companies. Large systems divisions emerge to maintain the hardware and software of mainframe computers and control access to computing resources. Used for transactions alone, the computer does not restructure the organization or change its core production. For example, payroll departments prepare coding sheets that are keypunched by the systems division to produce checks and pay records, but the payroll department's basic function and its relationship to other divisions are unchanged. Computer service companies emerge to function as "external" systems divisions for smaller companies that cannot afford to maintain the hardware and talent to process key administrative data. This is the period in which computer systems growth dominates the growth of personal computers. Employment in computing using firms grows; however, because the decline in costs enables firms to produce a much higher volume of transactions. That is why, for example, clerical employment grows throughout this phase of back-office automation.

In the second phase, the growth of personal, on-line, and interactive computing restructures the use of computer resources by changing core tasks and key relationships between divisions and functions. By subverting the batch process it creates a plentiful supply of access channels enabling numerous workers to use the computer to transform their work and the services they produce. For example, using the computer to facilitate ticketing, the airlines restructure the service itself so that customers have more flight and seating options, airlines can access one another's schedules and seating plans, and flight plans can be produced more rapidly. Similarly, using on-line access through an electronic cash register, retail managers can integrate the sales and inventory control functions at the point of sales enabling them to order new inventory in a more timely fashion. Finally, as we have seen in the case of the bank, in many sales settings, clerks use computers to facilitate the selling process and increase their sales productivity. Bank clerks with on-line access to financial records as well as bank product descriptions can check a customer's records, assess which services he or she may need or want, and so appropriately focus their sales talk. Finally, the spread of personal computers in the front office enables numerous managers, professionals, and secretaries to produce higher-quality documents, do more exhaustive research, and consider a broader range of information when "producing" or supporting decisions. In this phase, employment grows because computers are used to add new value to products and services rather than simply reduce the labor expense associated with old products. Thus as we have seen, the most computer-capitalized sectors and establishments tend to be the fastest growing in terms of employment.

Interactive computing combined with the reduced cost of computer circuitry decreases the power of the systems division in many companies. Unable to control access through the batch process and often unable to regulate computer hardware and software purchases, systems divisions are developing a new "customer orientation." They work with the end-user to design user-friendly systems and develop more helpful training programs to ensure that the total set of computer uses throughout a company creates a consistent set of protocols for access, coding, and classification.

By changing core tasks, interactive computing restructures the balance of power and influence in organizations. For example, as we have seen, as banks capitalize their branch clerks with computers, the prestige of the branch office division grows. No longer functioning as a mail box for deposit and applications, its clerks, tellers, and managers add to bank profits by actively selling loans and services. Staff personnel who once monitored branch managers find that they must support them instead. As computers migrate from back to front offices, the staff-line relationship is recalibrated. In this phase, distributed computing systems grow in significance as the front office is automated and personal computers are linked into mainframe or minicomputer systems.

Finally, while these two phases of computerization have been analytically and historically distinct, the twin processes of back-office and front-office automation now affect each other. As financial service companies, for example, continue to design and introduce new products, front-office personnel need new computer support systems to service customers. Information structures that supported prior product lines need to be redesigned, and a tension emerges between the back-office's needs for stability and continuity and the front office's needs for flexibility and change. Higher-level programming languages that enable front-office personnel to easily configure data in new ways and automated program writers that help back-office personnel write new codes quickly can help ameliorate this tension. Table 9–7 highlights the critical dimensions of each phase.

THE COMPUTER'S AGGREGATE AFFECTS: A SYNOPTIC VIEW AND SOME SPECULATIVE THOUGHTS

Let me review the key elements of my argument thus far. Technology is helping to shape a new mode of production that changes manufacturing, services, and the relationship between the two while helping to create a new mode of production. The latter is based on the economics of high value-adding production systems, through which enterprises focus on quality rather

Table 9–7. Phases in the Introduction of Computers into Banking.

	(1) *Transactive:* *Back Office*	*(2)* *Interactive:* *Front Office*
Examples	Payroll	Bank clerks selling
Effect	Speed-up but no change in organizational structure	Better fit between customer needs and services provided, timeliness, quality choice become key
Computer technology	Primarily computer systems	Computer systems and personal computing leads to interactive computing
Obstacles	Level of resources	Training of personnel, particularly those with historically lower-level skills such as bank clerks
Effect on jobs	Grows due to effect of falling costs on volume	Grows due to changes in structure of output and quality improvement[a]

a. We are now entering the third stage of computer use, in which managers and computer engineers are working to develop and implement decision support and decision production systems. Innovators and product champions hope that expert systems may actually replace the professional or manager, but recent research and experience suggest that the diffusions process will be slow, difficult, and quite uneven. When knowledge engineers first produced decision support systems, they hoped to amplify the manager's ability to use data, solve problems, and consider alternatives. But field studies of managers revealed that managers rarely "think analyze and decide" in three separate steps. Rather they combine these stages of problem framing and solving by enmeshing themselves in broad-scale *social* processes. The informal meeting in the hall, rumors from corporate headquarters, the politics of corporate programming, fads in the field, as well as hard analytic data all shape the decision process. The successful manager is active, listening talking, conferring, telephoning but rarely simply thinking and writing. Decision support systems have their place, particularly in helping lower-level managers prepare the briefings and reports they need to convince a superior that a certain decision is good or bad. But the social process that shapes decisions dominates the analytic one.

There is some hope nonetheless that experts systems for professional rather than managerial work may nonetheless transform certain categories of white-collar work. The underwriter, the loan rater, the paramedic, the drafter, who all make semi-structured decisions based on regular though implicit protocols may be able to effectively use such expert systems. Although this system may increase the professional's rate of decision production, it will more likely pay off by reducing the gap between the most and least qualified professional. Decisionmaking across professionals will become more consistent and the average quality of decisions will rise.

than costs, see labor as an asset rather than as an expense, and reduce their exposure rather than their costs by minimizing rather than maximizing the value of their long-term assets. The outlines of this new mode of production emerged in the 1980s as computing technology was rapidly diffused throughout the economy. Firms for the first time have the information power they need to erect flexible production systems. Data suggest that the business and professional services sectors and the small firm are playing a critical role in propelling this development. The former provides the services to help transform older manufacturing enterprises, while the latter has the flexibility to absorb the new computing technologies and use them in new ways. Although in the earlier phase of computerization (which began in the mid-1950s and ended in 1980) computers were used primarily to transform back-office operations, in the current phase computers restructure front-office operations creating interactive systems that link computer systems to personal computers.

The empirical evidence suggests overall that computers have not replaced workers or destroyed jobs; if anything, they have created jobs. In the first phase of automation, despite extensive back-office automation, the number of clerks, typists, secretaries, and bookkeepers grew, while in the second phase the fast-growing business and professional services sector is also the among the most computer intensive. Similarly, the data suggests that small firms that contribute a disproportionate number of new jobs have also capitalized their workers with computing power at a faster rate than large firms. Indeed, the dual role that both business and professionals services and small firms play, as both users of the new technologies and as important job creators, strongly suggests that computer usage is directly linked to the creation of jobs. Finally, aggregate forecasting methods typically project increases in white-collar workers in all occupational categories while one in-depth input-output study actually predicted an absolute labor shortage by the year 2000. All this suggests that over the next few decades employment fluctuations will be determined by the familiar dynamics of the business cycle rather than by technological change. Indeed, the later will in general provide a continuing secular boost to employment.

Contingent Employment

This does not suggest that computers pose no employment problems whatsoever. Because in the new prototypical enterprise labor is an asset rather than an expense, the level of firm employment is more stable over the course of the business cycle (thus dampening the business cycle itself). Firms are more reluctant to fire employees they have trained at great costs, but they are similarly careful before they hire full-time employees as well. Although creating more stability for full-time core employees, such firms may similarly rely more on contingent labor to meet unpredicted or apparently short-term increases in the demand for different kinds of workers. Indeed, data suggest that the number of contingent workers is growing rapidly, while their educational level is also rising. The temporary help services industry, though relatively small (about 735,000 workers representing less than 1 percent of the total wage and salary employment in nonagricultural establishments in 1985), has doubled in the three years since 1982 and now accounts for 3 percent of total job growth (Carey and Hazelbaker 1986). This trend is projected to continue into the 1990s with a 5 percent average annual projected growth rate. The temporary help industry reports a dramatic expansion of their payroll base between 1970 and 1983, from $547 million to over $4 billion, with an average annual growth rate of 17 percent (Dennis 1983).

Companies hire temporaries to control labor costs and manage peak load production periods, but they increasingly have a need for workers with special skills due to the rapid and radical changes in office technology. Employers may have difficulty filling certain positions due to a chronic shortage of technologically skilled clerical and technical employees. Thus Manpower, one of the largest temporary help service companies in the world, has found that from 1980 to 1985, the proportion of Manpower temporaries with a college degree or college background has increased from 30 to 65 percent because companies are asking for more skilled secretarial and office help. Moreover, specialty Temporary Helps Service firms have emerged to place accountants, doctors, and engineers in companies looking for people who are willing to work on short-term projects.

To respond to the demand for skilled labor, Manpower has developed a $15 million computer program called "skillware," a self-pacing computer program that teaches recruits the basics of word-processing for nine different languages. Moreover, the company is currently developing similar skillware software for data-based management and telecommunications skills. Contingent workers are no longer simply a "pair of hands" needed in the warehouse but skilled clerical and technical help.

This suggests that as the distinction between core and periphery workers is reproduced *within* rather than *between* firms, the new mode of production may not create problems of unemployment but rather problems in the working conditions for the temporary worker: How contingent will their relationship to the labor market be? What benefits might they receive? Will they receive adequate training? Are they working in temporary positions for voluntary reasons?

Moreover, the growth in contingent workers highlights a more general though more diffuse trend. Career ladders within companies are less secure. Production flexibility means that jobs and roles change more frequently, and workers are laid off not because they are too expensive but because their skills no longer match firm needs. Consequently, managers can no longer expect a career for life in a large company, where their upward mobility is based simply on their seniority. Similarly, as educational qualifications becomes important for many lower-level management, technical, and paraprofessional jobs, workers must obtain more education during their working lives to be promoted. Internal labor markets that once connected the shop and office floor to management and supervisory positions have been broken, and workers interested in upward mobility must increasingly post to positions outside their departments and divisions. Upward mobility is shaped by lateral as well as vertical steps.

Rather than destroying jobs, the new mode of production may be destroying the institutional links that once tied jobs together in an orderly sequence. Rather than displacing workers it is making it harder for people to create a satisfactory work history, to articulate work, nonwork, and education together in satisfying ways.

In Sum

A new mode of production is emerging based on a new paradigmatic enterprise and the new computing technologies. It is restructuring services, manufacturing, and the links between the two. Far from creating unemployment, the sectors and firms that use technology most intensively are expanding employment at the fastest rate. Nonetheless, because the new mode of production changes the employment relationship between workers and firms, treating core workers as an asset rather than as an expense, the level of contingency and uncertainty in the careers market is growing. We are not facing a problem of jobs but a problem in the institutional system that links jobs to the life-cycle.

APPENDIX 9A. COMTEC

Comtec researchers contact executives by phone and ask them a series of questions about past purchases, installed base, and planned acquisitions. For example, a simplified version of the protocol for assessing personal computer uses reads as follows

Do you have any PCs? (How many? When purchased?)
Do you have plans for the future acquisition of PCs? (How many?)
Do you have PCs that communicate? (How many?)

Comtec data appears to be statistically reliable (see Table 9A–1). It has been produced for three years now, and each year the sample size has grown and the data collection process has been checked and refined. Statistical error is measured by computing a coefficient of variation for any single statistic. Conceptually, a statistic computed from the 8,000 establishments is then checked by computing it from a randomly drawn subset of 4,000 businesses thirty-two times to produce a bell-shaped curve and thus a measure of the coefficient's stability. Typically, sample sizes of 100 or more produce reliable statistics. Comtec claims statistical error in the range of plus or minus 3 percent for widely used computer based products and plus or minus 10 percent for narrowly used products.

Table 9A–1. Comparison of Comtec and BLS Industry Groups.

	COMTEC			BLS		
	Group Name	Two-digit SIC Codes	Employees (1985)	Group Name	Two-digit SIC Codes	Employees (1984)
Agricultural/Mining/Constructions		07–17	6402827	Agriculture	07–09	356881
				Mining	10–14	974285
				Construction	15–17	4171763
Manufacturing durable		24–39	11592339	Manufacturing	20–39	19325352
Manufacturing nondurables		20–31	7829110			
Transport/utilities		40–49	6079697	Transportation/utilities	40–49	4675385
Retail		52–59	18245738	Retail	52–59	16060830
Wholesale		50,51	5837649	Wholesale	50,51	5387724
FIRE		60–67	6612027	FIRE	60–67	5783225
Business/professional services		73,81,89	7339240			
Miscellaneous services		70,72,75–79,83–85	8538917	Services	70–89	20349322
Health services		80	7870820			
Educational services		82	8937061			
Government		—	6228847			

APPENDIX 9B. SECTORAL GROWTH IN PERSONAL COMPUTERS AND COMPUTER SYSTEMS, 1980–85.

Figure 9B–1. Personal Computer Growth (all industry groups).

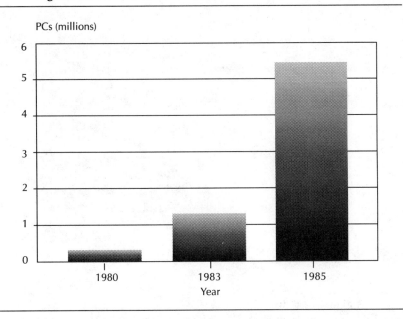

Figure 9B–2. Computer Systems Growth (all industry groups).

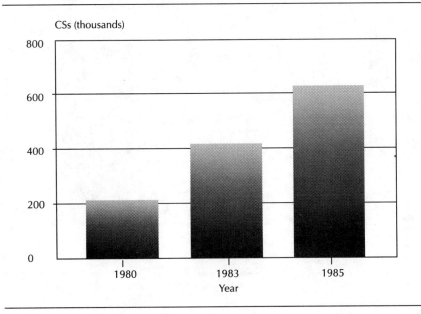

Figure 9B–3a. Personal Computer Growth: Retail Trade.

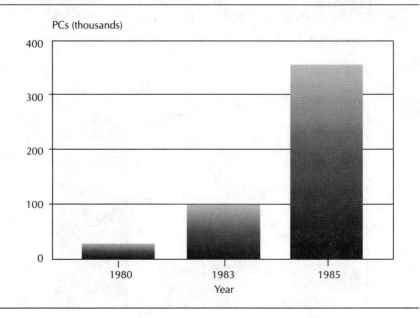

Figure 9B–3b. Personal Computer Growth: Wholesale Trade.

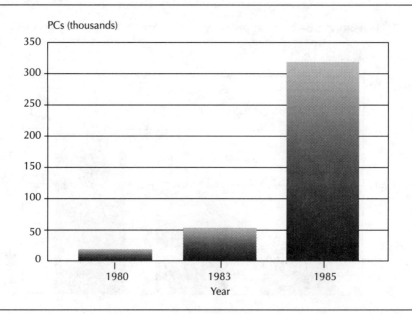

Figure 9B–3c. Personal Computer Growth:
Finance / Insurance / Real Estate.

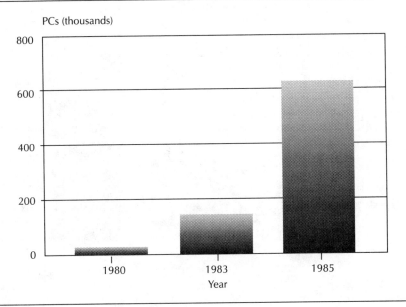

Figure 9B–3d. Personal Computer Growth:
Business and Professional Services.

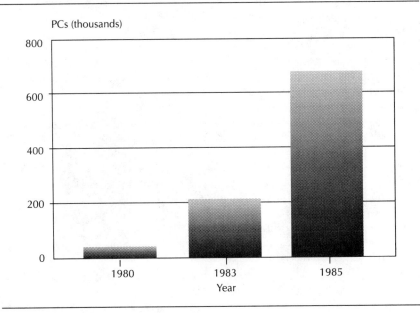

Figure 9B–4a.   Computer Systems Growth: Retail Trade.

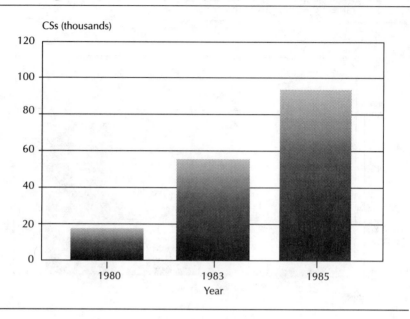

Figure 9B–4b.   Computer Systems Growth: Wholesale Trade.

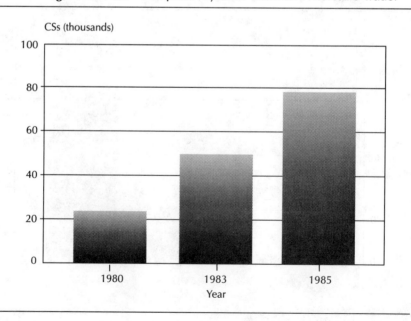

Figure 9B–4c. Computer Systems Growth:
Finance/Insurance/Real Estate.

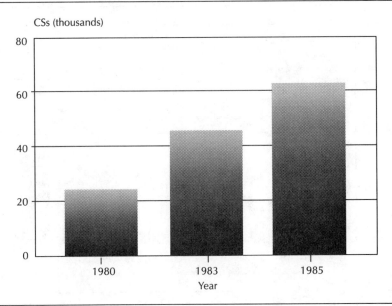

Figure 9B–4d. Computer Systems Growth:
Business and Professional Services.

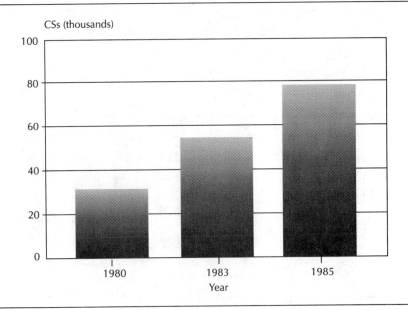

Figure 9B–5a. Personal Computers per Employee, 1985: Retail Trade.

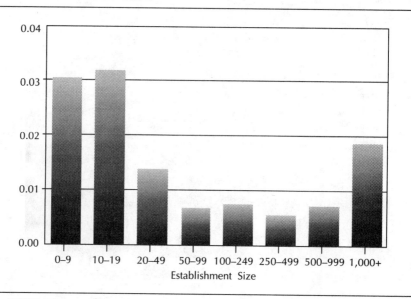

PCs per employee

Figure 9B–5b. Personal Computers per Employee, 1985: Wholesale Trade.

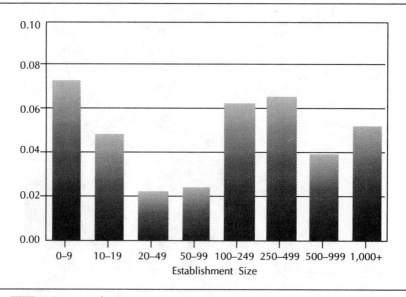

PCs per employee

Figure 9B–5c. Personal Computers per Employee, 1985: Finance / Insurance / Real Estate.

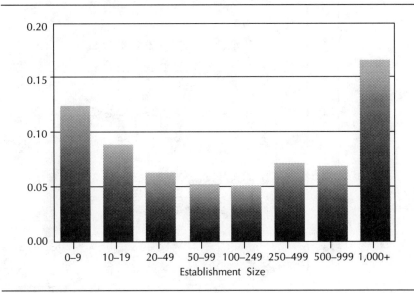

PCs per employee

Figure 9B–5d. Personal Computers per Employee, 1985: Business and Professional Services.

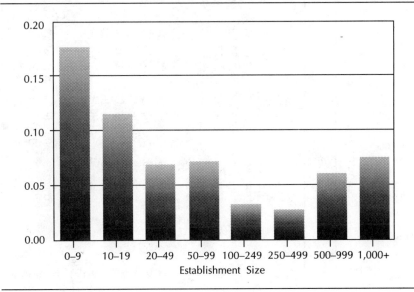

PCs per employee

Table 9B–1. Computer Systems, 1985.

	Universe	0–9	10–19	20–49	50–99	100–249	250–499	500–999	1,000+
Total CSs in 1985									
Sample	7096	72	125	335	450	830	849	1145	3290
Universe	610479.9	130228.7	81940.52	127617.7	83627.2	65078.47	34512.6	22762.69	64711.9
AgMinCon	35785.06	12536.07	4068.24	7317.9	5560.54	3091.84	1210.87	385.52	1614.08
MnfDur	53490.23	2139.47	3788.58	7566.41	8395.7	8449.09	7632.76	5129.98	10388.24
MfrND	39119.54	2846.92	975.79	7381.99	6395.16	8020.69	5204.4	3276.55	5018.04
TransUtil	44407.1	6560.64	5901.59	7629.28	3217.99	5722.25	2729.59	3175.13	9470.63
Retail	93756.61	25774.88	16528.94	24577.98	16608.19	7740.52	1854.13	474.4	197.57
Whlsle	78127.45	20519.55	13485.16	27865.18	9316.17	4896.34	1029.84	924.23	90.98
Finance	61446.65	14841.04	10800.84	13633.54	9401.73	4877.28	2389.9	1907.35	3594.97
BusProf	78473.73	28566.59	10515.15	16183.54	9200.47	8365.75	3313.13	1215.35	1113.75
MiscServ	34514.22	4376.29	11292.34	7100.82	5489.87	3189.16	1604.43	714.33	746.98
Health	28161.58	6903.43	3776.74	3842.93	2293.6	2545.04	1676.75	2215.48	4907.61
EdServ	28103.46	704.26	0	1552.96	4381.97	3092.74	3241.67	1861.2	13268.66
Govt	35094.32	4459.65	807.15	2965.25	3365.81	5087.77	2625.13	1483.17	14300.39
CSs per establishment in 1985									
Universe	0.092059	0.026054	0.106969	0.237158	0.434099	0.696049	1.372739	2.204650	8.852747
AgMinCon	0.049664	0.021048	0.060419	0.185731	0.454743	0.699376	1.178749	1.271839	6.850352
MnfDur	0.232221	0.017957	0.099881	0.208847	0.512667	0.674788	1.738859	2.595802	5.844692
MfrND	0.233546	0.034769	0.038597	0.274249	0.425308	0.676768	1.247997	1.944470	7.369716

TransUtil	0.155855	0.032818	0.163272	0.258911	0.322681	0.909462	1.505415	4.203354	16.50970
Retail	0.052711	0.019220	0.071780	0.169301	0.357187	0.586825	0.870655	1.432453	3.157078
Whlsle	0.150989	0.055348	0.160032	0.625953	0.731116	1.150696	1.497644	4.343186	4.098198
Finance	0.105868	0.031394	0.203037	0.386278	0.829656	0.880693	1.797782	3.765224	7.510173
BusProf	0.118023	0.052711	0.166510	0.452834	0.711103	1.062209	1.505236	1.704056	3.291126
MiscServ	0.035659	0.005517	0.112386	0.139763	0.386203	0.444149	1.127545	1.697229	4.881903
Health	0.069143	0.019909	0.137483	0.300286	0.264060	0.369733	0.870148	1.540474	3.648563
EdServ	0.168855	0.017411	0	0.023684	0.172769	0.383229	1.883531	2.659161	26.23249
Govt	0.241985	0.046959	0.049099	0.183002	0.463748	0.939032	1.130235	1.156657	12.58283

CSs per employee in 1985

Universe	0.006013	0.007387	0.007993	0.007638	0.006553	0.004664	0.004036	0.003353	0.004349
AgMinCon	0.005588	0.006414	0.004631	0.006224	0.007231	0.004569	0.003349	0.001966	0.004124
MnfDur	0.004614	0.004443	0.007260	0.007072	0.007755	0.004602	0.004913	0.003898	0.002783
MnfND	0.004996	0.009547	0.002911	0.008947	0.006392	0.004636	0.003794	0.003057	0.004191
TransUtil	0.007304	0.009436	0.012702	0.008720	0.004701	0.006189	0.004156	0.006172	0.007487
Retail	0.005138	0.005109	0.005401	0.005581	0.005705	0.004033	0.003057	0.002300	0.002069
Whlsle	0.013383	0.014086	0.011718	0.019836	0.011905	0.007919	0.004442	0.006736	0.001633
Finance	0.009293	0.009509	0.014996	0.012720	0.012695	0.006140	0.005225	0.005295	0.003963
BusProf	0.010692	0.016396	0.012348	0.015380	0.010771	0.006984	0.004596	0.002598	0.002459
MiscServ	0.004041	0.001687	0.008595	0.004618	0.005630	0.003021	0.003359	0.002470	0.002513
Health	0.003577	0.005645	0.010746	0.009663	0.003747	0.002538	0.002443	0.002259	0.001874
EdServ	0.003144	0.004055	0	0.000649	0.002363	0.002293	0.005007	0.003936	0.007905
Govt	0.005634	0.011002	0.003464	0.005891	0.006803	0.005995	0.003371	0.001911	0.006531

NOTES

1. Shifts in the rank order between the two forms of computing power highlight some dimensions of computer use in a particular sector. Thus, for example, wholesale trade ranks sixth in personal computer use but first in computer system use, suggesting that it uses computers primarily for the complex tasks of inventory control, warehousing, booking, and ordering rather than for increasing the productivity of the individual worker. It has, in other words, fewer work stations and more "back-office" systems. Similarly, educational services rank the highest on per capita personal computer investments and lowest on computing systems, highlighting of course, the distinctive role that personal computers play in education and research. Finally, the business and professional service sectors' significant investments in both kinds of computing power suggests that it uses computers to both process large databases while also amplifying the information work of the individual employees.

2. In the case of computer systems, this pattern may simply reflect the fact that since a computer *system* must be bought "in bulk" a small establishment must buy a minimally sized system to get the computing power it needs, regardless of the number of employees it has. Thus, for example, as Table 9B–1 in Appendix 9B shows, the largest establishments in all sectors have the smallest ratio of computer systems to employees.

    But this argument holds less for the purchase of personal computers, and as the second set of figures in Appendix 9B shows, the penetration of personal computers through the system of firms creates a distinctive *U*-shaped curved. Rates of penetration are high for small establishments, low for intermediate ones, and high again for the large establishments. The striking exceptions to this pattern, education and government services, suggests that a very different decision process, shaped by the goals of not-for-profit organizations are at play.

3. I am indebted to Thierry Noyelle for this conception.

4. This is for a sample of sectors. Trends displayed are similar for other sectors.

REFERENCES

Carey, M., and Hazelbaker, K. 1986. "Employment Growth in the Temporary Help Industry." *Monthly Labor Review* 109: 37–44.

Dennis, M. 1983. "Vital Statistics of the Temporary Help Industry." *Contemporary Times*, vol. 2, no. 6: 5–6.

Feldberg, Roslyn L., and Glen, Evelyn N. 1983. "Technology and Work Degradation: Effects of Office Automation on Women Clerical Workers." In *Machina ex dea: Feminist Perspectives on Technology*, Joan Rothschild, ed. New York: Pergamon Press.

Hicks, Donald A. 1985. *Advanced Industrial Development: Restructuring, Relocation and Renewal.* Boston: Oelgeschlager, Gunn and Hain (in conjunction with the Lincoln Institute of Land Policy).

Hirschhorn, Larry. 1984. *Beyond Mechanization: Work and Technology in a Post-Industrial Age.* Cambridge, Mass.: MIT Press.

Hunt, Allan H., and Hunt, Timothy L. 1986. "Clerical Employment and Technological Change: A Review of Recent Trends and Projections." Research Report Series 86-14, National Commission for Employment Policy, Washington, D.C., February.

Leontieff, Wassily, and Duchin, Faye. 1984. *The Impacts of Automation on Employment, 1965–2000.* Monograph, New York University, April.

Noyelle, Thierry. 1986. *Beyond Dualism.* Boulder: Westview Press.

Roesnner, J. David. 1985. "Forecasting the Impact of Office Automation on Clerical Employment, 1985–2000." *Technological Forecasting and Social Change*, 28: 203–226.

IV TRADE, TAX, AND DIFFUSION POLICY ISSUES

10 TECHNOLOGY, STRUCTURAL CHANGE, AND TRADE

C. Michael Aho

For most of the postwar period, the U.S. economy led the world. By almost any measure, the United States was number one. It was richly endowed with natural resources. It had the world's best-educated and best-trained labor force, the most modern plant and equipment, the leading edge technologies in almost every industry, and superior management. As a result, our citizens enjoyed the world's highest living standard—twice that of the nearest European country.

The relationship between the United States and the rest of world was asymmetrical: Foreigners depended on U.S. technologies to advance productivity—new products were invented in the United States and diffused abroad. They also depended on U.S. direct foreign investment and aid for capital, on the U.S. dollar for the international medium of exchange, and on U.S. leadership for maintaining the international economic institutions. The United States, on the other hand, had its economic fate determined at home. Trade comprised only a small share of U.S. economic activity and foreign capital flows accounted for just a small share of financial activity.

For most of the 1950s and 1960s, Americans felt secure in their economic relationship with the rest of the world and in the performance of the economy at home. U.S. living standards seemed

419

to rise inexorably for all, lifting large numbers of Americans from poverty.

From these economic heights, Americans looked at the world with confidence. Sure of their lead, business and labor agreed that free trade was in the nation's interest. U.S. economic policy was geared almost completely toward domestic objectives. U.S. foreign economic policy was aimed at political rather than economic goals. U.S. concerns about Europe and Japan focused on their economic weakness. The United States aided economic development abroad as a means of keeping the world safe from Communism.

But today almost everything has changed. While the United States continues to have the world's largest GNP and to occupy a leading position in the global economy, in several respects the U.S. economy is no longer clearly preeminent. It is not obvious that U.S. workers are trained in the world's best schools. Nor do they necessarily have the highest skills or work with the most modern equipment. U.S. management may not lead in quality control, in motivating its workforce, and in making decisions for the longer term. U.S. technology in several areas is no longer the world's best. The United States may still provide its citizens with the world's highest living standards—but the lead has narrowed substantially.

The channels linking the United States with the global economy have become deep and wide, and they transmit shocks in both directions. Exports and imports have doubled as a share of GNP. About one-third of U.S. farmland grows crops for export. Almost 70 percent of U.S. manufacturing competes directly with foreign firms, who often are equipped with similar or better technologies. Today vast electronic networks integrate U.S. capital markets into a global financial network, and Americans depend on these capital channels for investment opportunities and as a source of funds. The major U.S. banks have large multiples of their capital at risk in foreign loans. The United States has become the world's largest debtor nation and the major foreign investment host. Direct foreign investment in the United States is now outpacing such investment by U.S. corporations abroad. Shocks—such as OPEC pricing decisions, Japanese automotive achievements, and developing country debt problems—send major ripples across the U.S. economy.

In short, the United States has become less important in the world economy just as the world economy has become more important for the United States. We have become much more vulnerable to global economic developments. But both the government and private business have not developed the capability to respond to them.

The increased global integration of the economy has been associated with a period of much weaker domestic economic performance. During the past fifteen years the nation has been wracked with inflation and deep recessions. Although inflation has been dramatically reduced in the 1980s, unemployment has been extremely high by historical standards. U.S. budget and trade deficits have been mushrooming to record levels. Productivity growth has slowed to a crawl, and average real weekly wages in 1986 were almost 14 percent below their 1973 levels. Major parts of the nation's industrial base and key regional economies have been wrenched by major plant closings and layoffs.

The U.S. economy and the world economy are undergoing a wave of rapid industrial change caused by technological advance and heightened international competition. This wave of change is creating new wealth and higher living standards worldwide. Now that we no longer have a monopoly on technological advance, we benefit from being able to borrow from foreigners. Advanced telecommunications and computer-assisted manufacturing are creating huge increases in productivity and new products for work and leisure.

These advances have also quickened the pace of change. To survive in this rapidly evolving competitive environment, U.S. businesses must change with the world around them.

Rapid change is being most painfully felt in the labor market. Old jobs are disappearing and workers must move on to new jobs and skills. For many years to come, problems of worker adjustment will be among the most difficult economic and political challenges facing all countries. The temptation to resist change will be very great. But change is inevitable and it is the key to growth and to a better future. Countries that resist it condemn themselves to falling behind. The United States, to remain number one, will have to adapt to these changes but will also need an adjustment policy to cope with their consequences and social costs.

Many Americans today worry that the increasing competition facing U.S. firms here and abroad is hurting our standard of living.

The coincidence of widespread layoffs, slowing real income growth and the deterioration in U.S. trade balances, coupled with burgeoning foreign debt, has raised questions about our ability to sustain high living standards in the future. The nation's support for free trade has wavered and foreign economic development is seen increasingly as a threat to U.S. welfare rather than a help. These concerns have recently reached a fever pitch in the national debate about U.S. competitiveness.

But the issue of competitiveness is mixed up with structural change and is often confused by economists and policymakers who make different assumptions about the world and about the linkages between technological progress and trade. This chapter is a critical examination of the usefulness of classical economic models in analyzing competitiveness and structural change.

For many years, widespread concern that the United States may be losing its economic competitiveness has prompted many experts to examine the underlying causes of the sharp decline in U.S. productivity growth (since the late 1960s or early 1970s) as well as the causes of the ongoing productivity growth rate advantage enjoyed by most other industrial countries over the United States. Such examinations have usually concluded that a great deal of the apparent loss in U.S. competitiveness cannot be explained by standard classical models that impute productivity growth as an economic "output" determined by standard and easily measurable economic "inputs" (that is, factors of production such as labor, capital, and natural resources).

There seems to be a consensus, in fact, that only about two-thirds to one-half (or even less) of all historical productivity change can be explained by conventional growth accounting.[1] The rest, if it is not discarded as an inexplicable "residual," is usually attributed to immeasurable "structural change" or more specifically to innovation in productive technology. Of all immeasurable structural changes in the economy, economists often assume that technology is the most important single contributor to aggregate productivity growth.

It is thus understandable that over the last decade economists have grown interested in the productivity role played by structural change in general and by technology in particular. It is also understandable, given the more recent debate over the efficacy of "industrial policy" and the rapid recent growth of the U.S. deficit in net goods and service exports, that this interest is tending to focus increasingly on possible links between innovation and trade policy on the one hand, and structural changes that affect productive technology on the other.

The debate over "industrial policy" has, according to some experts, called attention to the long-term, structural enhancements to technological innovation and national income growth that some nations (such as Japan) have been able to create for themselves through "directed" or "targeted" economic policy. These experts reject the conclusion (or at least what is often thought to be the conclusion) of classical theory that public policy cannot improve on "market-determined" outcomes and therefore cannot create, through structural or technological change, significant competitive advantages. To the contrary, they argue that in many countries, with the notable exception of the United States, public policies are achieving this result. Japan, as well as the so-called newly industrializing countries or NICs (such as Taiwan, South Korea, Hong Kong, and Singapore), are often cited as examples.

Meanwhile, the unprecedented magnitude of the current U.S. net export deficit makes many U.S. observers worry whether the costliest U.S. failure in economic policymaking may not be in the specific area of trade policy. It has been estimated that, between 1979 and 1984, the effect of trade on the composition of output demand may have led to the loss of 1.7 million jobs in U.S. manufacturing.[2] Though most of biggest losses, in absolute employment terms, have been in heavy industries such as metalmaking, machinery, motor vehicles, and construction, trade also seems a threat to newer, technology-intensive industries. In 1986, the United States suffered its first trade deficit in high-technology goods since the category was invented back in the 1960s, and the recent Japanese challenge to U.S. trade production of standard RAM semiconductors now raises the specter of our losing yet one more "strategic" technology—just as we lost leadership in autos, CRTs, machine tools, and robotics in earlier years. If structural change induced by trade is "neutral" with regard to long-term competitiveness, we might reasonably conclude that public policy ought to be "neutral" with regard to structural change. If, however, trade policy can indeed create comparative advantage for one country and destroy it for another, we might want to reexamine our neutrality assumptions.

What follows is an overview of the basic "schools of thought" on the linkages between economic competitiveness and trade policy, with special attention to linkages involving technology and structural change. Obviously, since this is a vast subject area teeming with competing models, hypotheses, and sociopolitical assumptions, we can attempt here only the most summary treatment of the issues involves. In the first part we discuss generally how the structural change and technology can influence economy-wide productive efficiency, both in the basic classical model and in variations from that model. In the second part we discuss the special problems introduced into this picture by international trade. In the third part we review specific policy recommendations from the recent literature. Finally, the appendix reviews the recent literature on the subject more thoroughly.

TECHNOLOGY AND STRUCTURAL CHANGE: THE GENERAL POLICY ISSUES

The Classical Model

Although the starting point for most current investigations of competitiveness remains the classical model, it must be emphasized up front that this model, in its pure form, has precious little to say about precisely how structural or technological change affects economywide productive efficiency. The classical approach to the production of goods and services is to conceive of it as a variable dependent on a "production function" that transforms all measurable inputs ("factors of production") into measurable outputs. Ordinarily the factor inputs are thought of as exogenous or "independent" economic endowments, even though some "general equilibrium" models show that they too can be determined over time (and thus "dependent") if we know enough about all of the production functions and consumer preferences. Numerous assumptions about the competitive nature of all markets, moreover, lead to the conclusion that the outcome automatically guarantees the productive efficiency of the system as a whole. Consumer welfare, in other words, is "optimized" in the model, meaning that no public policy can affect a different result without making someone worse off.

Our purpose, of course, is not to investigate all of the implications of the classical model. Instead, we want to focus specifically on its treatment of production. How does the classical model identify and categorize the so-called factors of production? Basically, by whichever means are convenient. Labor is nearly always considered to be one factor, partly because it seems easy to quantify (just count heads) and partly because the returns to labor determined by the production function can be roughly equated with wages. A second convenient and often-used factor is natural resources—a given and limited set of inputs into the production process that cannot, by definition, be duplicable by man (for example, one can think simplistically of a fixed quantity of land or minerals). A third popular factor is capital. This can be regarded as a catch-all category—all inputs other than labor that are artificially duplicable without limit, or at least without practical limit and without inherently rising long-term costs.

Given a definition of factors, the classical production function tells us what quantifiable output result we can obtain in our economy by different combinations of quantifiable input factors. How do economists know what the production function is? A priori, of course, they cannot know. The best they can do is to measure the output from different combinations of factor inputs where they can actually observe the process and to make reasonable hypotheses about the combinations they cannot observe directly. With skill and mathematics, they can then formulate a simple, smooth, and idealized function that roughly describes how well, quantitatively, our economy transforms all different combinations of factor inputs into final outputs. It is usually assumed that this function reflects our economy's level of "technical" of "technological" efficiency. But any economist will admit that this is just a convenient manner of speaking. In fact, the production function reflects every characteristic about the individual or collective productive process that cannot be isolated and quantified in terms of factor inputs—everything from human skills, managerial efficiency, and nifty ideas to prevailing attitudes toward work, cooperation, and trust.

Limitations of the Classical Model

Productivity Measurement Problems.   The first point to be emphasized about the classical model, therefore, is that changes in efficiency due to changes in the supply of identifiable factor inputs are the only developments that economists can in some sense "predict" or "explain" via the classical model. All other changes will appear as surprise alterations in the underlying production function. One way economists can minimize surprise is to try to define and include in their function as many factor inputs as they think feasible. Accordingly, investigations into what has caused change in U.S. productivity growth rates often include many ingenious attempts to quantify education, experience, age, sex, and so forth, as measurable types of additional factor inputs. To be more precise, sometimes they are likened to uncontrollable natural resource endowments (such as, the age composition of the labor force), and at other times they are likened to a form of capital investment (such as, labor skills as human capital).

We now arrive at the most formidable problem that appears to confront the classical approach to explaining productivity. Most economists who follow it quickly reach the conclusion that—no matter how widely they define their factor inputs—the most important causes of productivity change cannot be explained, and in fact are usually underestimated, by changes in factor supply. They are well aware, on the basis of historical and empirical evidence, that most actual upward shifts in productivity seem to be the result of improvements in some sort of technical know-how. But it is obviously difficult to quantify "technical know-how" as a factor input—which "explains" why they can't explain it.

Economists, to be sure, are a sturdy lot and don't give up easily. What is most interesting about their attempts to solve or finesse the technical know-how problem, however, is not that they have been notably successful, but rather that they reflect widespread disagreement over just what the x-factor is. Some economists, for instance, tend to minimize technical innovation as an independent source of productivity advantage. The key variable, they feel, is the speed with which such innovation can be embodied in new physical capital, and they try to capture this effect by enhancing the value of "new" physical capital as opposed to "old" physical capital. Other economists, taking a somewhat broader view, try to redefine technological innovation as itself an output dependent on research input. Here we find attempts to quantify R&D and education as stocks of human capital that have productive returns just like physical capital. Finally, there are economists who think all such attempts are doomed to failure. The x-factor, they conclude, is the result not just of formal R&D but of all sorts of informal research, managerial changes, and social and attitudinal transformations that we cannot hope to quantify directly.

Market Failure. The second point to be emphasized about the classical model is the assumption that, ordinarily, the value of all productive output can be contractually "captured" by the agent or agents (that is, the firm, worker, partnership, joint-venture, and so forth) involved in the production. Technically, it is said that all gains are "internalized" within the contract. The importance of this assumption cannot be overemphasized, for on it stands or falls the efficiency and rationality of the entire model. With it, we can conclude that labor and capital will all move to those employments that maximize their social utility, the reason being that a higher or lower return realized ("owned in contract") by the worker or capitalist will then correspond to a greater or lesser advantage to the economy as a whole. When a firm, for instance, builds a new coal plant in Virginia rather than a new oil refinery in California, we may conclude on the basis of this assumption (together, of course, with other competition assumptions) that greater rewards will accrue not just to the firm but also to all workers and all other firms as a group.

Not even the nineteenth-century classical economists, of course, believed that all "economies" of production are necessarily "internalized." Ever since Adam Smith, they have left us with a long and growing list of exceptions, special cases where certain activities generate "external" returns to society. The classic example is the lighthouse, which today has come to symbolize public infrastructure in general. Without state subsidies for infrastructure, it is assumed that private actors, left to themselves, will underinvest in it because they cannot "own" all of the capital returns on such investment. The lighthouse cannot charge every ship for precisely the value of the help it renders to them in safety, nor can the road charge every user for precisely the value it renders to them in transportation. This so-called public good argument has long been applied not just to physical investment but also to various forms of human investment. The allocation of public resources toward education, research, labor-skill training, public health, welfare benefits, and much more has often been defended on the basis of alleged "positive externalities" (just as environmental regulation has been defended on the basis of alleged "negative externalities").

For our purposes, what is important about this aspect of the classical model is that it tells us very little about the presence or absence of externalities if we don't really understand the productive process in question. Thus we return to the mysterious x-factor input mentioned earlier. In the case of the lighthouse, for example, we know what is being provided to ships even if it is impractical for ships to be individually charged. Or at least we could make a good guess—say, by estimating the expected value of shipwreck costs, or the additional insurance costs, that shippers might incur in the absence of a lighthouse. But what about the economic value of inventing the transistor? Or the economic value of three CEOs being won over by a first reading of *Zen and Art of Management?* A classical approach would be to assume that the returns on such products can be fully appropriated by the entrepreneurs in question, perhaps with the help of a sympathetic legal system. The inventor, for example, can take out a patent and sell an idea for royalties; and the writer can differentiate his product and hold expensive seminars for business executives. But in this case there is no way to prove the classical assumption. A critic could equally well claim that the value of these products has external benefits that are entirely disproportionate to the return that is privately earned. And since there is no consensus on how much such products in turn "produce" as economic inputs, there can be no definitive winner in the debate.

It is easy to see why this issue underlines much of the controversy over "industrial policy." Those who are impressed by how *well* the classical model explains production will tend to restrict the x-factor to a minimal "technical" role and resist the call to direct industrial policies (that is, public efforts to improve on market outcomes in economic terms) toward activities where the results are not quantifiable. But those who are impressed by how poorly the classical model explains production may, on the contrary, embrace a very broad definition of the x-factor and call it "structural change." They may also advocate industrial policies, even where the results are unquantifiable, with the argument that otherwise we will have to resign ourselves to performing far beneath our economic potential.

It might be imagined that defenders of the former argument lean politically to laissez-faire and that defenders of the latter argument lean politically to socialism or state planning. But that would be a distorting simplification. Those in the former school may indeed advocate state remedies to externalities when and where they can be quantified according to the classical model. Some indeed may advocate so many such remedies (such as the British Fabians or today's so-called neoliberal economists) that some may consider them socialists. It is worth noting that all of the major efficiency- and innovation-stimulating public policies now practiced in the United States (such as patent law, R&D tax credits, ESOPS, federal research institutes) have been extensively defended on quantifiable "externality" grounds. The most that can be said about the former school is that it views state policy in the essentially liberal role of an impartial referee unless special circumstances pertain.

Similarly, those in the latter school may oppose activist state policy even though they are entirely unpersuaded by the classical model. Just because they believe that unpredictable innovation or structural change is the mainstay of economic progress, after all, does not necessarily mean that they doubt that the returns on such elusive "inputs" are not largely internalized by their creators. And even if they do doubt it, they still may doubt that state intervention will be able to improve on the market-determining outcome. Here we might cite Joseph Schumpeter, the Austrian economist, or even some of today's so-called supply-side theorists. The distinguishing feature of the latter school is that all of its adherents tend to reject the view that economic progress is somehow predictable or rational. They view the structure of government and society, moreover, not as a "nightwatchman" or as an incidental supplier of "labor" and "capital" factors but rather as a set of forces organically connected to every aspect of economic activity. When these two schools do advocate state intervention, perhaps a further distinguishing feature is worth mentioning: the former school prefers supporting generic, economywide inputs (such as R&D or education), while the latter school prefers supporting institutionally defined sectors (such as steel or telecommunications).[3]

Returns to Scale and Imperfect Competition. The third and final observation to be made about the classical model is that it ordinarily assumes both "replicable output" and "decreasing return to scale" for a competitive number of firms in every productive activity or sector. Again, these are assumptions that most classical theorists admit do not hold universally. There always seem to be outputs, such as the service value of real estate owned in downtown Manhattan, that are not easily duplicable. And there always seem to be certain activities, such electricity generation or local phone service, where the returns to scale keep declining at a sufficiently large volume of output that a competitive market equilibrium finds one or a few large producers dominating the activity. The former case can be called a rent or scarcity monopoly and the latter case has been called a scale monopoly. In either situation, classical theory agrees that we have to watch out for the possibility of market-power inefficiency. Specifically, if the demand for the output is small and relatively inelastic, a single monopoly or a noncompetitive oligopoly may find it easy to earn above-market rates of return by suboptimal restrictions on output. Classical theory agrees that such an outcome is inefficient and that special policy intervention may be necessary when it occurs.

Both types of monopoly are frequently at issue in discussions of technology, structural change, and industrial policy. Firms that profit from technological innovation, for instance, are often likened to rent monopolists who, if they are able to prevent others from benefiting from the innovation, can enjoy abnormally high rates of return. But again, whether or not economists consider this a true case of monopoly is likely to depend on their understanding of the innovative process. If they see it, in classical terms, as a return on a factor investment (such as on the cost of R&D, on the time of trial and error, or on various forms of human capital such as experience, skill superior organization, and so forth), they are likely to disagree that such returns constitute monopolistic rents. If, on the other hand, they see it as a purely serendipitous event or the result of "externalities" elsewhere in the economy (such as a return on skills developed at the expense of other firms that failed), they may well view it as a true monopoly. The classical view warns us never to mistake ex-post rents with successes that never could be anticipated in advance; the structural view rejoins that such "probablistic" calculus is often self-serving and leads to an infinite logical regression.

Much of the debate over the "innovation rents" on so-called product-cycle goods centers on these distinctions. And it leads to policy recommendations that are as murky as the distinctions. If we agree that there are positive externalities associated with innovation-oriented investments, does that also mean that there must be negative externalities associated with innovation rents? If we subsidize the former (say, skill and research inputs), must we also tax the latter (the successful innovative output)? Is it worthwhile trying to strike this sort of balance? Or is it better just to make a small across-the-board subsidy and allow everything else to cancel itself out?

There is also the possibility of "scale monopolies," which become an issue here not when the economies of scale can be internalized within a single firm but rather when the scale economies are themselves "external." Alfred Marshall was the first economist to identify external technological scale economies, which refer to instances when, for technological reasons, costs decline for all firms in an industry when their total output is grown even though they do not decline for a single firm when its own output is grown. Marshall seemed to have in mind "lumpy" efficiencies that cannot be achieved incrementally by a single firm but can be achieved by a "critical mass" increase in aggregate industry size. His examples include the advantages of a region filled with skilled labor or a single industry publication or information service.[4]

Marshall's concept of "external scale economies" appears to be quite similar to what today's "structural change" theorists have in mind when they imply that impressive efficiencies of scale can result from infrastructure, skilled communities, and shared knowledge. It also appears to be similar to the "learning curve" economies of scale so often described by Japanese trade experts. To the extent these efficiencies can be created by public policy, they would of course be economically beneficial. We should be careful, moreover, to distinguish this argument from another often-made claim that certain innovative projects should get public support because uninsurable risk, investor myopia, consumer reluctance, or sheer size prevents private agents from acting on their appropriable future return. The former argument points to true, industry-specific externalities. The latter argument points to market-failure in other institutions.[5]

TECHNOLOGY AND STRUCTURAL CHANGE: THE TRADE POLICY ISSUES

Classical trade theory is best known for its strong conclusions that free trade between nations maximizes world welfare just as free trade within a nation maximizes national welfare. It is also well known for its famous Ricardian Law of Comparative Advantage, which asserts that every nation benefits from trade, no matter what the absolute advantage or disadvantage of its productive factors, as long as the domestic relative prices of its goods and services differ, in the absence of trade, from world prices. (If the prices are the same, then the nation does not trade and thus gains no advantage.)

In the early twentieth century, the appearance of the so-called Heckscher-Ohlin Factor Price Equalization Theorem seemed to point to a yet more wondrous result of free trade. Even assuming complete immobility of factors between nations, the Heckscher-Ohlin Theorem declares on the basis of classical free-competition assumptions that factor prices in all nations must tend toward equality. This implies, for instance, that the wages of labor and the profits on capital in Burma should tend toward equality with wages and profits in Switzerland. The logic of Heckscher-Ohlin stems from a demonstrated property of most well-behaved production functions: that a single set of economywide *product* prices can only be consistent with a singly set of economywide *factor* prices. Because the worldwide arbitrage prompted by free trade automatically equalizes all product prices, so must it also equalize all factor prices. The production adjustment by which this equalization takes place is national specialization on products demanding relatively more of the same factor (say, labor-intensive versus capital-intensive goods) that is relatively more plentiful in that country.

The patent unreality of Heckscher-Ohlin's conclusions has led in recent decades to a long and often highly technical academic debate on how its classical assumptions may not apply to the real world. Although this is no place to even attempt a summary of this debate, it is worth pointing out how it fits into our prior discussion. The nonclassical response is simply to say that production functions differ between nations, perhaps due to technological or structural differences between societies. This violates the classical assumption and thus ends all discussion. The "classical" response is to say that we can always assume an identical production function in all nations if we simply recognize that we don't know what all the "input" factors are. This assumption indeed saves the theorem but makes it of questionable usefulness. It is one thing to observe, for instance, that U.S. workers earn twenty times as much as Indonesian workers because, in addition to the return on "pure" labor, U.S. workers also get a return on various types of human capital inputs such as skill, organization, education, and so forth. It is, however, another thing to say that this is a useful observation when the technological or structural inputs generating these returns are both unknowable and immeasurable.

It is popularly believed that classical trade theory denies that it is possible for one nation to gain material advantage over others by means of trade policy. This popular notion is false. At least as early as the late nineteenth century, neoclassical economists formally demonstrated, under the rubric of optimal tariff theory, what even the earliest classical economists knew intuitively: that by means of an import tax (or an export tax) a nation can usually improve its terms of trade at the expense of its trading partners. The extent of maximum possible advantage is determined by the relative price elasticity of the trading partners' offer curve. A country facing a high-elasticity offer curve (usually a small country) can gain little advantage; a country facing a low-elasticity offer curve (usually a large country) can gain a great deal of advantage. Optimal tariffs work by exercising one country's market power advantage over another (just as a monopoly works within a country). Because large nations generally possess more market power, they have the most to win. It is perhaps remarkable, given the lack of official enthusiasm expressed by Americans toward trade policy, that classical theory suggests the United States could gain more by it than any other country.

It is correct, however, to observe that the classical tradition has consistently advised against using tariffs—optimal or otherwise—since they diminish world economic welfare and invite retaliation in kind by other countries. Further tariffs may win back some advantage, but further retaliation always takes it away again. Eventually, tariff wars result in putting an end to all trade, and leave all countries much worse off than they were to begin with. In theory, large economies ordinarily gain relatively less from trade than small economies—which might strengthen their rationale for risking a tariff war.

In practice, however, large and wealthy economies historically tend to place more importance on absolute levels of productive efficiency than do smaller and poorer economies. One reason may be strategically military. Large and wealthy economies tend to be

defense leaders who benefit the most from stronger trading partners to the extent they are also allies (a rationale that might well
apply to U.S. trade policy during the Cold War). Another reason
may be strategically political. Large and wealthy economies tend
to benefit most from the political stability and convergent values
(political, cultural, and so forth) that are said to accompany rising
standards of living and trade ties (a rationale often expressed by the
British during the nineteenth century).

There is, of course, another possibility that has often been raised
by critics of the ideology of free trade: that leading economies
promote free trade as a principle in order to preserve their world
economic hegemony. Free trade, say critics, helps the leader by
destroying any foreign competition in the leading economy's export industries and by preserving for itself the special productivity
and terms-of-trade advantages of such industries. This line of reasoning leads, in turn, to the famous infant industry defense of
selective trade protection. Infant industries are those for which
comparative advantage cannot appear but can be created with the
help of trade protection and state encouragement. What sorts of
industries are these? As we have already seen, they may be characterized as industries that have external economies of scale or structural innovation rents. In the context of world trade, the argument
in favor of their existence becomes stronger since the market for
their product is (especially for relatively small economies; one
thinks of Japan in the 1950s) much larger. Again, the argument
does not rest on the inability of new firms to obtain investment on
the basis of risky future returns that it will then be fully able to
appropriate (that would be an infant firm argument).

The infant industry argument is usually associated with anti-
classical, or at least nonclassical, economists who have generally
employed organic, structural, and partisan language to characterize the relationship between the economy and the state. We think
of ardent republicans (Hamilton), romantic nationalists (List), cor-
poratist reactionaries (Schmoller), and militant socialists (Lenin).
But it is not often appreciated that most classical economists
conceded the issue. John Stuart Mill, to take a notable example, is
worth quoting at length:

> The only case in which, on mere principles of political economy,
> protecting duties can be defensible, is when they are temporarily

imposed (especially in a young and rising nation) in hopes of naturalizing a foreign industry, in itself perfectly suitable to the circumstances of the country. The superiority of one country over another often arises only from having begun it sooner. There may be no inherent advantage on one part, or disadvantage on the other, but only a present superiority of acquired skill and experience. A country which has this skill and experience yet to acquire, may in other respects be better adapted to the production expected that individuals should, at their own risk, or rather to their certain loss, introduce a new manufacture, and bear the burden of carrying it on, until the producers have been educated up to the level of those with whom the processes are traditional. A protecting duty, continued for a reasonable time, will sometimes be the least inconvenient mode in which the nation can tax itself for the support of such an experiment.

In Mill's suggestion that, absent protection, "individuals" would have to "bear the burden" until "producers have been educated," it is not hard to infer Marshall's later notion of "external economies of scale." At the end of this passage, however, Mill brings up another important question: Is it indeed true that "a protecting duty" is "the least inconvenient mode in which the nation can tax itself" for this purpose? Most later economists have disagreed with Mill. Classical theory unambiguously supports subsidies, not import duties, as the most efficient means of protection. The reason is that a subsidy, by revealing to domestic consumers the true opportunity cost of their purchases, enables the nation to attain a higher level of economic welfare with the same degree of protection. As a practical matter, on the other hand, subsidies require the expensive, troublesome, politically vulnerable mechanism of public taxation and public spending. Even if they had been aware of the efficiency cost, most infant industry proponents would probably still have favored tariffs.

We have already mentioned that public support for industries with external economies of scale may have special advantages in the context of world trade, namely, further efficiencies of scale. But what about the possible problem of strategic firms with innovation rents? To the extent that the market for such firms is foreign rather than domestic, it seems clear that such a problem is instantly transformed into a benefit. The monopolistic or oligopolistic market power possessed by such firms, which would create diseconomies domestically, instead creates a favorable terms-of-trade advantage internationally. A relatively new field, strategic trade theory, is now investigating how these market-power trade advantages can be exploited in the presence of foreign policies that have the same goal.[7]

One might be skeptical of the teachings of this new field, however, on two counts. First, as we have already seen, it is still an open question whether firms with innovation rents are true monopolies at all, or whether, from an ex ante perspective, their returns are no better than average. Even if dumb luck might be considered a legitimate basis for a monopoly, most strategic trade thinking works from a contrary premise that such situations can be planned for in advance. The key theoretical issue, in short, is whether it is possible to predict any *future* monopoly that does not presently exist.[8] Second, even if it is possible, the fact that such strategies already anticipate foreign retaliation raises the question why policymakers do not simply use their tariff powers to exploit international market power even in industries that are now competitive. (This could be done by means of an import or export tax, which is just the opposite of the export subsidy usually proposed for creating monopolistic power by strategic trade theorists—an odd result suggesting that trade policy will backfire if the competitive nature of an industry is misjudged.)

Throughout the discussion of trade policy, we have assumed that all goods and services flowing between borders are pure or contractually internalized. The possibility that externalities may also flow between borders, however, raises some additional problems worth mentioning. If the spillover from R&D or a strategic industry is significant only within a nation, the nation may gain by publicly supporting it.[9] If, on the other hand, the spillover goes just as easily to other nations, such support may no longer make sense. If the public good is global, a global tax must be created to support it; a national tax simply becomes a subsidy to other nations. The practical significance of these considerations has often been raised in international comparisons of the institutional openness of publicly funded know-how. If it is true, as is sometimes said, that the United States is open while Japan is closed, one wonders to what extent the difference is the result of policy (as opposed to culture) and whether closure may actually be far more cost-effective for the United States than the generation of more public knowledge. Very little economic analysis has been done on this subject.[10]

POLICY RECOMMENDATIONS: INDUSTRIAL POLICY REVISITED

The decline in the U.S. position in the world combined with the quickening pace of change has spawned numerous policy recommendations from various analysts. The significant recent analyses are critically reviewed in the appendix. But there is a strong element of *déjà vu* in all of these discussions. They are all remarkably similar to the discussions of industrial policy of a few years ago.

The concept is ill defined, but we have an industrial policy. It is unarticulated and *ad hoc*, but what is the total effect of our myriad of policies for antitrust, government procurement for defense, agricultural subsidies, and so forth, on the structure of industry and our ability to compete internationally? What about the nonuniform effects of our uniform tax policies?

I believe in letting the market allocate resources as much as possible, but if the market does not work well—in providing infrastructure, investments, and education, or in providing for our national defense—then we should not use it.

The phrase *industrial policy* seems to conjure up visions of the government picking winners and losers, and this offends our market sensibilities. But we should not forget that the government does have a role to play in setting and administering our laws— that is, the rules by which the game is played.

What will the objectives of the government be in setting and administering those rules? Three broad possible orientations come to mind: (1) Will government policy be defensive or protective, oriented toward preserving the status quo? (2) Will policy be adaptive in allowing the market to work and in facilitating adaptation and flexibility in response to technological change and import competition? (3) Will policy be initiative as in Japan where industries are targeted as the next sources of industrial growth and success?

I prefer the second, by which I mean adopting policies that enhance flexibility to cope with the manifold changes in economic environment that we face today and will increasingly face in the future. But that does not mean the government has no role to play in determining our revealed industrial policy.

Proposals that fall under the rubric of industrial policy do lie along a spectrum; one which stretches from process improvements, to better coordination of existing policies, to policies to correct for market failures, to, finally, proposals for direct government intervention to allocate resources to assist private industry— what some partisan observers have labeled as planning.

Improvements in labor force quality and opportunity include better information on labor market shortages and surpluses; job bank; new incentives like merit pay for teachers to meet national goals.

Improvements in existing policies include tax reform; adjustment assistance for workers permanently displaced from their old jobs; deregulation.

Better coordinated existing policies include, at a macroeconomic level, improved coordination of fiscal and monetary policies; at a microeconomic level, improved coordination between different tax, regulatory, and trade policies.

Policies to correct market imperfections include R&D support; education support; waiver of antitrust rules for joint R&D projects; infrastructure improvements (bridges, highways, and so forth).

More explicit government interventions include tripartite (labor, management, and government representatives) national industrial boards, which can be useful in problem industries where excess capacity exists internationally (for example, steel) or that are undergoing massive technological or demand changes (for example, autos). But such deliberations or negotiations must have explicit objectives (international competitiveness), should only recommend changes in policy that deal with the source of the problems, and entail offers of mutual sacrifice in exchange for changes in government policies.

Investment banks (a resurrected Reconstruction Finance Corporation, RFC) would explicitly allocate resources to industries. Given that resources are scarce, on what basis would the industries be selected? Who would select them and who would those people represent? Can such an organization be kept free of partisan political influence? Why isn't the market providing sufficient resources? If it is because of a market failure (such as infrastructure) or to achieve a noneconomic objective (such as national security), then there is some possible scope for an investment bank. But in general, the provision of an RFC raises more questions than it is intended to answer. We would be better off using existing statutes, some of which could be broadened—like our antitrust and trade relief statutes—to deal with exceptional industry problems.

So if what we mean by industrial policy is to improve our educational system, to reform our tax code, to increase R&D and innovation, to improve our policy coordination, and to rebuild our crumbling infrastructure, then I am for it. If it is only to be defensive and protective, resulting in investment distortions and impediments to change, then I am against it.[11]

Reasonable people will disagree on where to draw the line along this spectrum, but we should not forget that we live in a dynamic, interdependent world economy. If we try to resist change, our standard of living and future growth will suffer. Blocking change is equivalent to blocking growth. We need to enhance the flexibility of our resources—both human and physical—and invest in those resources in order to ensure a better world for our children and our children's children.

NOTES

1. See summary discussion at the beginning of Ralph Landau, "Technology, Economics, and Public Policy," Ralph Landau and Dale Jorgenson, eds., *Technology and Economic Policy* (Cambridge, Mass.: Ballinger, 1986); or the review of findings in Edward N. Wolff's "Comment" in John W. Kendrick, ed., *International Comparisons of Productivity and Causes of the Slowdown* (Washington, D.C.: American Enterprise Institute, 1984).

2. For an excellent analysis of trade-related U.S. employment changes—by sector, industry, region, and profession—see Charles F. Stone and Isabel V. Sawhill, "Labor Market Implications of the Growing Internationalization of the U.S. Economy," Research Report of National Commission for Employment Policy (Washington, D.C., June 1986).

3. In a recent book Richard R. Nelson, *High-Technology Policies: A Five*

Nation Comparison (Washington, D.C.: American Enterprise Institute, 1984), makes a distinction between "leading" and "strategic" industries. Leading industries are those that generate technological advances that lead to productivity leaps, not only for themselves but also via the spillover effect to other parallel and downstream industries. Nelson agrees with the Schumpeterian conclusion of many economic historians that leading industries seem to play a major historical role in economywide productivity growth. A strategic industry, on the other hand, is one in which the specific economywide productivity returns to technology investment are not fully captured by the firms making the investment. It makes sense, then, for public policy to remedy this "market externality" by encouraging (such as, through subsidies) technological investment in strategic industries beyond the amount justified by market returns. The categories of leading and strategic, therefore, are not identical though they overlap.

4. Nelson's general conclusion in that a strong underlying investment in scientific and research human capital is a necessary condition for successful public R&D policy (the United States), but it is not a sufficient condition (the United Kingdom). As for lessons from specific industries, he observes that the U.S. and Japanese successes in electronics had more in common than is generally thought. Both involved a broad, sectoral encouragement that allowed individual breakthroughs to take their own path (even though commercial application was a secondary consideration in the United States) and both depended on a large and captive domestic market for initial commercial development (even though the captive U.S. market merely reflected its leadership role). The more strictly targeted measures of France and the United Kingdom, which were not aided by captive markets, did not turn out well. Similarly, the single successes of European nations in aviation and nuclear power occurred when the government took an entrepreneurial rather than a directive role (in the case of Airbus) and followed consumer demand rather than engineering targets (in the case of West German and Japanese successes with nuclear-power development). Despite the U.S. success in electronics, "one clear lesson of the post–World War II experience is that trying to blend commercial and military procurement objectives is a mistake. If a program is aimed specifically at enhancing competitive strength, it should stand separate from procurement-oriented programs."

5. Nelson also observes that faster international transmission of ideas and techniques makes it increasingly less likely that leading industries will also be strategic industries. The enduring margin in Japa-

nese and West German productivity growth over the United States, he observes, is apparently unhindered by the fact that leading exports as a share of all exports is much smaller in Japan and West Germany than in the United States. "Increasingly, technological knowledge and capability are international rather than national . . . The argument that leading industries are strategic nationally because they feed into national industries downstream is, to a considerable extent, vitiated by the growing strength and breadth of the international networks and the export orientation of the strongest firms in these industries."

6. John Stuart Mill, Chapter X, "Of Interferences of Government Based on Erroneous Theories," *Principles of Political Economy* (Toronto: University of Toronto Press, 1965).

7. For an introduction to this new field, see Paul R. Krugman, ed., *Strategic Trade Policy and the New International Economics* (Cambridge, Mass.: MIT Press, 1986).

8. Stephen Cohen and John Zysman in their recent book, *Manufacturing Matters: The Myth of the Post-Industrial Economy* (New York: Basic Books, 1987), set out six hypotheses:

> First, technological developments can provoke rapid market shifts. Second, technologies are shaped by the needs and arrangements that exist in the nations from which they emerge. Third, some critical technologies can affect the competitive position of a whole range of industries; and if one nation uses these technologies to gain a lead in a vital product, it can forge an important trade advantage for itself. These are strategic transformative technologies, characterized by imperfect competition and with powerful interindustry spillovers. Fourth, continued technological development depends heavily on the connections between producing firms, their suppliers, and their customers. A web of structural and operating arrangements supports technological development, and that web can unravel. Fifth, this reshuffling of market position in a period in which important new strategic transformative sectors are emerging is powerfully influenced by government policy. Sixth, the reshuffling can result in new international hierarchies of wealth but also of power.

9. Cohen and Zysman, note 8 above, review a miscellaneous assortment of policy recommendations—such as cutting the deficit, raising capital formation, devaluing the dollar—and conclude that they are insufficient to meet America's coming challenge. They prefer, instead, a series of recommendations made on the basis of their underlying analysis that (in their words) "investment rests in peo-

ple, in human capital, and not just narrowly in engineers or scientists—though certainly that—but broadly in the community as a whole." Accordingly, they suggest more publicly mandated investment in education, civilian R&D, in standard infrastructure, in the "new" infrastructure of telecommunication. They believe that targeted industrial strategies, including strategic trade policy, can work in specific situations. Altogether, their book is provocative, but does not give the reader a clear idea of how to choose those specific industries.

10.   See, however, the insightful observations made in Gary R. Saxonhouse, "Why Japan Is Winning," in *Issues in Science and Technology* (Spring 1986): 50–62.

11.   In a recent survey Sven W. Arndt and Lawrence Bouton, "Competitiveness: The United States in World Trade," Working Paper No. 8 (Washington, D.C.: American Enterprise Institute, 1987) ask whether state policy can "create" comparative advantage. Following classical logic, the authors look for market externalities and find it in generic productive inputs such as R&D, job retraining, and education. They take a dim view of sectoral targeting, however, and they point out the many instances where the fabled MITI of Japan has made bad mistakes either in choice of sectors or in timing: such as steel, shipbuilding, cars, textiles, and aircraft.

11 THE EFFECT OF TAX POLICY ON THE CREATION OF NEW TECHNICAL KNOWLEDGE
An Assessment of the Evidence

Joseph J. Cordes

In the late 1970s considerable attention was paid to an apparent slowdown both in private spending for industrial R&D and in the rate of industrial innovation by U.S. firms. As a consequence questions were raised about whether U.S. businesses faced adequate incentives to invest in creating the new technical knowledge that is indispensable to industrial innovation.[1]

The U.S. tax system was identified as one of several institutions with a potentially significant effect on incentives for innovation. On the one hand, it was claimed that some features of the tax system—inadequate incentives for investment in general and for investments in innovation in particular—contributed to the observed slowdown. On the other, it was argued that tax policy could play a useful role in reversing the observed downward trend by providing more incentives for investment both in physical capital and in the creation of new knowledge.

Tax Policy and the Creation of New Technical Knowledge in the 1980s

Since that time, there have been several significant changes both in the trend of R&D spending by U.S. industry and in U.S. tax

443

policy. First, as is seen in Figures 11–1 and 11–2, after declining through 1978, the share of resources committed to industrial R&D began to increase again in 1979 and continued to grow through 1984.

Second, shortly after growth in R&D spending resumed, major changes were also made in the corporate income tax through pas-

Figure 11–1. R&D Sales Ratio In Manufacturing.

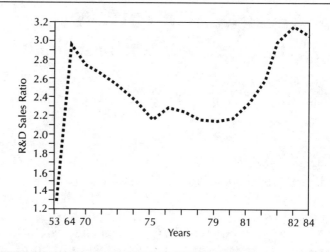

Figure 11–2. Ratio of R&D to GNP.

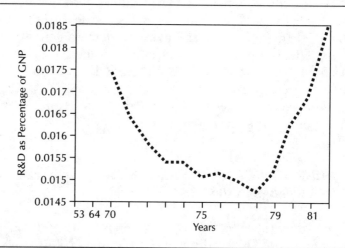

sage of the Economic Recovery Tax Act of 1981 (ERTA), which was subsequently modified by the Tax Equity and Fiscal Responsibility Act of 1982 (TEFRA). These were noteworthy changes because they affected features of the tax code that were believed to significantly affect incentives for industrial innovation.

For example, prior to 1980, the option to expense rather than to capitalize R&D spending was the principal tax incentive extended to R&D. However, as part of ERTA, firms were also allowed to claim a tax credit for R&D spending in excess of some base amount. In addition, capital equipment used in R&D became subject to different depreciation rules than those applied to equipment used in other production activities. Lastly, firms were generally allowed to claim much more liberal depreciation allowances under the Accelerated Cost Recovery System (ACRS) than they had previously been allowed under the Asset Depreciation Range (ADR) system.

The adoption of these changes in the tax treatment of U.S. businesses in the early 1980s raises a number of issues. Because the turnaround in R&D spending occurred *before* the enactment of ERTA, and because the passage of ERTA was generally not expected to take place until after the election of President Reagan in 1980, it is unlikely that the 1979 turnaround in R&D spending can be attributed to subsequent changes in tax policy. Based on this observation one might conclude that spending for industrial R&D has a life of its own that is little affected by tax policy.

However, as Baily, Lawrence, and DRI (1985) have observed, growth in R&D spending remained strong in the midst of one of the most severe recessions in recent history. It is therefore possible that, though changes in tax policy were not responsible for the initial turnaround in R&D spending, they contributed to continued strong growth in R&D spending in the early 1980s.

Objectives of the Chapter

This chapter has two main objectives. One is to assess the validity of these competing interpretations of the recent record. To do so it is necessary to address two questions: (1) How have recent tax changes affected incentives to invest in the creation and the use of new technical knowledge? (2) How have U.S. businesses responded to any changes in such incentives? The other main objective is to draw some policy lessons from the answers to each of these questions.

TAXES AND INNOVATION: A SIMPLE CONCEPTUAL FRAMEWORK

To provide a framework for discussing these issues it is useful to begin by highlighting the different ways in which the tax system impinges on the process by which new technical knowledge is created and used.

Taxes and the Propensity to Innovate

The decision to commit resources to the creation of new technical knowledge is first and foremost an investment decision involving the outlay of time and money in the present in order to create something of value in the future. As with any other investment, the decision to make such a commitment should be based on a comparison of the expected return on the resources invested with the cost of the investment.

In a world in which taxes are collected on income and deductions are allowed for the costs of earning such income, there are two related, yet distinct, investment criteria that might be used by a profit-maximizing firm when deciding whether to invest in creating new technical knowledge. Such a firm could decide to invest an additional $1 in R&D only if the total or gross cash flow per dollar was greater than or equal to the *user cost* of the investment—the amount needed to both return the capital invested plus pay a competitive return on that capital, taking into account the tax treatment of the investment. An alternative criterion would be to invest an additional $1 in R&D only if the pretax return net of any depreciation in the commercial value of the new knowledge equaled or exceeded some threshold or *hurdle rate* of return. This hurdle rate would be the minimum pretax return, net of depreciation, required to pay investors a competitive return on the resources invested in R&D after taking into account the value of any taxes owed on the future income produced by the knowledge created by the R&D.[2]

Changes in tax policy can therefore affect incentives to invest in R&D through their effect on the user cost or on the hurdle rate. Other things being equal, tax measures that lower either the user cost or the hurdle rate should be favorable to the creation of new technical knowledge because prospective R&D projects need to pass a less stringent investment test. This will be particularly true if such tax measures not only reduce the user cost or hurdle rate for R&D in absolute terms, but also relative to user costs and hurdle rates for investments that are alternatives to investments in developing technological know-how. Tax measures that raise the user cost or hurdle rate for investments in R&D either in absolute or relative terms will conversely be less favorable to the creation of new technical knowledge.

How favorable or unfavorable they are will depend not only on how much the user cost or hurdle rate is actually changed by changes in tax policy—the *size* of the incentive—but also on how responsive the firm or investor is to changes in the user cost or hurdle rate—the *behavioral response* to the incentive. The more responsive a firm or investor is to such changes, the greater will be the ultimate impact "per dollar" of change in tax policy.

Thus, if one is to assess empirically the effects of changes in tax policy on the propensity to undertake R&D, one needs to address two distinct questions. First, by how much is the user cost or the hurdle rate for investments in R&D affected by any particular change in tax policy? Second, given the change in the user cost or the hurdle rate, by how much are firms and investors likely to change their investment behavior?

The Role of Cash Flow.   Some researchers have also argued that spending on R&D is sensitive to the firm's cash flow. If this contention is correct, then changes in tax policy can affect the firm's propensity to spend on R&D through their effects on the firm's cash flow. Specifically, tax measures that increase (decrease) the firm's cash flow would also increase (decrease) its level of spending on R&D. However, as is discussed in Charles River Associates (1985) there is little empirical evidence to support the proposition that changes in cash flow as distinct from changes in corporate profits have a significant effect on R&D spending.

Taxes and the Propensity to Adopt New Technical Knowledge

It is also important to examine the effects of tax policy on the propensity of firms to adopt innovations. On the one hand, the total economic benefits of innovation depend on the rate at which innovations are diffused throughout the economy. On the other, the level and pattern of some innovative activities may be sensitive to changes in the demands for certain types of goods that embody technological innovations.

Two important ways in which new technical knowledge is adopted are through the purchase of patents and new capital goods. As in the case of investments in innovation, user costs and hurdle rates of return for patents and for different types of capital goods will play a role in determining how much is spent on acquiring patents from others as well as on different types of capital equipment. Tax policy changes that affect user costs and hurdle rates will therefore affect the financial payoff to acquiring patents from others as well as the overall amount and the mix of spending on capital goods.

As before, several issues are involved. The first is the effect of tax policy changes on the relevant user costs or hurdle rates. A second is the behavioral response of firms to these changes. A further issue pertaining to induced innovation is the sensitivity of innovative activity to changes in demands for capital goods embodying innovation.

Cash Flow and Capital Spending.   Finally, it has also been argued that policy measures such as investment tax credits and accelerated depreciation affect the demand for new capital goods by affecting the firm's cash flow. However, unlike the case of R&D spending, where there is little empirical support for the hypothesis that firms' behavior is affected by cash flow, there is more evidence to support the hypothesis that changes in cash flow can affect spending on capital goods.

A TYPOLOGY OF TAX MEASURES

There are three principal administrative ways in which tax policy can affect the return to investing in R&D either as defined in terms of the user cost or the hurdle rate of return. First, receipts attributable to the creation of new technical knowledge could be taxed at a statutory tax rate different from that applied to other receipts. Second, the costs of creating new technical knowledge could be treated more or less favorably than other costs. Third, the cost of creating new technical knowledge could be indirectly subsidized through tax credits.

In the United States, as in other industrialized countries, distinctions have not been made, as a general rule, between receipts attributable to the creation of new technical knowledge and other receipts.[3] However, the cost of creating new technical knowledge is treated differently than other costs of doing business, and in some circumstances, spending on R&D is eligible for tax credits.

Expensing of R&D

The conduct of industrial research is one step in the process of creating commercially valuable technical knowledge. Because such knowledge will enhance the profitability of the firm over some period of time, the research therefore creates a commercially valuable though intangible asset.

By this logic, a firm that invests in industrial research is in a position analogous to that of a firm that invests labor and materials to construct its own equipment or buildings, rather than buying these assets from other producers. As a general rule, the costs of labor and materials used to "self-construct" an asset with a useful life that extends beyond the taxable year must be *capitalized* into the value of the asset that is self-constructed and then deducted over several years once the asset is put in service.

In contrast, current expenditures on R&D may be deducted *immediately* or deferred and deducted over a period of at least sixty months, beginning with the month in which the taxpayer first realizes benefits from the R&D. Hence, the costs of creating new technical knowledge may be deducted more rapidly than the costs of comparable activities. This reduces the effective tax rate on any income produced by R&D.

The option to expense rather than to capitalize R&D costs has been a long-standing feature of the U.S. corporate tax system. It appears that the provision was originally enacted for administrative reasons, rather than to provide incentives *per se* to R&D. However, notwithstanding the original intent, the effect of the provision is to reduce the user cost as well as the hurdle rate of return to investments in the creation of new knowledge.[4]

Depreciation of Capital Used in R&D

The after-tax cost of creating new technical knowledge can also be affected by tax depreciation rules that apply to the equipment and buildings used in the R&D process. Prior to 1980, such capital was subject to the same tax rules as capital used in other activities. Beginning in 1981, under both ERTA and TEFRA, R&D equipment was treated differently than equipment used in other activities. Specifically, all R&D equipment, regardless of its useful life, was grouped in the three-year ACRS recovery class. However, the difference in actual tax treatment of R&D equipment was more apparent than real, inasmuch as the effective tax rate on R&D equipment under ERTA/TEFRA was virtually the same as that on equipment used in other activities. As a consequence, this feature of pre-1987 tax law will not receive further attention.

Tax Credits for R&D

Lastly, tax credits can reduce the cost of creating new technical knowledge by providing a rebate, through the tax system, of some portion of the expenses of R&D. Prior to 1980, R&D spending was not eligible for any tax credits. However, beginning in 1981, firms were eligible to claim a 25 percent credit for qualified research and experimental expenses incurred in any year if such expenses exceed a base period amount. This amount was generally defined to be the average of the three prior years' spending on R&D. Firms were allowed to claim such credits through December 31, 1985, at which time the credit was to expire. At the end of 1985, the Congress had failed to extend the R&D credit. However, the credit was renewed in the Tax Reform Act of 1986, at a rate of 20 percent rather than 25 percent, from January 1, 1986 through December 31, 1988.

EFFECTS OF SPECIAL TAX PROVISIONS ON USER COSTS AND HURDLE RATES

How have expensing and the 25 percent incremental R&D tax credit enacted under ERTA affected incentives to invest in the creation of new technical knowledge? Table 11-1 provides some insight about the effects of both measures on (1) the user cost of R&D—that is the gross receipts that would have to be earned by $1 invested in R&D in order to pay investors a required return of 4 percent, plus return the amount of capital invested, taking into account the tax treatment of the receipts from and costs of R&D, as well as the decline in the commercial value of technical knowledge once it is created; (2) the hurdle rate of return to R&D—the pretax return net of depreciation that must be earned to pay investors some assumed competitive rate of return; and (3) the effective tax rate on the income earned from an additional dollar of R&D. The mathematical expressions used to calculate these magnitudes are discussed in Cordes, Watson, and Hauger (1986). The basic data

Table 11-1. Effects of Expensing and the R&D Tax Credit on Incentives to Invest in R&D: 1985 Tax Law.[a]

	No Expensing, No Credit	Expensing No Credit	Expensing plus Credit
User cost	.161	.143	.136
Hurdle rate of return	.061	.042	.036
Effective marginal tax rate	.349	.043	−.118

a. The calculations assume (1) that each $1 of additional R&D is financed by shares of debt, new equity, and retained earnings equal to .337, .049, and .614, respectively, (2) that the combined federal-state corporate tax rate is .495, (3) that the tax rates on holders of debt, new equity, and "old" equity are .235, .295, and .052, respectively, (4) that the required after-tax return is .04, (5) that the inflation rate is .05, and (6) that the rate at which new technological knowledge declines in value is .1.

underlying each calculation are based on the characteristics of a "typical" R&D project conducted by a "typical" technology-oriented firm.[5]

Expensing

The first column of Table 11–1 presents estimates of what the user cost, hurdle rate, and effective tax rate *would have been* if the costs of R&D had to be capitalized and then deducted over a five-year period in a manner comparable to assets grouped in the five-year ACRS class under 1986 tax law. The second column of Table 11–1 presents estimates of these magnitudes when such costs can be expensed, but do not qualify for the incremental R&D tax credit. Thus, a comparison of columns 1 and 2 provides some indication of the effect of expensing alone, during the period from 1981 through 1985, assuming that no tax credit for R&D had been enacted in 1981. The results suggest that expensing by itself substantially reduced both the user cost and the hurdle rate of investments in R&D, as well as the effective marginal tax rate.

Incremental R&D Tax Credit

Assessing the actual effect of the incremental R&D tax credit on incentives for R&D is somewhat more complex. Some of the data suggest that the tax credit has had a small effect on incentives. It is well-known by now that the 25 percent incremental R&D tax credit did not actually reduce the *marginal cost* of qualified R&D spending by 25 percent. Because of the way in which the base period of the credit was defined, an additional $1 of R&D spending in the present allowed the firm to claim a $.25 tax credit, but at a cost of not being able to claim a comparable tax credit in the future because the additional $1 reduced the amount of future R&D eligible for the credit by $1. The net effect was to reduce the effective rate of the tax credit to something on the order of 4 to 5 percent. This estimate of the marginal effect of the credit is consistent with the calculations in Table 11–1 that show that the R&D credit reduced the user cost of R&D during the period 1981 through 1985 by 4.2 percent.

In a similar vein, if the estimated revenue cost of the 25 percent incremental R&D credit is used to gauge the size of the incentive, the amount—on the order of $1.6 billion—pales in comparison to the revenue cost of other business incentives, such as the investment tax credit. Moreover, whatever its potential incentive, the incremental R&D credit is not refundable and therefore could be claimed only if the firm had sufficient tax liability. Based on data compiled by the U.S. Treasury Department, roughly 70 percent of all R&D tax credits earned in 1981 were actually claimed against current year taxes by U.S. firms. In 1982, the comparable figure ranged between 43 and 59 percent.

However, these conclusions should be tempered by several considerations. First, as can be seen from Table 11–1, though the 25 percent incremental R&D credit had a much smaller effect on the hurdle rate than did expensing, it nevertheless reduced the hurdle rate to R&D by seven-tenths of 1 percent, or by roughly 16 percent. Given the assumptions that underlie the calculations in Table 11–1, this was sufficient to actually *subsidize* the hypothetical project depicted in Table 11–1 at a rate of 11.8 percent.

Second, one should note that relatively few firms undertake significant amounts of R&D. It is, therefore, to be expected that revenue losses from the credit will tend to be modest. However, for firms whose competitive strategy depends in a critical way on the creation of new technical knowledge, investments in R&D may be at least as large as, say, investments in tangible plant and equipment. Hence, the R&D credit may provide substantial reductions in tax liabilities for such firms. Indeed, recent research by Cordes, Watson, and Hauger (1986) indicates that the 25 percent incremental R&D credit reduced the simulated tax liabilities of nine hypothetical technology-intensive firms by amounts ranging from 1.0 to 56.0 percent. In five of the nine cases these percentage reductions were equal to or greater than the percentage decrease in tax liabilities attributable to the investment tax credit.

Third, though firms eligible to receive the credit in a given year may have insufficient tax liabilities in that year, they can carry such credits back for up to three years against past tax liabilities. Though there are no hard data on the extent to which firms are able to avail themselves of such carrybacks, estimates by Cordes and Sheffrin (1983) and Eisner, Albert, and Sullivan (1984) suggest that perhaps half of tax credits that cannot be claimed in a given tax year can be carried back. Thus, perhaps 85 to 90 percent of the tax subsidy provided by the R&D credit in 1981 and 72 to 80 percent of the subsidy provided in 1982 was extended to firms who were ultimately able to use it fully.

On the other hand, the calculations in Table 11–1 may overstate the net effect of introducing the R&D credit in 1981 because the liberal depreciation provisions of ACRS also reduced the user cost and hurdle rate for investments in depreciable assets. As noted by Barth, Cordes, and Tassey (1985), the ratio of both the user cost and the hurdle rate for R&D relative to that for other alternative investments was roughly the same in the period following passage of ERTA as it was before 1981.

Thus for firms that were able to claim and use the 25 percent incremental R&D credit, the effect of the credit on their incentives to invest in the creation of new technical knowledge may be summarized as follows.

1. By itself, the incremental R&D credit translated into a small reduction in the user cost and a larger reduction in the hurdle rate for R&D.
2. The 25 percent incremental R&D credit is likely to have had a major effect on the amount of taxes paid by high-technology firms.
3. But the adoption of ACRS also reduced the user costs and hurdle rates for all business investments, so that R&D may have been no more attractive in relative terms in the years following adoption of the incremental R&D credit than it was in prior years.

Comparative Tax Treatment of R&D

In addition to these effects, it is also interesting to note that the package of R&D tax incentives embodied in the U.S. corporate income tax prior to the Tax Reform Act of 1986 were quite generous by comparative international standards. For example, Cordes, Watson, and Hauger (1986) have shown that application of 1985 U.S. tax rules resulted in both a lower user cost and a lower hurdle rate for an R&D project than did application of either Japanese or West German tax corporate tax rules (holding the discount rate faced by the firm and the return required by savers fixed across all three countries). Some comparisons are presented in Table 11–2 for the same R&D project that underlies the calculations presented in Table 11–1. However, as may also be seen in Table 11–2, because of the relatively less generous tax treatment of depreciable capital in Japan and Germany as compared to the United States, the user cost and hurdle rate of R&D relative to the user cost and hurdle rate of depreciable capital may have been higher between 1981 and 1985 in the United States than it was in either Japan or West Germany.

BEHAVIOR OF R&D SPENDING

How have tax incentives for R&D in place from 1981 through 1985 affected the propensity of U.S. firms to invest in the creation of new technical knowledge? Evidence bearing on this question comes in several forms. First, indirect evidence is available from

Table 11–2. Tax Treatment of R&D in the United States, Japan, and West Germany (1985 Tax Law).[a]

	United States	Japan	West Germany
User cost of R&D	.136	.146	.152
Hurdle rate of return to R&D	.036	.046	.052
Effective marginal tax rate on R&D	−.118	.147	.230
User cost of equipment	.256	.328	.341
Hurdle rate of return of equipment	.049	.121	.134
Ratio of R&D user cost to equipment user cost	.533	.445	.445
Ratio of R&D hurdle rate to equipment hurdle rate	.734	.380	.388

a. Calculations are based on the model described in Cordes, Watson, and Hauger (1986). The calculations of the user costs and hurdle rates for equipment and structures assume depreciation rates of .207 and .036, respectively. The calculations of the user cost and the hurdle rate for patents assume that patents are depreciated on a straightline basis over eight years and that the commercial value of the patent declines at a rate of .1. The values of the other parameters are the same as those assumed in the calculations in Table 11–1.

econometric studies of the relationship between spending on R&D and the user cost of R&D. Second, direct evidence is available from a variety of sources on how firms have actually behaved since enactment of the incremental R&D tax credit.

Indirect Evidence

Econometric Estimates of the Price Elasticity of R&D.   The calculations in Table 11–1 suggest that expensing reduced the user cost of R&D by roughly 11.1 percent and the hurdle rate by roughly 31 percent while the incremental R&D tax credit reduced the user cost by an additional 4.2 percent and the hurdle rate by an additional 16 percent. A rough estimate of how these two incentives have affected the propensity to spend on R&D can be obtained by multiplying these percentage price changes by the price elasticity of R&D—the ratio of the percentage change in R&D to the percentage change in its price.

Though there are no econometric estimates of of the elasticity of R&D with respect to changes in the hurdle rate, such estimates have been made of the elasticity of R&D with respect to the user cost. As noted by Charles River Associates (1985), and most recently confirmed by an econometric analysis presented in Baily and Lawrence (1985), these estimated elasticities have generally ranged between −.2 and −.5. That is, a one percentage point drop (rise) in the user cost of R&D is estimated to increase (decrease) R&D spending by between two-tenths and one-half a percentage point.

Based on the user cost estimates presented in Table 11–1, the range of estimated price elasticities for R&D suggests that R&D spending would have been between 2.2 and 5.6 percent percent below its observed level if R&D expenses had to be capitalized rather than expensed. Based on 1984 levels of spending on R&D, this implies that expensing stimulated between $1.5 and $3.9 billion in additional R&D spending.

Similarly, the user cost calculations in Table 11–1 suggest that R&D spending would have been between .8 and 2.1 percent below its observed level in the absence of the 25 percent incremental R&D credit. This implies that a permanent 25 percent R&D credit would stimulate between $560 million and $1.5 billion in additional R&D spending. When compared to an estimated revenue loss of $1.6 billion, this means that as little as 35 cents and perhaps as much as 93 cents of the implicit subsidy provided by the tax credit was translated into additional R&D.

Econometric Estimates of the Effects of Corporate Cash Flow. As has been noted above, there is at best weak evidence that corporate R&D spending is significantly affected by changes in corporate cash flow. To the extent that cash flow does exert an effect on R&D spending, Mansfield (1985) has suggested that $1 of additional cash flow would increase R&D spending by no more than $.10. Based on this estimate, Mansfield concludes that the cash flow effects of the R&D tax credit would have increased R&D spending by .5 percent. This estimate is consistent with results presented in Charles River Associates (1985) that imply that the increase in R&D spending that could reasonably be attributed to cash flow effects would be roughly one-half as large as the estimated increase attributable to tax-induced changes in the price of R&D.

Direct Evidence

In addition to the rather indirect evidence presented above, more direct information about the impact of the incremental R&D tax credit is available from (1) corporate tax return data, (2) aggregate time series of R&D spending by U.S. firms, (3) data on R&D spending collected at the firm level, (4) responses of firms to survey questions about their behavior in response to the credit, and (5) studies of the impact of similar tax credits in other countries. The first two sources seem to provide some evidence that the 25 percent R&D credit stimulated a significant amount of additional R&D. In contrast, the remaining sources uniformly indicate that a relatively small amount of additional R&D was stimulated by the credit.

Evidence from Tax Returns.   Because the R&D credit is incremental, firms must report the amount by which current R&D spending exceeds the average amount of R&D spending in the previous three years. The ratio of the amount of qualified R&D to base period R&D thus provides one measure of the increase in R&D spending relative to the base.

Based on data from 1981 tax returns, summarized in Table 11–3,

Table 11–3. Distribution of R&D Tax Credits, 1981 ($ amounts in 1,000).

Industry	Total Base Quality R&D (1)	Total 1981 Quality R&D (2)	Percent Change in Quality R&D (3)	Tentative Credit (4)
Agriculture, forestry, and fisheries	$17,343	$24,604	41.87%	$1,739
Mining	$4,886	$8,218	68.19%	$685
Crude petroleum and gas	$71,859	$127,338	77.21%	$12,253
Contracting; construction	$11,423	$24,820	117.28%	$2,580
Food and tobacco	$231,890	$314,607	35.67%	$20,157
Textiles and apparel	$39,650	$65,080	64.14%	$5,248
Paper and printing	$25,474	$53,689	110.76%	$5,606
Petroleum refining	$420,743	$782,697	86.03%	$73,779
Chemical and rubber	$1,681,586	$2,210,235	31.44%	$131,254
Wood and stone products	$301,055	$408,111	35.56%	$25,643
Machinery and miscellaneous manufacture	$3,613,314	$4,920,423	36.17%	$297,701
Transportation equipment	$396,021	$523,864	32.28%	$29,516
Motor vehicles	$1,161,974	$1,416,135	21.87%	$63,110
Transportation, communications, utilities	$1,232,984	$1,848,175	49.89%	$134,756
Trade	$69,238	$125,144	80.74%	$11,353
Financial, insurance, real estate	$172,148	$292,313	69.80%	$25,470
Services	$130,733	$294,066	124.94%	$31,177
N.A.	$369	$587	59.08%	$55
Total	$9,582,690	$13,440,106	40.25%	$872,082
Agriculture, forestry, fisheries	$17,343	$24,604	41.87%	$1,739
Mining and extracting	$76,745	$135,556	76.63%	$12,938
Contracting; construction	$11,423	$24,820	117.28%	$2,580
All manufacturing	$7,871,707	$10,694,841	35.86%	$652,014
Transportation, communications, utilities	$1,232,984	$1,848,175	49.89%	$134,756
Trade and services	$372,488	$712,110	91.18%	$68,055

Source: Office of Tax Analysis.

Share of Tentative Credit (5)	Share of Base Quality R&D (6)	Credit Claimed (7)	Share of Credit Claimed (8)	Percentage of Credit Claimed (9)
0.20%	0.18%	$1,115	0.18%	64.12%
0.08%	0.05%	$288	0.05%	42.04%
1.41%	0.75%	$10,065	1.59%	82.14%
0.30%	0.12%	$1,910	0.30%	74.03%
2.31%	2.42%	$12,906	2.04%	64.03%
0.60%	0.41%	$4,678	0.74%	89.14%
0.64%	0.27%	$4,967	0.79%	88.60%
8.46%	4.39%	$71,376	11.28%	96.74%
15.05%	17.55%	$95,385	15.08%	72.67%
2.94%	3.14%	$20,392	3.22%	79.52%
34.14%	37.71%	$199,693	31.57%	67.08%
3.38%	4.13%	$15,196	2.40%	51.48%
7.24%	12.13%	$36,366	5.75%	57.62%
15.45%	12.87%	$114,907	18.17%	85.27%
1.30%	0.72%	$8,672	1.37%	76.39%
2.92%	1.80%	$16,036	2.54%	62.96%
3.58%	1.36%	$18,546	2.93%	59.49%
0.01%	.00%	$55	0.01%	100.00%
100.00%	100.00%	$632,553	100.00%	72.53%
0.20%	0.18%	$1,115	0.18%	64.12%
1.48%	0.80%	$10,353	1.64%	80.02%
0.30%	0.12%	$1,910	0.30%	74.03%
74.77%	82.15%	$460,959	72.87%	70.70%
15.45%	12.87%	$114,907	18.17%	85.27%
7.80%	3.89%	$43,309	6.85%	63.64%

firms that claimed the 25 percent incremental R&D credit increased their 1981 R&D expenses by an average of 40.3 percent over base period spending. This ratio is surprisingly large and, taken at face value, might lead one to conclude that the R&D credit led to a substantial increase in R&D spending.

However, this conclusion is not warranted for several reasons. First, only a portion of R&D spending qualifies for the incremental R&D credit. Thus the ratio of qualified R&D to base R&D overstates the growth in total R&D if nonqualified R&D spending grew at a slow rate or actually declined. This could, for example, occur if firms increased their spending on labor and supplies, which qualify for the R&D credit, but not on R&D capital, which does not qualify. Second, the observed increase may be attributable to reclassification of existing activities as R&D in order to qualify for the incremental R&D credit.

That the Treasury data reflect an increase in *R&D qualified for tax credits* rather than *total R&D* is supported by a comparison of the average growth rate reported in Table 11–3 with the 1981 increase in total R&D relative to the same three-year base period, as reported by the National Science Foundation. This latter figure is 19.9 percent, rather than 40.3 percent, suggesting that much of the increase in R&D reported for tax purposes may have been due either to increased spending for some R&D inputs, offset by decreased spending on other inputs, or by reclassification of existing activities as R&D, rather than to an increase in total R&D.

Though the evidence in Table 11–3 should be interpreted with care, it does provide information on some of the initial effects of the incremental credit. For example, three of the four industries reporting the largest increases in qualified R&D—contract construction, services, and trade—are in the nonmanufacturing sector of the economy that in the three years prior to 1981 had a ratio of R&D to sales that was well below that of firms in manufacturing. The fourth industry—paper and printing—is in the manufacturing sector, but had an R&D to sales ratio that was roughly a third of that reported by other manufacturing firms. Hence, the firms that were most responsive to the credit in 1981 were among those devoting relatively few resources to R&D in the years prior to 1981. By comparison, most manufacturing firms, who have traditionally been the most R&D-intensive, reported relatively modest increases in qualified R&D.

Moreover, among manufacturing firms, there is additional evidence that firms in high-technology industries reported increases in qualified R&D that were no larger, and in some instances smaller, than did other firms. Specifically, while high-technology firms accounted for 41 percent of R&D credits claimed in 1981, they also accounted for 45 percent of qualified R&D in the base period. In addition, in 1981 high-technology firms reported increases in qualified R&D relative to base period R&D of 34.6 percent while the comparable figure for other firms was 40.3 percent.

Based on a series of studies financed by the National Science Foundation, there is ample historical evidence that (1) almost 90 percent of all major innovations have been developed by manufacturing firms, (2) that within the manufacturing sector, high-technology, R&D-intensive industries have been responsible for a significant share of these innovations, and (3) that a smaller share of the cost of financing major innovations has been subsidized by public grants or contracts in manufacturing than in the non-manufacturing sector. It is therefore at least arguable that to be judged successful, the incremental R&D credit should have provided proportionately more benefits to manufacturing firms, especially high-technology firms.

The fact that this did not happen in the first two years of the credit's existence can be interpreted in two ways. First, it is possible that manufacturing and high-technology firms were more likely than other firms to have classified activities as R&D before the credit was enacted, thereby making it more difficult for them to "reclassify" activities after the credit was enacted. If this interpretation is correct one would expect differences between manufacturing and other firms to diminish over time as the range of activities that could be reclassified narrowed. There is some evidence that this occurred in 1982, inasmuch as Treasury data show that the share of credits earned by firms in high-technology industries increased from 41 percent in 1981 to 47 percent in 1982.

Alternatively, it may be that the price elasticity of demand for R&D is lower among firms for whom R&D is an important element of their corporate strategy—that is, manufacturing and high-technology firms—than it is for firms for whom R&D is less important—that is, nonmanufacturing firms. If this interpretation is correct, one would expect the R&D credit to continue to have a greater effect on the behavior of firms who have devoted relatively few resources to R&D in the past, than on the behavior of traditional R&D-intensive firms.

Evidence from Aggregate Time Series. As shown in Figure 11–1 above, R&D as a percentage of company sales averaged a little over 3 percent during the period 1982 to 1984 as compared to 2.3 percent in 1980. Similarly, R&D as a percentage of GNP averaged 1.75 percent during 1982 to 1984, as compared to 1.6 percent in 1980. Thus, the share of resources devoted to R&D grew rapidly in early 1980s, even though the economy experienced a severe recession in 1981–82.

As may be seen from Table 11–4, observed values for total R&D spending are greater in 1981 to 1984 than would have been predicted from projections based on a simple linear regression of current R&D spending in manufacturing on current sales and R&D spending in the previous year. These results are comparable to those obtained for 1981 and 1982 from a similar analysis performed by Edwin Mansfield, as well as to findings reported by Baily, Lawrence, and DRI (1985) of the trend and cyclical behavior of the R&D to sales ratio in several manufacturing industries. The results reported by Mansfield and Bailey, Lawrence, and DRI are reproduced in Tables 11–5 and 11–6, respectively.

Table 11–4. Comparison of Projected and Actual R&D Spending, 1975–1984.

Year	Projected R&D[a] (billion $)	Actual R&D (billion $)	Difference (percentage)
1975	$25.7	$23.5	−8.5%
1976	28.0	26.2	−6.4
1977	30.9	28.9	−4.9
1978	34.2	32.1	−6.1
1979	38.1	36.7	−3.6
1980	41.9	42.7	+1.9
1981	45.9	49.9	+8.7
1982	48.4	56.8	+17.4
1983	51.3	64.4	+25.5
1984	55.1	69.9	+26.8

a. Projection derived from the regression: $RD = 303.98 + .0065 \text{ Sales} + .778 \, RD_{-1}$.

Table 11–5. Comparison of Projected and Actual R&D Spending for Selected Industries in 1981 and 1982 (in percentage).[a]

| Industry | Average Difference between Projected and Actual Spending | |
	1981	1982
Aerospace	11%	23%
Chemicals	12	22
Drugs	10	24
Electrical	8	29
Oil	19	35
Computers	0	8
Average	10	23

Source: Mansfield (1985).

a. Average percentage increase in reported 1981 and 1982 R&D over what would have been expected on the basis of naive forecasts generated by regressing current R&D spending on sales and time.

Evidence on R&D Spending at the Firm Level. Robert Eisner, Steven Albert, and Martin Sullivan (1983, 1984) have conducted extensive analyses of corporate R&D spending patterns using three different sets of data: (1) the 1981 Treasury data discussed above, (2) Standard and Poor's Compustat tapes containing information contained on corporate financial reports, and (3) the McGraw-Hill surveys of R&D corporate R&D spending plans. Among the issues considered were (1) whether there was any observable difference in behavior between firms that could not actually claim the R&D credit and therefore derived no incentive from it and firms that could claim the credit, and (2) whether there was any evidence that reported increases in R&D spending were attributable to reclassification of existing activities or substitution of qualified for nonqualified R&D.

Evidence bearing on the first issue was obtained in two different ways. With regard to 1981 tax return data, information was available by industry on (1) the growth in 1981 reported R&D relative

Table 11–6. Differences between Actual R&D Spending and Trend-Cycle Projections for 1982–83 in Selected R&D-Intensive Industries.[a]

Industry	Average Difference in Percentage
Industrial chemicals	13.9%
Other chemicals	12.0
Petroleum	−3.6
Steel	10.0
Nonferrous metals	24.7
Fabricated metals	0.0
Machinery	2.4
Electrical equipment	19.8
Autos	−13.7
Aircraft	9.6
Scientific and engineering instruments	7.5
Surgical and other instruments	7.6
Weighted average	7.3

Source: Baily, Lawrence, and DRI (1985).

a. Projections are based on historical regressions of R&D as a percentage of sales on time, time-squared, time cubed, and industry sales divided by five-year average sales.

to base period R&D, and (2) the amount of unclaimed R&D credits—that is the difference between tentative credits earned in each industry and credits actually claimed. Other things being equal, the incentive effects of the credit would be strongest in those industries in which the ratio of unclaimed credits to industry assets was low (indicating that a substantial number of firms in the industry actually received the tax benefits of the incremental credit). Accordingly, one would expect that the growth in qualified R&D would be higher in industries with a relatively low ratio of unclaimed credits to assets.

To test this hypothesis, Eisner, Albert, and Sullivan (1983) estimated a regression equation with the growth in qualified R&D as the dependent variable, and the ratio of unclaimed credits to total industry assets as one of the independent variables. The coefficient on the unclaimed credit variable was statistically significant, indicating that industry growth in qualified R&D spending was positively related to the ability of firms to fully avail themselves of the R&D credit. However, though the differences are statistically significant, they are quantitatively small.

In the case of the Compustat data, Eisner, Albert, and Sullivan (1983) compared the difference between current R&D spending and either the previous year's R&D spending, or the average of R&D spending in two previous years. The results of these comparisons when the base is defined to be the previous year are summarized in Table 11–7. Comparable results are obtained when the base is defined as the average of the two prior years' spending.

Several patterns emerge from these comparisons. First, as may be seen by examining the second column of Table 11–7, there does not appear to be any significant difference in behavior in 1981 and 1982 as compared to previous years. For example, in 1982, firms

Table 11–7. Additional R&D as a Percentage of Previous Year's R&D Spending.

Year	RD < Base[a]	RD > Base[b]	Difference
1982	−9.3 %	18.0 %	27.3 %
	(0.8)[c]	(1.3)	(1.5)
1981	−8.6	23.4	32.0
	(0.8)	(2.2)	(2.4)
1980	−9.5	21.7	31.2
	(0.9)	(2.2)	(2.4)
1979	−9.4	23.5	32.9
	(1.0)	(2.3)	(2.5)
1978	−10.0	18.6	28.6
	(1.1)	(1.4)	(1.8)
1977	−9.9	19.5	29.4
	(1.0)	(2.8)	(2.9)
1976	−11.2	21.4	32.5
	(1.0)	(3.8)	(3.9)

Source: Eisner et al. (1983).

a. Firms with R&D spending below the average of prior three years' R&D spending and hence ineligible for the credit.

b. Firms with R&D spending above the average of prior three years' R&D spending and hence eligible for the credit.

c. Amounts in parentheses are standard errors of the percentages.

able to take advantage of the credit increased their expenditures an average of 18.0 percent above the previous year. However, these same firms increased their R&D spending by comparable or larger amounts from year to year during the period 1976 to 1980 when there was no R&D credit. Second, the difference in behavior between firms that qualified for the credit in 1981 and 1982 and firms that did not is much the same in 1982 and 1981 as it was in years in which there was no credit. These results stand in rather sharp contrast to those obtained from analyses of aggregate time trends in that they imply that the credit had little or no effect on the behavior of firms receiving the credit.

Moreover, insofar as firms were observed to change their behavior, the change may have involved substituting R&D spending that qualified for the credit for R&D spending that did not. Evidence to support this contention is obtained from the McGraw-Hill survey. In particular, a supplementary survey administered by McGraw-Hill on behalf of Professor Eisner provided important information on the breakdown between spending on qualified and nonqualified R&D. An analysis of the response to these questions revealed that in 1981 respondents increased their spending on qualified R&D by 21.3 percent while decreasing their spending on nonqualified R&D by 4.4 percent. The reported differences in 1982—8.6 percent for qualified R&D, and 2.1 percent for nonqualified R&D—are not as dramatic but are still consistent with the hypothesis that the R&D credit may have had a discernible effect on the type of R&D undertaken, if not on the total amount of R&D undertaken.

Evidence from Questionnaires. A different source of evidence is available from an extensive questionnaire administered by Mansfield (1985) to 110 companies, accounting for some 30 percent of company-financed R&D. In the questionnaire, the companies surveyed were asked to estimate how much lower their R&D spending would have been without the R&D credit. The responses given are reproduced as Table 11–8. The magnitudes reported in Table 11–8 represent considerably lower behavioral responses than would be inferred from analysis of time series data. However, unlike the results reported by Eisner, Albert, and Sullivan the findings from the questionnaire imply that the credit had a discernible effect, albeit a small one. Indeed, as is observed by Mans-

Table 11–8. Estimated Percentage Reduction in Company-Financed R&D Spending in the Absence of The R&D Tax Credit: Based on Responses to Mansfield Questionnaire.

Industry	Percentage of Reduction		
	1983	1982	1981
Chemicals and drugs	1.7%	1.5%	0.7%
Electrical and electronics	0.4	0.3	0.0
Machinery (including computers)	1.0	0.7	0.0
Instruments	5.9	3.7	0.4
Metals and steel	4.9	2.4	4.2
Oil and oil supply	0.3	0.6	0.1
Aerospace	0.0	0.0	0.0
Telecommunications	1.4	1.4	1.5
Rubber	2.2	0.6	0.6
Paper	0.7	0.4	0.0
Other	0.3	0.2	0.2
Total	1.2	1.0	0.4

Source: Mansfield (1985).

field, the results are quite consistent with those obtained by simply multiplying the estimated change in the user cost of R&D resulting from the R&D credit by the estimated price elasticity of demand for R&D.

In addition, a randomly selected subset of firms in Mansfield's sample was asked to what extent the R&D credit had induced them to redefine activities as R&D. The responses indicate that taken as a whole, R&D expenditures reported for tax purposes increased by about 4 percent per year during 1982–83 simply because existing activities were reclassified as R&D.

Evidence from Other Countries. In addition to surveying the experience of U.S. firms, Mansfield also conducted surveys of Swedish and Canadian firms because Canada currently provides special tax incentives to R&D and Sweden did so until 1984. In the Swedish case, R&D may be expensed as in the United States, and R&D capital depreciated over its normal economic life. Moreover though no R&D tax credits have been available to Swedish firms, until 1984 they were allowed to claim special additional deprecia-

TRADE, TAX, AND DIFFUSION POLICY ISSUES

tion deductions currently equal to 5 percent of the firm's R&D expenditures plus 30 percent of the increase in R&D over the previous year.

The current mix of Canadian tax incentives for R&D is the outcome of a series of measures that have been adopted beginning in 1961. During the early 1980s, in addition to being able to expense R&D and claim depreciation deductions for R&D capital expenditures, Canadian firms received a tax credit for R&D equal to 10 percent of both current and capital expenditures for research for large firms and 25 percent for small firms. Unlike the U.S. tax tax credit, the value of the Canadian tax credit was included in taxable income. In addition, between 1978 and 1984 firms could also claim a special research allowance as a deduction against income. The amount of the special allowance was equal to 50 percent of any increase in qualifying current and capital expenditures for R&D over the average of qualified R&D spending in the previous three years.

In the Swedish survey, forty companies, accounting for 80 percent of company-financed Swedish R&D, were asked to estimate by how much R&D spending would decrease if the special tax allowance were not in existence. Based on the responses to this question, Mansfield estimates that in the absence of the special tax allowance, R&D spending would have declined in 1981 by between .4 percent and 1.6 percent, or by between 23 to 98 million Swedish kroner. In that same year, the estimate revenue loss from the special allowance equaled some 176 million kroner. Thus, between .13 and .56 kroner of additional R&D spending was encouraged per 1 kroner of forgone tax revenue.

In the Canadian survey, fifty-five companies, accounting for 30 percent of company-financed Canadian R&D, were asked to estimate the effect of both the Canadian tax credit and the special research allowance. The responses indicate that R&D spending was increased 2 percent and 1 percent by the tax credit and the special research allowance, respectively. Based on these responses, Mansfield estimates that Canadian R&D tax incentives, which resulted in a revenue loss of some $130 million, stimulated between $50 million and $87 million of additional R&D spending. Thus, between $.38 and $.67 of additional R&D spending was encouraged per $1 of forgone tax revenue.

Assessing the Empirical Evidence

It seems clear from most of the evidence presented above that the R&D tax credit *has* had some effect on the behavior of U.S. firms in the early 1980s, with comparable results observed in other countries. However, questions remain about both the *nature* and the *size* of the effect.

It is not as yet clear how much of the initial response of firms to the credit represents actual increases in total R&D as opposed to increases in R&D that is qualified for the R&D credit. The Treasury tax return data indicate that there was a significant increase in the amount of R&D reported for tax purposes in 1981. However, findings reported by Eisner, Albert, and Sullivan as well as comparable growth rates in R&D, as measured by the National Science Foundation, suggest that much of this increase may reflect the substitution of qualified for nonqualified R&D spending rather than an increase in total R&D. Similarly, evidence reported by Mansfield suggests that much of the initial increase in R&D reported for tax purposes reflects a reclassification of existing activities as R&D.

In addition, different empirical approaches produce substantially different estimates of the size of the effect of the R&D credit. In particular, based on time series data one would infer that the credit has had a significant effect on R&D spending. However, indirect evidence based on estimates of the price elasticity of demand for R&D, as well as the more direct evidence reported by Eisner, Albert, and Sullivan and by Mansfield imply that the effect of the credit has been considerably more modest.

There are several ways in which these divergent estimates of the size of firms' responses can be reconciled. However, it is noteworthy that each suggested reconciliation points toward using the more modest estimates to gauge the effects of a permanently enacted incentive for R&D.

First, some of the increased R&D spending observed in the time series data may be due to reclassification of activities as R&D. Whether this is the case can be determined more definitively as more data become available on the response of firms to the credit over time. But in any event, as long as reclassification takes place, time series comparisons will result in upwardly biased estimates of the true effect of the credit.

Second, and perhaps more fundamentally, like all analyses that rely on extrapolations from time trends, there is no way of knowing whether the structural shift observed in R&D spending is due to changes in tax law, or to other factors. There is, for example, at least casual evidence that a number of U.S. firms began to dramatically change their business operations in response to pressures from foreign competition sometime around 1980. It is plausible that part of this change may have involved a renewed and heightened commitment to research and development. If this is the case, for example, respondents to Mansfield's survey may have implicitly, and correctly, "factored" out any increases in R&D spending that reflected a shift in corporate strategy from their subjective estimate of the effect of the credit.

Some evidence bearing on this issue is presented in Figure 11–3 in which both the projected and the actual time trends in R&D spending are shown. The projected time trend is obtained from the regression of R&D spending on sales and lagged R&D spending that is used to calculate the estimates presented in Table 11–4.

Figure 11–3. Projected Versus Actual R&D Spending, 1970–1984.

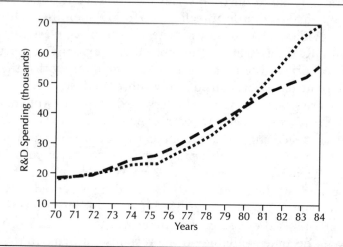

- — Projected R&D
▪▪▪▪▪ Actual R&D

It seems clear from Figure 11–3 that a structural shift in R&D spending of some sort did take place, beginning around 1980. However, from Figure 11–4, it is at least arguable that this structural shift had taken place *before* the enactment of new incentives for R&D in 1981. If so, much, if not all, of the observed difference between projected and actual R&D may be attributable to factors other than changes in tax policy.

Third, as is noted by Baily, Lawrence, and DRI (1985), estimates obtained from time series analyses are often interpreted as measuring short-run behavioral response. In contrast, estimates obtained from multiplying the price change in R&D by the estimated price elasticity of R&D may be interpreted as measuring the longer-run response. Under this interpretation, estimates of the effects of the credit based on estimates of the elasticity of demand would provide the most reliable indication of the lasting effects of tax incentives on the propensity of firms to devote additional resources to the creation of new technological knowledge.

In the same spirit, it should also be noted that the R&D tax credit was enacted as a temporary rather than as a permanent tax incentive. As Barth, Cordes, and Tassey (1985) have shown, the effect of doing so is to reduce the after-tax price of conducting R&D only during the period in which firms can claim the credit. This could, in theory, encourage firms to take advantage of a temporary drop in the after-tax price of R&D by accelerating planned spending on R&D into the years in which the credit was known to be in effect. If firms responded in this way, the initial response, which is apt to be reflected in simple time series comparisons, could be significantly greater than one that would be observed if the credit were permanent. Once again, however, this latter response is precisely the one measured by applying econometric estimates of the price elasticity of R&D to the estimated change in the user cost of R&D.

TAX POLICY AND THE ADOPTION OF NEW TECHNOLOGIES

There is much less evidence on how tax policy affects the adoption of new technical knowledge than there is about how tax policy affects the propensity to create such knowledge. However, some statements can be made about the way in which recent tax changes have affected the incentive of firms either to acquire such knowledge directly through acquisitions of patents from others or indirectly through acquiring the new technology that is embodied in capital goods.

Patents

By purchasing a patent from another firm, one acquires an intangible asset. Moreover, as in the case of tangible capital, the commercial value of this asset will depreciate over time as the new technical knowledge protected by the patent declines in commercial value. This fact is recognized in the tax codes of most industrial nations, including the United States, in that firms are allowed to claim depreciation deductions for the diminution in the value of the patents they have acquired.

However, under 1985 U.S. tax law, these depreciation allowances were far less generous than those allowed on tangible capital, and less generous still than the depreciation allowance—that is, expensing—allowed on self-created technical knowledge. For example, Cordes, Watson, and Hauger (1986) estimate that a patent acquired by the type of firm underlying the calculations in Tables 11–1 and 11–2 faced an effective marginal tax rate under 1985 U.S. tax law of 43 percent as compared to a marginal tax rate of -11.8 percent on R&D and 18 percent on equipment. As a consequence, firms faced a tax disincentive to acquire new technical knowledge through purchases of patents from others.

Acquisition of Embodied New Technologies

Insofar as much new technology is embodied in capital goods, there is some indirect evidence about how tax incentives for capital spending have affected the acquisition of new technologies. The first source of evidence is from the econometric literature on the effects of the investment tax credit on corporate investment spending. The second is from data on the demand for capital goods in the period following the enactment of ERTA/TEFRA.

The Investment Tax Credit. From 1962 through 1967 and again from 1973 through 1985, U.S. firms have been able to claim a tax credit for new investment spending. Whether the investment tax credit (ITC) has been effective in stimulating substantial new investment has long been a controversial subject in the econometric literature.

Chirinko and Eisner (1983) have shown that the various points of view about the effectiveness of the ITC are embodied in the investment equations in each of the major large-scale econometric models of the U.S. economy. From a set of simulations based on modified versions of each of these models, Chirinko and Eisner conclude that the ITC may have increased annual investment spending above its baseline value by between .9 and 6.4 percent, or by as little as $.12 and as much as $.86 per dollar of revenue lost.

Thus, it appears that the ITC has had a small to modest effect on new investment spending by firms. Accordingly, the role played by the ITC in encouraging the adoption of new technologies embodied in new capital goods is also likely to have been a modest one.

Effects of ERTA/TEFRA. Two main patterns are evident from data on capital spending during the period following passage of ERTA as subsequently modified by TEFRA. First, as is documented by Bosworth (1985), the overall level of capital spending increased by substantially more during the recovery from the 1981–82 recession than it did in previous recessions. Second, the investment recovery was not widely distributed among all capital goods but was concentrated in producers durable equipment generally, and more specifically in nonindustrial computers and automobiles.

This resurgence in investment has no doubt encouraged the diffusion of new technologies embodied in new capital goods. However, Bosworth concludes that, though the unprecedented revival in capital spending followed closely on the heels of the enactment of ERTA and TEFRA, it is unclear whether the observed increase was actually due to changes in tax policy or to other factors.

This conclusion is based on a disaggregated analysis of patterns of capital spending observed in 1980–84. Bosworth first uses a statistical model to predict how much capital would have been invested in each of twenty-two capital goods between 1980 and 1984 based on pre-1980 trends. These predictions are then compared to the actual amount of invested in each capital good between 1980 and 1984. The cumulative difference between the predicted and the actual amount of capital was then computed for each capital good. These estimated deviations between the predicted and the actual capital stock in each of the twenty-two capital goods were then correlated with changes in the user cost that were attributable to ERTA and TEFRA, as well as with changes in the user cost caused by other factors, notably changes in the pretax acquisition cost of the capital goods.

There was no significant correlation between tax-induced changes in the user cost and the deviation between the predicted and actual capital stock. However, there was a significant correlation between changes in the pretax acquisition cost and the amount of capital invested. Based on these results, Bosworth concludes that while increased capital spending was caused by a decline in the user cost of capital, any effects of tax-induced changes were overwhelmed by changes in other components of the user cost, notably the acquisition cost of capital.

The instance of industrial computers is a case in point. While Bosworth finds evidence that increased demand for such computers was linked to an overall drop in the user cost, it is also the case that, by themselves ERTA and TEFRA actually *increased* rather than decreased the user cost. Thus, the observed increase in demand for nonindustrial computers could not have been stimulated by a tax-induced reduction in the user cost. Instead, demand appears to have been stimulated by a sharp decline in the acquisition cost of computers.

To be sure, there are numerous limitations of Bosworth's analysis, not the least of which is the small size of his statistical sample (that is, twenty-two assets). Nevertheless, it is striking that fully 93 percent of the increased spending on equipment between 1979 and 1984 occurred in two categories, computers and automobiles, which were *not* favored by ERTA and TEFRA. This suggests that one keep in perspective the role of tax policy in influencing the diffusion rate of a new technology. At least in the case of personal computers, increased demand was stimulated not by any tax advantages associated with acquisition of the new technology but, rather, by a decline in its price.

SUMMARY AND CONCLUSIONS

There are a number of policy lessons to be learned from recent experience about tax incentives for the creation of new technical knowledge. While time series data appear to suggest that firms significantly changed their behavior in response to the R&D credit, there are several reasons for placing more reliance on other evidence that implies that the credit has had an effect, but a small one.

This in turn has two related implications for public policy. First, since the R&D tax credit is a form of price incentive, the findings reported above imply that it may be extremely difficult to permanently increase the amount of private resources devoted to the creation of new technical knowledge by means of *any* type of price incentive. Second, insofar as price incentives are used, it is highly likely that such incentives will not be "target efficient" by one frequently used standard. Namely, the amount of additional R&D spending induced by the incentive may be well below the budgetary cost of the incentive.

It should, however, be noted that this commonly used yardstick may not be entirely appropriate. In particular, if the additional R&D that is induced by the R&D credit has significant commercial benefits that are not captured by the firm conducting the R&D, it is easily shown that a price subsidy to R&D could be efficient in one very important sense, and at the same time be inefficient in the budgetary sense described above. Specifically, if such nonappropriable or external benefits to R&D were present, a subsidy to R&D could stimulate additional R&D whose economic benefits exceeded the economic costs of conducting the R&D, even though the budgetary cost of the subsidy exceeded the amount of additional spending on R&D.[6]

Moreover, the budgetary target efficiency of tax incentives for R&D is comparable to what has been observed for other investment tax incentives. For example, Chirinko and Eisner (1983) estimate that the investment tax credit may have stimulated between

$.12 and $.86 of additional investment per each $1.00 of forgone tax revenue. These amounts may be compared to results presented above that imply that the R&D credit may have stimulated between $.35 and $.93 of additional R&D spending per each $1.00 of revenue loss.

Although these last two considerations may provide a qualified case for keeping an open mind about renewal of the incremental R&D tax credit beyond 1988, there are also several reasons for placing less emphasis on tax policy as an instrument of technology policy in the near term. First, the existing mix of tax incentives (expensing plus the R&D credit) reduces the after-tax price of R&D substantially. Second, the estimated increase in the the amount of resources devoted to the creation of new technical knowledge in response to tax incentives in place from 1981 through 1985 has been modest. Thus, even if the R&D tax credit is renewed, future efforts should focus on mechanisms other than the creation of additional price incentives to foster the creation of new technical knowledge.

Effects of Tax Reform.   Table 11–9 summarizes the principal channels through which the Tax Reform Act of 1986 is likely to affect the tax treatment of R&D. As may be seen from Table 11–9, the creation of new technical knowledge will continue to enjoy relatively favorable tax treatment. Firms can continue to expense the costs of R&D, and the R&D tax credit has been retained, though in a slightly less generous form. Thus, the statutory tax treatment of the costs of creating new technical knowledge has been relatively unaffected by otherwise sweeping changes in the income tax. This is in sharp contrast to the tax treatment of investments in plant and equipment, which has been substantially altered, most notably, through abolition of the investment tax credit. Moreover, limiting the range of activities that qualify for the credit should increase the likelihood of subsidizing R&D that will have at least some external benefits.

On the other hand, some aspects of tax reform may be relatively unfavorable to investments in R&D. First, insofar as debt finance has enjoyed a subsidy under 1985 tax law, reducing the corporate tax rate from 46 to 34 percent reduces the value of this subsidy. Since an asset such as R&D (which can be expensed) benefits relatively more from tax subsidies to borrowing than does an asset

Table 11-9. Effects of 1986 Tax Reform on The Corporate Tax Treatment of Investments in the Creation and Acquisition of New Technical Knowledge and in Depreciable Capital.

Expensing of R&D	No change
Incremental R&D tax credit	Rate of incremental credit reduced from 25 to 20 percent; definition of R&D eligible for the credit tightened to include only R&D undertaken for the purpose of "discovering information (a) that is technological in nature, and also (b) the application of which is intended to be useful in the development of a new or improved business component of the taxpayer."
Depreciation of R&D capital	Loss of 6 percent investment tax credit; R&D capital included in the five-year depreciation class, to be depreciated on a double declining balance basis with switchover to straightline to maximize depreciation allowances.
Amortization of patents	No change
Corporate tax rate	Reduced from 46 to 34 percent
Investment tax credit	Abolished effective January 1, 1986
Tax depreciation of physical capital	Assets currently in three-year, five-year, and ten-year ACRS classes to be placed either in revised three-year, five-year, or ten-year classes, or in new seven-year, fifteen-year, or twenty-year classes; assets included in the three-year, five-year, seven-year, or ten-year classes to be depreciated on a double-declining balance basis with switchover to straight-line; assets in the fifteen-year and twenty-classes to be depreciated on a 150 percent declining balance basis with switchover to straight-line; commercial buildings to be depreciated over 31.5 years on a straight-line basis.

(which must be capitalized), the former type of asset will also be relatively more affected by any decline in the value of such subsidies. Second, as is shown in Barth, Cordes, and Tassey (1985), the value of the R&D credit will decline with the corporate tax rate. Accordingly, the direct effect of reducing the incremental credit rate from 25 to 20 percent will be magnified by the effect of reducing the corporate tax rate.

The calculations in Table 11–10 illustrate the effects of the changes described in Table 11–9 on the user costs and hurdle rates of R&D, equipment and structures. At least for the hypothetical company considered in Tables 11–1 and 11–2, it appears that tax reform will increase the user cost of R&D by 7.4 percent and reduce the user cost of acquiring a patent by 3.0 percent. By comparison, the user costs of equipment and structures and equipment are estimated to increase by 13.2 and 7.3 percent, respectively.

Insofar as the case considered is representative of high-technology firms as a whole, the results in Table 11–10 suggest that the relative tax treatment of the creation and acquisition of new technical knowledge will be at least as favorable after tax reform as it was before. This would seem to offer further support for a strategy

Table 11–10. Estimated Effects of 1986 Tax Reform on the Tax Treatment of the Creation and Acquisition of New Technical Knowledge and of Investment in Physical Capital.[a]

	1985 Law	Tax Reform	Percentage Change
R&D user cost	.136	.146	7.4%
R&D hurdle rate	.036	.045	27.8
Patent user cost	.170	.165	−3.0
Patent hurdle rate	.070	.065	−7.1
Equipment user cost	.256	.290	13.2
Equipment hurdle rate	.049	.083	69.4
Structures user cost	.110	.118	7.3
Structures hurdle rate	.074	.082	10.8

a. These calculations assume that tax reform will increase the firm's discount rate from .084 to .09, decrease the combined federal-state corporate tax rate from .495 to .383, and that all of the firm's R&D will qualify for a 20 percent incremental R&D credit. The values of the other parameters in the calculations of user costs and hurdle rates are the same as those assumed for the U.S. firm in Table 11–2.

of deemphasizing the use of additional tax or other price incentives to stimulate the creation of new technical knowledge.

NOTES

1. For example, see Cordes (1980), Landau and Hannay (1981), Fullerton and Lyon (1983), Collins (1983), and Hulten and Robertson (1984).
2. See Fullerton and Lyon (1983) for a discussion of different ways of measuring the cost of capital. For an interesting discussion of the use of hurdle rates to set R&D budgets, see Ellis (1984a, 1984b).
3. See Cordes, Watson, and Hauger (1986).
4. See the testimony by Ronald Pearlman, Assistant Secretary of the Treasury, in U.S. House of Representatives (1985).
5. See Cordes, Watson, and Hauger (1986). The user costs, hurdle rates, and marginal tax rates presented in Table 11–1 are calculated for a hypothetical firm whose characteristics are designed to be representative of the characteristics of a firm producing computers and computers components.
6. This assumes, as is most often the case, that the alternative to a subsidy for R&D would be no governmention action.

REFERENCES

Baily, Martin, Lawrence, Robert, and Data Resources Incorporated. 1985. "The Need for a Permanent Tax Credit for Industrial Research and Development." Commissioned by The Coalition for the Advancement of Industrial Technology, February.

Barth, James, Cordes, Joseph, and Tassey, Gregory. 1985. "Taxes and Incentives for R&D Spending." In Bozeman, Barry, and Link, Alfred, eds., *The Strategic Management of R&D* (Lexington, Mass.: D.C. Heath).

Bosworth, Barry. 1985. "Taxes and the Investment Recovery." in Brainard, William, and Perry, George, eds., *Brookings Papers on Economic Activity*, Vol. 1 (Washington, D.C.: Brookings Institution).

Charles River Associates. 1985. "An Assessment of Options for Restructuring the R&D Tax Credit to Reduce Dilution of its Marginal Incentive." Prepared for National Science Foundation, Division of Policy Research and Analysis, February.

Chirinko, Robert, and Eisner, Robert. 1983. "Tax Policy and Investment in Major U.S. Macroeconomic Models." *Journal of Public Economics* (March): 139–66.

Collins, Eileen. 1983. "An Early Assessment of Three R&D Tax Incentives Provided by the Economic Recovery Tax Act of 1981." National Science Foundation, PRA Report 83-7, April.

Cordes, Joseph. 1980. *The Impact of Tax and Financial Regulatory Policies on Industrial Innovation* (Washington, D.C.: National Academy of Sciences).

Cordes, Joseph, and Sheffrin, Steven. 1983. "Estimating the Tax Advantage of Corporate Debt." *Journal of Finance* (March): 95–105.

Cordes, Joseph, Watson, Harry, and Hauger, Scott. 1986. "An Analysis of Domestic and Foreign Tax Treatment of Innovation and High Technology Firms," Vol. 1, Prepared for National Science Foundation, Division of Policy Research and Analysis, October.

Eisner, Robert, Albert, Steven, and Sullivan, Martin. 1983. "Tax Incentives and R&D Expenditures," paper presented at the Colloque sur L'Econometrie de la Recherche, Paris, France, and at the 16th CIRET Conference, Washington, D.C.

Eisner, Robert, Albert, Steven, and Sullivan, Martin. 1984. "The New Incremental Tax Credit for R&D: Incentive or Disincentive?" *National Tax Journal* (June): 171–83.

Ellis, Lynn. 1984a. "Viewing R&D Projects Financially." *Research Management* 27 (March/April): 29–34.

———. 1984b. "Viewing R&D Projects Financially," *Research Management* 27 (May/June): 35–40.

Fullerton, Don, and Lyon, Andrew. 1983. "Differential Effects of the Accelerated Cost Recovery System on Investment: A Survey With Implications for Understanding Effects on High Technology Investment," Prepared for National Science Foundation, Division of Policy Research and Analysis, October.

Hulten, Charles, and Robertson, James. 1984. "The Taxation of High Technology Industries," *National Tax Journal* (Sept.): 334–345.

Landau, Ralph, and Hannay, N. Bruce. 1981. *Taxation, Technology and the U.S. Economy* (New York: Pergammon Policy Studies).

Mansfield, Edwin. 1985. "Studies of Tax Policy, Innovation, and Patents: A Final Report." Prepared for National Science Foundation, Division of Policy Analysis and Research, October.

U.S. House of Representatives, Subcommittee on Oversight of the Committee on Ways and Means. 1984. *Hearings on the Research and Experimentation Tax Credit* (Washington, D.C.: U.S.G.P.O., August).

12 THE DIFFUSION OF NEW MANUFACTURING TECHNOLOGIES

David C. Mowery

The economic impact of innovation, whether revealed in productivity growth, employment creation and destruction, or changes in wages and profits, is realized only through the adoption of innovations. The development of computer-based manufacturing processes, for example, can have little or no effect on U.S. economic performance if these technologies are not widely adopted. Analysis of the effects on economic performance and employment of new manufacturing technologies therefore is incomplete without some consideration of factors that affect the speed and extent of adoption of these new technologies. Despite its importance for an assessment of the effects of technological change, however, the economic theory of diffusion does not provide a strong basis for predictions of the rate of diffusion of computer-based manufacturing technologies. Nonetheless, a review of the theoretical and empirical literature on this topic yields a number of important insights.

One of the most robust conclusions of this literature concerns the lengthy nature of the diffusion process. Adoption of an innovation by all of the members of a given industry, firm, or sector can take decades, for a number of reasons that are considered below in detail. This characteristic of the diffusion process, however, means that the employment consequences of new technologies are real-

481

ized gradually, rather than abruptly—especially by comparison with such other sources of economic and employment change as exchange rate flutuations. The development of policies to address the employment consequences of technological change therefore should not be an insurmountable challenge.

Computer-based manufacturing technologies exhibit a number of characteristics that may impede their diffusion. These technologies have consequences for both managers and labor in daily production operations that frequently are not easily predicted in advance of their adoption. In addition, the "systemic" character of many new manufacturing technologies distinguishes them from such earlier manufacturing technologies as "hard-wired" numerical control, which could be adopted in a more incremental fashion. This systemic character is likely to result in a rate of adoption of computer-integrated and other manufacturing processes that will at best not exceed that of previous manufacturing technologies. The very rapid diffusion of these technologies that would be necessary to yield the substantial job displacement predicted by Ayres and Miller (1983) seems unlikely within the United States.

Slow adoption of these manufacturing process technologies within U.S. industry, however, could have a destructive effect on manufacturing employment, reflecting the consequences of lagging productivity growth for the international competitiveness of U.S. industry. In view of the tendency for federal policy toward science and technology to favor the generation, rather than the adoption, of new technologies, serious consideration should be given to some form of support for the diffusion of new processes for manufacturing.

The section immediately below reviews the theoretical literature on diffusion, contrasting the "investment-centered" view developed primarily by David (1966, 1969, 1985a) with the "uncertainty-centered" model of Mansfield (1968). This is followed by a brief survey of a number of studies of the diffusion of new industrial processes. The characteristics of new manufacturing technologies, their rates of adoption within the United States, and a comparison of the utilization of these new technologies within the United States and other industrial nations are considered next. The next section examines impediments to the diffusion of these technologies, and the conclusion discusses some policy implications.

THEORETICAL CONTRIBUTIONS TO THE
STUDY OF DIFFUSION

Much of the theoretical literature on the diffusion of innovations is concerned with explaining the sigmoid or S-shaped curve that describes the adoption of an innovation over time within a population.[1] This characteristic pattern implies that the rate of adoption of an innovation is dependent on the proportion of the potential adopter population that has adopted the innovation.

Economic studies of the diffusion of innovations drew on a lengthy research tradition within rural sociology (see Rogers 1983 for a survey). One of the earliest and most influential studies of diffusion was Ryan and Gross (1943), who analyzed the pattern of adoption by individual farmers of hybrid seed corn in Iowa. Ryan and Gross focused on the characteristics of early adopters of these new hybrid strains within Iowa, in contrast to Griliches (1957), who compared the rate of diffusion of hybrid corn within different regions of the United States.

Although the different approaches of these sociological and economic studies contributed to a scholarly debate (see Griliches 1960, 1961, 1962; Rogers and Havens 1962), the conclusions of the economic and sociological studies of this diffusion process are not incompatible. To argue, as did Griliches, that the profitability of the innovation exerted a powerful influence on interregional differences in rates of adoption does not explain why adoption is not instantaneous within each region. The Griliches conclusions, however, support the hypothesis, embedded in economic models of diffusion, that the profitability of an innovation is among the most important influences on its diffusion. Ryan and Gross stress the cosmopolitan characteristics of early adopters, but this hypothesis does not explain the rate of diffusion within a population—Stoneman (1983: 97) concluded that the analysis of individual characteristics of early and late adopters "tells us little more than that early adopters adopt early and late adopters adopt late." Nonetheless, the emphasis by sociologists on the role of uncertainty and information is reflected in one group of economic models of diffusion.

Economists have developed two broad approaches to the explanation of the characteristically gradual pattern of adoption of innovations. The first emphasizes the role of uncertainty and imperfect information in the decisions of firms and individuals to adopt innovations, and the second stresses the role of differences in the payoff to adoption within the potential adopter population.

The model of diffusion developed by Mansfield (1961, 1963, 1968) assumes that uncertainty concerning the performance and other characteristics of a new technology decreases as the population of adopters increases. Although the profitability and costs of adoption of the innovation also are important in Mansfield's model, they are constant over time and across all potential adopters within the population. A critical element in the diffusion of innovations is the level of knowledge among potential adopters concerning the characteristics of the innovation. The fact that such knowledge is not instantly available to all members of the potential adopter population is fundamental to the characteristic pattern of diffusion. Mansfield's model of uncertainty reduction is based on an "epidemic" model of knowledge transmission, in which initial adopters "infect" other firms with their knowledge and operating experience, lowering the perceived uncertainty associated with the adoption of an innovation. The Mansfield model of diffusion is a model of sustained disequilibrium, as the adoption of a profitable innovation is impeded by uncertainty.

An alternative model of the determinants of the sigmoid pattern of diffusion over time was developed by David (1966, 1969, 1985a) and is a model of a moving equilibrium. Arguing that the diffusion of an innovation among a population of firms is the outcome of a series of individual investment decisions, David ascribes the non-instantaneous pattern of adoption to the heterogeneity of the adopter population. The initial version of an innovation, for example, may be profitable only for firms of a particular size. As the number of firms within this size class increases over time, or changes in the technology make the innovation profitable within other portions of the population, one observes the characteristic sigmoid curve of cumulative adoption.

Information asymmetries play no role in the David model. The distribution within the adopter population (as well as changes over time in this distribution) of characteristics that are critical to the profitability of the innovation (such as firm or farm size) yields the intertemporal diffusion path. The David model, similarly to the Mansfield model, emphasizes profitability and cost as key parameters within the diffusion process. Unlike Mansfield, however, David assumes that the differential profitability and costs of the innovation for various members of the adopter population, rather than imperfect information and uncertainty, are responsible for the extended duration of the typical diffusion process.

The Mansfield and David models of diffusion differ in their conceptualization of the determinants of expected profitability, but both models emphasize the central role of profitability. The implicit assumption in the David model that the diffusion of innovations is an equilibrium process, in which all potential adopters are equally well-informed concerning the costs and profitability of an innovation, conflicts with anecdotal evidence concerning the role of uncertainty and imperfect knowledge in diffusion of innovations. The acknowledgment within the David model that the adopter population for any innovation is heterogeneous, as well as the ease with which this model can accommodate incremental improvements in a specific innovation, are important strengths, however, and are consistent with other data on the diffusion of innovations.

The Mansfield model's emphasis on uncertainty and the role of prior adoption in the diffusion of an innovation also contains important insights. In most cases, including the manufacturing technologies discussed in this chapter, both information concerning the potential of innovations and the expertise necessary to evaluate the innovations are scarce. Accordingly, uncertainty over the payoff to adoption is pervasive within the potential adopter population.

Both the Mansfield and David models of diffusion are also limited in their focus. Neither model explicitly incorporates the determinants of the profitability of an innovation, which makes it difficult to predict changes over time in the rate of adoption of an innovation. Neither model considers the general determinants of the costs of adoption, an omission that is carried over into the empirical literature. The focus of these and other models on cross-sectional differences in the adoption of a single innovation also restricts their ability to predict diffusion rates for multiple innovations over time. Most economic models of diffusion cannot account for changes over time in diffusion rates within an economy. While a small literature dealing with "long waves" in economic growth purports to analyze changes over time in the diffusion behavior of entire economies, much of this work lacks microeconomic foundations and empirical support (Mensch 1979; Rosenberg and Frischtak 1984; and Freeman, Clark, and Soete 1982 provide reviews of the "long wave" literature).

Recent surveys by Rosenberg (1976) and David (1985a) develop additional hypotheses concerning the profitability of innovations. Rosenberg emphasizes the changes in an innovation during the diffusion process, similarly to David's theoretical model. The initial version of a specific innovation is rarely an optimal design, in this view—debugging and perfecting an innovation is a lengthy and gradual process and accounts for much of the delay in adoption. Diffusion of an innovation thus might better be conceptualized as a sequence of brief processes of adoption of different designs of an innovation. The cost savings realized by early adopters of an innovation may be less than those associated with the adoption of later, improved designs. The accumulation of knowledge from the operation of a new technology ("learning by using," as Rosenberg 1982 has denoted it) may increase the efficiency and profitability of use of the technology by later adopters. Rosenberg also emphasizes the importance of complementary investments that support adoption.

David considers two factors that affect the profitability of investments in new process technologies. The first concerns the minimum efficient scale of output associated with the new technology, which is closely related to the fixed costs of adoption. If this minimum output exceeds the attainable output for a production plant, adoption may be difficult or impossible. A second factor concerns the relative production costs of the new and old technologies. Although new process technologies may offer savings in operating costs over existing technologies, these may be insufficient to offset the high capital costs of replacing fully depreciated plant and equipment.[2]

Another hypothesis about the development of new technologies that has important implications for their diffusion was proposed by Nelson and Phelps (1966). These scholars argue that in the initial stages of the diffusion process, the skills required to operate a technology are high, declining over the course of the modification and diffusion of the technology. A comparison of the skills needed to operate the mainframe computers of the 1950s with those required to operate a desktop personal computer illustrates this point, which applies to a number of other technologies as well. As a result, higher levels of basic and job-related skills within the workforce should support more rapid diffusion. Unfortunately, due to the difficulties of measuring both diffusion and skills over time, this hypothesis has not been directly tested, although studies have found that educational attainment is associated with earlier adoption in a number of cases (see Rogers 1983).

EMPIRICAL STUDIES OF DIFFUSION

The empirical literature on diffusion processes has grown enormously during the past twenty years. This literature retains an anecdotal quality, however, since most empirical studies examine the diffusion of a single innovation. Several studies, however, have compared the diffusion of different technologies. Enos (1962) compiled estimates of the length of the period between the invention of a new process or product and its *initial* application (in other words, substantially prior to extensive diffusion of the innovation), finding that for a sample of forty-six such innovations, this lag averaged nearly fourteen years. Other important comparative studies are Mansfield (1961) and the collection of studies assembled by Nabseth and Ray (1974).

Mansfield's study (1961) analyzed the diffusion of twelve manufacturing process innovations among large firms in the coal mining, railroad, brewing, and iron and steel industries. Consistent with the results of Enos and others, Mansfield found that the adoption of these innovations by all of the largest firms within these industries required more than ten years for nine of the innovations, while five of the innovations required more than twenty years to be adopted by all of the large firms. Moreover, the rate of diffusion of these innovations varied widely: "The number of years elapsing before half the firms introduced an innovation varied from 0.9 to 15" (Mansfield 1966: II-122). The key determinants of the rate of diffusion in Mansfield's 1961 study, which were incorporated in his theoretical model, were the profitability and cost of the innovations, as well as the number of firms that have adopted the innovation.

Nabseth and Ray (1974) examined the diffusion of six process innovations (numerically controlled machine tools, float glass, basic oxygen steelmaking, continuous casting, tunnel kilns in brickmaking, shuttleless weaving, gibberellic acid in malt manufacture, and papermaking presses) in six European nations (France, the United Kingdom, Sweden, Austria, West Germany, and Italy). The comparative analysis, spanning as it did both technologies and nations, proved to be extremely difficult, because of changes in the character of the innovation during its diffusion, as well as the heterogeneity of the adopter population.[3]

The study concluded that the profitability of an innovation was the most important single factor supporting rapid diffusion, although "managerial attitudes" toward new process technologies and firm liquidity and capital costs also influenced the adoption of innovations (Ray 1984 summarizes the findings of the project). With the possible exception of "managerial attitudes," a variable that includes a multitude of sins (many of which are associated with specific characteristics of the technology, as Davies 1979 notes), the findings of this study are broadly consistent with Mansfield's empirical results and support the predictions of the David and Mansfield models.

The empirical studies in Nabseth and Ray also support the conclusions of Enos and Mansfield that widespread adoption of manufacturing process innovations takes considerable time (as much as twenty years were needed for the adoption of an innovation in 50 percent of the relevant population), although the duration of this process varies across industries and nations.[4] The nations in which innovations were first introduced exhibited the slowest rates of diffusion, ascribed by Nabseth and Ray to the difficulties of applying new technologies that also were noted by Rosenberg (1976).[5]

A small number of empirical studies have examined characteristics of firm and industry structure, in an assessment of how such variables as firm size, demand growth, and industry concentration affect diffusion. Typically, these studies analyze the differences among industries or firms in the adoption of a single innovation. Mansfield (1963) found that larger, faster growing firms tended to adopt process innovations more rapidly. He rejected the hypotheses that firm liquidity, profitability, the age of managers, and profit trends significantly influenced the speed of adoption of new processes.[6]

Romeo (1975, 1977) examined the influence of industry structure on the adoption of numerically controlled machine tools in a large sample of industries, concluding that adoption rates were highest in industries with lower levels of concentration and less skewed firm size distributions. The Romeo studies provided broad support for the hypothesis, not considered in most theoretical models of diffusion, that competitive pressure within the user industry supports more rapid diffusion of innovations. Romeo's 1977 study also found, consistent with the conclusions of Nabseth and Ray (1974) for different nations, that the industries in which the first adoption of an innovation was relatively late exhibited the most rapid rates of adoption.

ADOPTION OF COMPUTER-BASED MANUFACTURING PROCESSES IN U.S. INDUSTRY

This section examines evidence on the rate of adoption and utilization by U.S. firms of new manufacturing processes, which include robotics, computerized numerically controlled (CNC) machine tools, and computer-integrated manufacturing technologies. Despite rapid recent growth, the utilization by U.S. industry of these new manufacturing technologies does not appear to be increasing significantly more rapidly than previous process innovations, such as "hard-wired" numerically controlled machine tools. Moreover, and of greater importance for the long-run employment effects of technological change in U.S. manufacturing, adoption rates and utilization within U.S. industry appear to be lagging behind those observed in many industrial competitors, notably Japan and Sweden. As the role of international trade within the U.S. economy continues to grow, lagging adoption of productivity-enhancing innovations may reduce employment growth in U.S. manufacturing.

Longitudinal data are available only for robotics technologies within the United States. An examination of forecasts and actual growth rates in shipments of robotics equipment supports two conclusions. Until 1986, robotics industry sales grew at roughly 45 percent per year, including some years in which sales grew by 50 percent. Moreover, sales of these capital goods grew (albeit at a more modest rate) throughout 1980–83, a period that included a severe recession in the United States.

Nonetheless, it is important to note that even these rates of growth for sales and shipments are well below many of the forecasts made during the past decade. According to the U.S. International Trade Commission, shipments of robotics equipment in the early 1980s accounted for no more than 19 percent of the $1 billion market for such equipment forecast in the 1970s (1983: 50). Conigliaro's forecast (1983) that total U.S. industry sales in 1983 would be at least $205 million contrasts with the estimates of the U.S. International Trade Commission that total U.S. industry shipments in 1983 were no more than $169 million. Subsequent estimates (Prudential-Bache Securities, cited in U.S. Department of Commerce 1985: 19) that sales of robotics equipment would amount to $375 million in 1984 also appear to have overstated growth considerably.[7] The rate of growth of the industry has dropped sharply in the past eighteen months; 1986 shipments of robots in the United States declined slightly from the 1985 level (U.S. Department of Commerce 1987; Mearman 1987).

Comparable data are unavailable for other such process technologies, but forecasts of their growth also appear to have been consistently optimistic. The next section discusses several factors that contribute to the tendency for the diffusion of these technologies to fall below forecast levels.

Although the rate of growth of the robotics market is below projections sales of these devices have grown rapidly during the past five years. Does this growth represent an unprecedented "explosion" of new technology? The evidence does not support such a claim. Hunt and Hunt (1983) have argued that the rate of growth of robotics utilization in U.S. industry does not greatly exceed the rate of diffusion of other significant technological innovations of the postwar period. The annual rate of growth in the numerically controlled machine stock within U.S. manufacturing averaged 12 to 20 percent during 1965–81, while computers were adopted much more rapidly. Chow (1967) concluded that the stock of computers expanded at an average annual rate of more than 70 percent during 1954–65, while Hunt and Hunt estimated that this annual growth rate during 1961–79 was 25 to 30 percent.

The diffusion of robotics within U.S. industry also appears to be slower than the diffusion rate in other industrialized nations. A comparison of the U.S. Commerce Department's 1981 estimates of the robotics population in the United States and other nations with trade journal estimates of this population in 1984 (cited in

Technology Management Center 1985: 68) suggests that the U.S. robotics population is growing at a substantially lower rate than robots in five other industrial nations (Sweden, Japan, West Germany, the United Kingdom, and France). During 1981–84, the U.S. robotics population grew by roughly 270 percent, while the Japanese robot population (which was larger in 1981) grew by 460 percent. The robot populations of France, the United Kingdom, Sweden, and West Germany, all of which were substantially smaller than that of the United States or Japan in 1981, grew by 590, 365, 340, and 455 percent, respectively.[8]

The diffusion of this new process technology in the nation that first developed it thus seems to be lagging that of later adopters, consistent with Nabseth and Ray's (1974) findings. Relatively slow adoption of this technology in the United States may reflect the greater ease of late adoption, the poor performance of the U.S. economy during 1981–84, or differences in factor costs (a hypothesis rejected by Flamm; see Chapter 7). The indifference to diffusion exhibited by federal science and technology policy, however, also may be an important factor (see below).

Partly as a result of slower rates of diffusion within U.S. manufacturing, U.S. per capita use of robotics technologies in 1984, at 4.3 robots per 10,000 workers, lagged well behind that of other industrialized nations, including Japan (16.4 robots per 10,000 workers), Sweden (17.7), Belgium (6.4), West Germany (5.7), and the United Kingdom (4.8) (*Industrial Robot International Quarterly*, cited in Technology Management Center 1985). Both the rate of robotics diffusion and the level of utilization of robotics in U.S. industry appear to be lower than those of other nations.

Lags in U.S. adoption also appear to characterize advanced machine tools, including numerically controlled machine tools, based on comparisons of the age of the stock of machine tools employed in the manufacturing sectors of the U.S. and other industrialized nations. According to Ray (1984), Japan and Sweden ranked highest in 1982 in the percentage of their national machine tool stocks accounted for by numerically controlled machine tools, roughly 3.0 percent. This fraction stood at about 2.6 percent of the machine tool stocks in the United States and Great Britain, while West Germany was above the United States and below Sweden and Japan in utilization of numerically controlled machine tools. Slower rates of adoption of new machine tool technologies in U.S.

manufacturing are reflected as well in the advanced age of the overall stock of U.S. machine tools. Recent survey data published in the *American Machinist* suggest that the percentage of U.S. machine tools that are five to nine years old actually increased during 1982–83, the first increase since 1945 (survey data from the *13th American Machinist Inventory*, cited in Technology Management Center 1985: 72).

Data on the adoption of these manufacturing process technologies are fragmentary but support two broad conclusions. These technologies are not being adopted more rapidly than earlier process innovations within the U.S. economy. Moreover, and of greater importance for long-term performance and growth in U.S. manufacturing employment, both the rate of adoption and the level of utilization of such technologies as robotics and numerically controlled machine tools appear to be lower within U.S. manufacturing than in other industrialized economies. The growing role of international trade within U.S. manufacturing means that substantial differences between U.S. and foreign firms in the utilization of advanced manufacturing processes may be unsustainable at current levels of real wages. The real threat posed by new manufacturing technologies to U.S. employment results from their excessively slow, rather than rapid, adoption in the United States, relative to other nations.

OBSTACLES TO ADOPTION

What are the major obstacles to adoption of robotics and other computer-based manufacturing technologies? A number of factors appear to impede their adoption and utilization (Ettlie 1985a, 1985b, 1986; Graham and Rosenthal 1986; Miller and Bereiter 1985). A discussion of obstacles to adoption lends support to the conclusion that adoption of these technologies will be gradual, rather than explosive. The obstacles can be grouped conveniently into three broad and overlapping categories—those affecting the adoption cost of these manufacturing technologies, those affecting product standards, and those concerning the availability and evaluation of information relevant to these technologies.

These three groups of factors affect the adoption of virtually any new technology, but they are particularly important for computer-based manufacturing technologies, for several reasons. Adoption of these technologies, which often integrate numerous hitherto separate processes, requires extensive integration and applications engineering. As a result, the costs of adoption of these technologies often are greater than those associated with such prior innovations as numerically controlled machine tools and mainframe computers, discrete innovations with more modest requirements for integration. The computing technologies on which these innovations are based also are new to many potential adopter industries and firms, raising both the uncertainties surrounding evaluation and the costs of acquiring the necessary expertise. Finally, computer-based manufacturing technologies lack product standards for interconnection and hardware-software interfaces.

In contrast to mainframe computers and most varieties of "hard-wired" numerical control, the adoption of new computer-based manufacturing technologies requires the adoption of a new system for the control and performance of virtually all operations. The fixed costs associated with the adoption of these technologies therefore often exceed those associated with other process innovations. According to the U.S. International Trade Commission (1983: 42–43),

> About 90 percent of the robots sold today are being integrated with existing equipment which is often 10 to 20 years old. The integration of the robot with old equipment is one of the biggest problems for the industry, and this situation is expected to continue for the rest of this decade. Not surprisingly, the cost to adapt a robot to existing equipment along with necessary tooling and programming costs is often higher than the initial cost of the robot. According to responses to Commission questionnaires, purchasers reported that the median cost of making robots operational varied between 175 and 500 percent of the purchase price, depending on the type of robot acquired.

In addition to these costs, studies by Ettlie (1985a, 1985b) and others emphasize the need for substantial reorganization of the production process, including the redesign of products, to exploit fully the productivity and product quality benefits of these technologies.[9]

The costs of downtime in any single operation within computer-integrated or flexible production systems also is high, due to the greater interdependence among these operations and the lower levels of inventory and work in progress that can insulate the overall production process from the consequences of breakdowns. Investments in maintenance staff and procedures thus become more important, offsetting the putatively labor-displacing impact of adoption of these technologies.[10] The costs of adoption of computer-based manufacturing technologies also may increase as a result of the need for substantial reorganization of management, as well as production, operations.[11] Extensive investments in both retraining and personnel evaluation (in order to assess the new skills and aptitudes needed by many workers for operating these technologies) are necessary.[12] These costs may be particularly important impediments to the adoption of these technologies in the United States, due to the relatively high mobility of skilled and managerial workers.

A final dimension of the costs of adoption is the sustained period of suboptimal productivity and product quality performance that often follows the initial installation of an extensive computer-based manufacturing system, while the system is debugged, workers and supervisory personnel are retrained, and products are redesigned. The data in Miller and Bereiter (1985) are particularly sobering in this regard. Their study of the installation by General Motors of an extensive computer-integrated manufacturing system in its Baltimore truck assembly plant revealed that more than thirty weeks (rather than the twelve forecast) were required to reach target output levels, far more than the week normally required for converting the production line following minor changes in product design or the six weeks for major changes in product design.[13] The authors note (1985: 23) that GM management had planned to reach the target production rate within twelve weeks and concluded that

> The disparity between the management plan for the output acceleration and actual experience suggests that top management has a great deal to learn about planning the introduction of new computer-integrated systems. One manager pointed out that the company cannot build plants using complicated computer-integrated technologies and still keep the old style of planning and managing the installation and debugging phase.

All of these factors contribute to the costs of adopting computer-based manufacturing processes. To the extent that these costs are invariant with respect to the scale of output, adoption of computer-based manufacturing processes may be financially feasible only for larger firms. Financing for such investments may be unavailable for small firms due to capital market imperfections, further limiting the diffusion of these new technologies among smaller firms and production establishments.

Currently, there exist few if any technical standards for hardware or software in computer-based manufacturing technologies. Although the lack of standards is common in many young, high technology industries and pervasive in computers, it is particularly important for the adoption of computer-based manufacturing technologies. The use of customized hardware or software solutions to retrofit new and older manufacturing process equipment increases the costs of adoption considerably. In addition, the diversity of applications in which these process technologies are employed means that the products of several different vendors often must be combined within a single production establishment. The lack of standards for hardware and software interfaces among these devices may necessitate the development of costly customized solutions.

The engineering and design information needed for application of these technologies often requires the exchange of proprietary, firm-specific knowhow between the seller and buyer of the technology. One answer to such disclosure problems is the establishment of a product development joint venture between user and supplier firms, such as that between General Motors and Fanuc of Japan in the development and production of robotics equipment.[14] The attractiveness of such ventures, which provide nonmarket channels for the exchange of proprietary technical data and process knowhow, appears nevertheless to have diminished in recent years. Most of the original user-supplier joint ventures linked robotics producers with automotive firms. As markets for robotics equipment have expanded to include new industries, the importance of the automotive producer market has declined and the industry has not been replaced by any other single market, reducing somewhat the incentives of robotics producers to enter into joint ventures with auto firms or customer firms in other industries. (See Klepper 1988 for further discussion.)

The absence of standards has several other important implications for diffusion. By raising the fixed costs of adoption, the lack of standards further biases the choices of smaller firms and factories away from these manufacturing technologies.[15] A lack of standards also may confer considerable power on very large users and producers of these technologies to establish de factor standards based on their specific needs. Customers representing a large share of the user market are placed in a strong position to insist that producers conform to their product standards and protocols—the Manufacturing Automation Protocol developed by General Motors, one of the largest single customers for computer-based manufacturing technologies, is an excellent example. Alternatively, as occurred in the computer market, a major producer may resist the development of industrywide product standards and force the standardization of the products of other firms around its product architecture (Brock 1975). To the extent that the specific needs of smaller producers and consumers of the specific products differ from those of large consumers and producers, then, a *laissez-faire* approach to the establishment of hardware and software standards may retard the diffusion of innovations that depend on such standardization.

The implications of standardization, however, are even more complex than this discussion suggests. The inherently incremental, cumulative nature of technological change, as well as the important influence exercised over new product choice by the existing capital stock, mean that once established, a product or process standard may exercise a powerful influence over the future direction of technological change. Uninformed or hasty standardization thus may effectively "lock in" an inefficient technology, with costs that are apparent to anyone who has used the conventional typewriter keyboard (see David 1985b). The lack of standards thus retards diffusion, while the informal establishment of such standards confers additional market power on large firms and may increase the risks of technological suboptimization.

A final category of impediments to the diffusion of computer-based manufacturing technologies consists of informational problems. The adoption of computer-based manufacturing technologies in the face of diverse adopter needs and in the absence of product standards is a knowledge-intensive activity. Despite the frequent assumption in the economic theory of research and development investment (Arrow 1962; Stigler 1956) that the transfer of knowledge is a low-cost or costless activity, much evidence suggests that it is in fact quite costly. A great deal of the relevant knowledge necessary to adopt and debug a new technology is in fact not codified, but "tacit," consisting of knowhow and informal routines (Nelson and Winter 1982). As a result, the adoption, or even the evaluation, of technology from sources external to a firm may require considerable in-house expertise.[16]

Many of the computer and information hardware and software technologies that are central to the implementation of these process technologies have not previously been employed within prospective adopter firms. In-house expertise therefore is lacking, requiring staff additions or extensive retraining in advance of adoption. The costs of developing this expertise have obvious consequences for adoption costs.

The novel character of many computer-based manufacturing technologies, as well as their "systemic" character, also affect the ability of prospective adopters to evaluate the cost consequences of adoption. Numerous case studies of the adoption of these technologies emphasize the fact that many of the essential areas in which these process technologies yield significant cost savings are not incorporated in conventional investment analyses. Reductions in inventory or in work in progress, for example, often receive little attention in analytic methodologies developed for the evaluation of investments affecting only one or a few manufacturing processes (Technology Management Center 1985; Ettlie 1986; Kaplan 1986). Many of the most significant benefits from the adoption of these process technologies stem from such savings, however, and the heightened awareness of these benefits provides an excellent example of the reduction in uncertainty over adoption benefits that results from more widespread use of an innovation.

CONCLUSIONS AND POLICY IMPLICATIONS

The policy implications of the preceding discussion depend fundamentally on an assessment of the likely consequences of widespread adoption of computer-based manufacturing technologies. If one believes that adoption will have a massive labor-displacing impact, policies to increase the rate of diffusion of these technologies are of little interest. If, on the other hand, one believes that the aggregate job destruction effects of widespread adoption of these technologies are likely to be minor, and that the increasingly open character of the U.S. economy makes rapid adoption necessary to prevent even greater job losses, policies to support more rapid diffusion are appropriate. This survey takes the second view of the consequences of adopting computer-based manufacturing technologies.

Both the theory and practice of programs to support the diffusion of new technologies are underdeveloped within U.S. science and technology policy.[17] Economic theory has focused on the conditions necessary to offset an putative *undersupply* of research investment, due to difficulties of appropriating the returns to such private investments, while largely ignoring the requirements for effective *utilization of the results* of this research investment (Mowery 1983). Heavily influenced by this theoretical model, federal civilian science and technology policy in most areas has supported the supply of R&D, rather than concerning itself with the adoption and utilization of the results of R&D.[18]

Additional investments to improve the sophistication of currently available computer-based manufacturing technologies, however, appear unlikely to yield a social payoff comparable to that obtainable from public initiatives to support the adoption of these technologies. A central problem within U.S. manufacturing is the absorption of existing levels of computer-based manufacturing technology, rather than the absence of appropriate technologies. Indeed, several studies have suggested that the very sophistication of U.S. robotics systems, by comparison with less complex Japanese products, has impeded the diffusion of the U.S. technology (U.S. Department of Commerce 1983; U.S. International Trade Commission 1983; Flamm, Chapter 7).

The models of diffusion discussed above suggest that reductions in the costs and uncertainty associated with a new technology increase the rate of its diffusion. Two alternative policy strategies, which are not mutually exclusive, are implied by the David and Mansfield models of diffusion. One strategy, which reflects Mansfield's focus on uncertainty, emphasizes the provision of information, so as to overcome potential adopter uncertainties about the new technology. The other strategy, which is based on David's model, emphasizes the use of subsidies to defray the costs of adoption and increase the rate of diffusion.[19]

Policies that lower the costs of adoption could particularly facilitate the adoption by smaller firms of new manufacturing processes. What are examples of federal policies that could reduce the costs of adoption? Worker training and retraining in job-related skills is one example of an activity that may aid diffusion. In view of the fact that U.S. firms frequently are unable to appropriate the full returns from such retraining, the costs of such training might better be shared by the private and public sectors.

Several of the nations characterized by rapid diffusion of advanced manufacturing processes, including Japan, West Germany, and Sweden, have public policies or labor market practices and institutions that reduce the problems of nonappropriability faced by U.S. firms. Japan's largest corporations possess a strong incentive to invest heavily in worker retraining, due to the practice of lifetime employment that has been widespread during the postwar period. Within Sweden and West Germany public funds support extensive programs of technical education and apprenticeship, increasing the supply of skilled blue-collar workers. Technical training in job-related skills for U.S. blue-collar workers, however, appears to be far less extensive. Indeed, employer-financed training goes largely to white-collar, better-educated workers, exacerbating differences in educational attainment between blue- and white-collar workers (see Tierney 1983).

Worker training programs can support more rapid adoption of computer-based manufacturing technologies. An increased supply of production workers with skills in computer operation and maintenance reduces the costs for many firms of adopting these technologies, hastening the diffusion of advanced manufacturing processes. In view of the importance of rapid adoption of new technologies for the international competitiveness of this economy, as well as the contribution that a more skilled workforce may make to such adoption, stronger public and private support for training of the employed workforce could yield a substantial social payoff (see also Bartel and Lichtenberg 1987).

A more skilled workforce also can adjust more easily to changes in industrial and occupational structure resulting from new technology or other sources of economic change. These strategies to support diffusion and enhance competitiveness thus can reduce the severity of worker displacement. Initiatives in the training and retraining of skilled and unskilled workers are important components of diffusion policies aimed at improving both the quality of the workforce and the rate at which new technologies are adopted in this economy. A number of training programs recently have been established at the state level, but much remains to be done to assess the effectiveness of these programs and develop guidelines for program design.

Product standards play an important role in the adoption of these technologies, as was noted above, and are another area in which federal initiatives could lower the costs of adopting computer-based manufacturing technologies. Existing programs of research by the National Bureau of Standards in the technological requirements of standards and cooperation with private firms in the development of these standards both could be strengthened.

Current federal policy is relatively permissive with respect to the formation of joint ventures among erstwhile competitor firms that focus on precommercial research. The formation of consortia of smaller user firms might be encouraged through tax credits or other mechanisms, as a means of supporting standard-setting and the research necessary to develop standards among smaller firms. The federal government could mandate the use of specific hardware and software standards in a wide range of new technologies. Federal research programs for the improvement of defense-sector manufacturing processes, such as the U.S. Air Force ManTech program, could be modified to develop and test process technologies in a more comprehensive program that also involves smaller, "second-tier" defense supplier firms far more heavily.[20]

The development of advanced software and hardware for improved manufacturing processes in the United States will not improve industrial competitiveness, and thereby have a positive effect on employment, unless these innovations are widely adopted within industry. Theory and policy have nevertheless devoted little attention to the diffusion process. This chapter has described and applied insights from the economic theory of diffusion to the adoption of computer-based manufacturing technologies. A number of critical obstacles have contributed to diffusion rates and utilization levels for these technologies that are substantially lower in the United States than in a number of industrial competitors. A reorientation of federal policy away from an exclusive preoccupation with the generation of new technology to greater concern with the adoption of this technology might improve this performance, especially in the area of standards and worker training.

NOTES

Prepared for the Panel on Technology and Employment, National Academy of Sciences. Research for this chapter was supported in part by the Technological Innovation Program of Stanford University's Center for Economic Policy Research.

1. "Although the finding that technical change follows this pattern is not very surprising or new, . . . it is very useful. It allows us to summarize large bodies of data on the basis of three major characteristics (parameters) of a diffusion pattern: the data of beginning (origin), relative speed (slope) and final level (ceiling)" (Griliches 1960: 212–13).

2. "Thus, the legacies of past capital formation decisions, as well as the costs of new investment, may combine to create differences which will determine the timing and pattern of technology diffusion within firms and industries. Durable facilities surviving from earlier epochs may pose barriers to the introduction of best-practice methods" (David 1985a: 381).

3. Nabseth (1974:297) noted several difficulties in the measurement of diffusion rates and levels that recall the discussion of diffusion theories above:

   > what is a reasonable basis of comparison? In studying the diffusion of a new technique, in the paper or steel industry for instance, the data can be related either to the total production of paper or steel, or to the number of firms in the industry. But problems at once arise: either the new technique may never be suitable for certain types of paper or steel, or it may improve over time, so that while it is unsuitable for parts of the production or some types of firms initially, it will be suitable later on . . . A third problem arises with numerically controlled machine tools, which can be used to produce parts of many different products that are, however, not clearly definable, and a fourth (relating both to the numerator and the denominator) is how to fix a starting date for commercial operation of a new process which has been improved over a long period, but initially could be used only in some types of plant. When is it possible to say that the process has really become an innovation in the Schumpeterian sense of the word?

4. Ray (1984: 82–83):

In the case of BOP [the basic oxygen process in steelmaking], the only one among the technologies studied that has nearly reached total diffusion (if the other modern method, electric steel, is also taken into account), it took twenty years to approach saturation. But in the cases of CC [continuous casting of steel] and TK [tunnel kilns in brickmaking] the same time span was needed to reach only 50 percent diffusion. The cheaper technologies required even more time for a weaker penetration into their respective markets.

5. Nabseth and Ray (1974: 19):

This result is consistent with the the hypothesis that the pioneer faces all sorts of teething troubles—new problems associated with the new technique—which are likely to be solved, partly and gradually, by the time others adopt it. It is therefore not necessarily desirable to be the first to introduce a new technique.

6. Anticipating a hypothesis popularized by Mensch (1979), Mansfield also examined the influence of economic fluctuations on the appearance of innovations in different industries, finding no significant relationship:

Contrary to the opinion of many economists, there was no tendency for innovations to cluster at the peak or trough of the business cycle. Apparently process innovation at the trough was discouraged by the meagerness of profits and the bleakness of future prospects; at the peak, it was discouraged by the lack of unutilized capacity. For product innovations, there was no evidence that the rate of innovation varied significantly over the business cycle. (1966: II-12; the empirical results are reported in Mansfield 1968).

7. Shipments of robotics equipment, which include the extensive production of robotics equipment for in-house use, from U.S.-based producers were estimated by the Robotics Industries Association to equal $332.5 million in 1984 (U.S. Department of Commerce, 1987: 21–6).

8. Although the definition of a robot employed by these various nations and the two surveys of robotics usage may differ, the estimates of the robotics population in both 1981 and 1984 are well within the range of estimates published in several other sources.

9. Ettlie (1985a: 17):

One consequence of installing new processing technology consistently noted in these plant visits and talking to managers, engineers and shop floor personnel responsible for installation, is almost always the same: the part or parts had to be redesigned so that the systems could work or be designed to work in a cost-effective way. Standardization of parts cuts down on the amount of tooling a robot needs to fabricate, handle or assemble parts.

10. Miller and Bereiter (1985:12) reported in their study of the adoption of computer-integrated manufacturing technologies for the assembly of trucks that in the aftermath of adoption, "The maintenance workforce experienced the largest change in relative proportion [of the total workforce]. The number of maintenance people increased and the proportion of maintenance personnel more than doubled [from 4.8 to 10.6 percent of the total workforce]."

11. Reporting on a study of thirty-nine manufacturing plants, Ettlie (1985a 11, emphasis in original) stated that "22 (56%) of the 38 plants have adopted an administrative innovation *specifically* to facilitate the introduction of their new technology processing systems." An example of these administrative innovations is contained in another discussion by Ettlie (1986:12): "one plant completely reorganized its management hierarchy and reduced it to just three levels, including the plant manager. Foremen or first-line supervisors were completely eliminated and every employee became a salaried member of at least one team."

12. Reporting on their field studies of firms adopting flexible manufacturing systems, Graham and Rosenthal (1986: 14–15) noted that

> We were impressed by the extensive need for users either to prepare vendors for new demands on their training capabilities or for the users themselves to develop their own supplementary training programs for FMS operators. In almost all cases, vendor training was narrowly hardware oriented and took little or no account of the new forms of job design and workforce participation that certain companies had employed in designing and running their FMS [flexible manufacturing systems] operations.

13. "It appears that the majority of the eight months of acceleration in 1985 [to the target production rate] is the result of the major changes in process technology. . . . the length of the acceleration at this plant is comparable to the acceleration period of other vehicle assembly plants that have recently installed similar types of computer-integrated production systems" (Miller and Bereiter 1985: 23).

15. Indeed, the role of standards in lowering the costs of adoption

suggests one additional reason for the cumulative pattern of technological diffusion. As markets expand, de facto standards are established, further hastening the adoption process.

16.   Graham and Rosenthal (1986:7) argue that

> FMS [flexible manufacturing systems] project teams [located
> within the adopter firm] need to include (or at least have easy
> access to) a wide range of disciplines and staff expertise previ-
> ously seldom used for what would ordinarily be considered sim-
> ple facilities planning. The expertise needed may not be as great
> as for in-house development of such systems, but skills that are
> complementary to those of the vendor are vital if effective im-
> plementation and transfer are to take place. These may include,
> among other things, NC [numerical control] programming, soft-
> ware and systems expertise, fixtures and tooling, machine con-
> trol, labor relations and/or personnel.

14.   A recent account of the Ford-American Robot venture contained
the following comment by Romesh Wadhwani, chief executive of
American Robot, concerning the development of new manufacturing technologies for a Ford factory in Ontario, "'It's the kind of stuff
that's still experimental,' explains Wadhwani. 'To make it really
work for Ford, we would have to disclose our technology and that
would require something more than a vendor/customer relationship.'" (Business Week, May 27, 1985:44). Ettlie and Eder (1985:6) also
reported from their case studies that "Two of the respondents in
Case 20 even referred to the evaluation they make of vendors concerning the degree to which these suppliers can keep shared information secret."

17.   David (1985a:377) notes that "Innovation has thus become our cher-
ished child, doted upon by all concerned with maintaining competitiveness and renewing failing industries, whereas diffusion has
fallen into the woeful role of Cinderella, a drudge-like creature who
tends to be overlooked when the summons arrives to attend the
Technology Policy Ball."

18.   Two areas in which federal policies have combined support for
research with some form of support for the diffusion of results are
agricultural research and commercial aircraft (albeit inadvertently
in the latter case). Support for the rapid adoption of innovations in
commercial aircraft was a largely unintended consequence of federal regulation of domestic air transportation (Mowery 1985 discusses the costs and consequences of the postwar policy structure
in the commercial aircraft industry). It seems scarcely coincidental
that these sectors of the U.S. economy have exhibited both rapid
adoption of innovations and remarkably high rates of productivity

growth during the postwar period. Moreover, the products of these sectors are among the most important U.S. exports.

19. See David and Stoneman (1985) for a formal treatment. Japanese government policy toward robotics has emphasized financial incentives to support the diffusion of these technologies, through the establishment of the Japan Robotics Leasing Company (JAROL), as well as the allowance of accelerated depreciation on investments in robotics (see Fleck 1984). In view of recent moves to reduce the differential treatment of various forms of investment within the federal tax code, similar policy initiatives seem unlikely within the United States.

20. See the National Research Council (1983) for a critical evaluation of the impact on smaller machine tool firms of ManTech and other defense manufacturing technology programs, and National Research Council (1987) for a broad critique of the design of the ManTech program.

REFERENCES

Arrow, K. J. 1962. "Economic Welfare and the Allocation of Resources for Invention." In *The Rate and Direction of Inventive Activity* (Princeton, N.J.: Princeton University Press for the National Bureau of Economic Research).

Ayres, R., and S.M. Miller. 1983. *Robotics: Applications and Social Implications* (Cambridge, Mass.: Ballinger).

Bartel, A. P., and F. Lichtenberg. 1987. "The Comparative Advantage of Educated Workers in Implementing New Technology: Some Empirical Evidence." *Review of Economics and Statistics* 59: 1–11.

Brock, G. W. 1975. *The U.S. Computer Industry: A Study of Market Power* (Cambridge, Mass.: Ballinger).

Business Week. 1985. "Will Ford Beat GM in the Robot Race?" (May 27), p. 44.

———. 1986. "GM Throws a Monkey Wrench into the Robot Market." (August 25), p. 36.

Chow, G. C. 1967. "Technological Change and the Demand for Computers," *American Economic Review* 57: 1117–1130.

Conigliaro, L. 1983. "Trends in the Robotics Industry (Revisited): Where Are We Now?" In *Proceedings of the 13th International Symposium on Industrial Robots and Robots 7*, vol. 1, *Applications Worldwide* (Dearborn: Robot Institute of America).

David, P.A. 1966. "The Mechanization of Reaping in the Ante-Bellum Midwest." In *Industrialization in Two Systems: Essays in Honor of Alexander Gerschenkron*, edited by H. Rosovsky (New York: Wiley).

————. 1969. "A Contribution to the Theory of Diffusion." Center for Research on Economic Growth Memorandum 71. Stanford, Calif.

————. 1984. "The Reaper and the Robot: The Diffusion of Microelectronics-Based Process Innovations in Historical Perspective." Center for Economic Policy Research Policy Paper 23. Stanford, Calif.

————. 1985a. "New Technology Diffusion, Public Policy and Industrial Competitiveness." In *The Positive-Sum Strategy*, edited by R. Landau and N. Rosenberg. (Washington, D.C.: National Academy Press).

————. 1985b. "Clio and the Economics of QWERTY." *American Economic Review* 75: 332–337.

David, P.A., and P.L. Stoneman. 1985. "Adoption Subsidies vs. Information Provision as Instruments of Technology Policy." Center for Economic Policy Research Publication 49. Stanford, Calif.

Davies, S. 1979. *The Diffusion of Process Innovations* (Cambridge: Cambridge University Press).

Economist. 1986. "Wanted: New Jobs for Sad Robots" (July 26), pp. 59–60.

Enos, J. 1962. "Invention and Innovation in the Petroleum Refining Industry." In *The Rate and Direction of Inventive Activity* (Princeton, N.J.: Princeton University Press for the National Bureau of Economic Research).

Ettlie, J.E. 1985a. "The Implementation of Programmable Manufacturing Innovations." Working paper, Center for Social and Economic Issues, Industrial Technology Institute. Ann Arbor, Mich.

————. 1985b. "Management and Robotics." Working paper, Center for Social and Economic Issues, Industrial Technology Institute. Ann Arbor, Mich.

————. 1986. "Systemic Innovation." in *Strategies and Practices for Technological Innovation*, edited by D. Gray et al. (Amsterdam: North-Holland).

Ettlie, J.E., and J.L. Eder. 1985. "Managing the Vendor-User Team for Successful Implementation of Flexible Automation." Paper presented at the 22nd Conference and Exposition of the Association for Integrated Manufacturing, St. Louis. May 14–17.

Farrell, J., and G. Saloner. 1985. "Economic Issues in Standardization." Massachusetts Institute of Technology Economics Department Working Paper 393. Cambridge, Mass.

Fleck, J. 1984. "The Adoption of Robots in Industry." *Physics in Technology*, vol. 15: 4–11.

Fleck, J., and B. White. 1985. "National Policies and Patterns of Robot Diffusion." In *Proceedings of the 14th International Symposium on Industrial Robots*, edited by N. Martensson (Amsterdam: North-Holland).

Freeman, C., J. Clark, and L. Soete. 1982. *Unemployment and Technical Innovation: A Study of Long Waves* (London: Frances Pinter).

Graham, M.B.W., and S. Rosenthal. 1986. "Flexible Manufacturing Systems Require Flexible People." *Human Systems Management* 6: 211–222.

Griliches, Z. 1957. "Hybrid Corn: An Exploration in the Economics of Technological Change." *Econometrica* 25: 501–522.

———. 1971. "Hybrid Corn and the Economics of Innovation." *Science* 132: 275–280. Reprinted in *The Economics of Technological Change*, edited by N. Rosenberg (Harmondsworth: Penguin).

———. 1961. "Congruence versus Profitability: A False Dichotomy." *Rural Sociology* 25: 354–356.

———. 1962. "Profitability versus Interaction: Another False Dichotomy." *Rural Sociology* 27: 327–330.

Howell, D.R. 1985. "The Future Employment Impacts of Industrial Robots: An Input-Output Approach." *Technological Forecasting and Social Change* 28: 297–310.

Hunt, H.A., and T.L. Hunt. 1983. *Human Resource Implications of Robotics* (Kalamazoo: W.E. Upjohn Institute for Employment Research).

———. 1985. "An Assessment of Data Sources to Study the Employment Effects of Technological Change." In National Research Council, *Technology and Employment: Interim Report* (Washington, D.C.: National Academy Press).

James, J.A. 1984. "Perspectives on Technological Change: Historical Studies of Four Major Innovations." National Commission for Employment Policy *Research Report* 84-07. Washington, D.C.

Kaplan, R.S. 1986. "Must CIM Be Justified by Faith Alone?" *Harvard Business Review* 64: 87–95.

Katz, M.L., and C. Shapiro. 1985. "Technology Adoption in the Presence of Network Externalities." Working paper, Woodrow Wilson School, Princeton University.

Klepper, S.I. 1988. "Joint Ventures in Robotics." In *International Collaborative Ventures in U.S. Manufacturing*, edited by D.C. Mowery (Cambridge, Mass.: Ballinger).

Mansfield, E. 1961. "Technical Change and the Rate of Imitation." *Econometrica* 29: 741–766.

———. 1963. "Size of Firm, Market Structure, and Innovation." *Journal of Political Economy* 71: 556–76.

———. 1966. "Technological Change: Measurement, Determinants, and Diffusion." In Vol. II, *Studies of the Employment Impact of Technological Change*, prepared for the National Commission on Technology, Automation, and Technological Progress (Washington, D.C.: U.S.G.P.O.).

———. 1968. *Industrial Research and Technological Innovation* (New York: Norton).

———. J. Rapoport, J. Schnee, S. Wagner, and M. Hamburger. 1971. *Research and Innovation in the Modern Corporation* (New York: Norton).

Mearman, J. 1987. U.S. Department of Commerce: personal communication, September 3.

Mensch, G. 1979. *Stalemate in Technology* (Cambridge, Mass.: Ballinger).

Miller, S.M., and S.R. Bereiter. 1985. "Modernizing to Computer-Integrated Production Technologies in a Vehicle Assembly Plant: Lessons for Analysts and Managers of Technological Change." Paper presented at the NBER conference on Productivity Growth in the U.S. and Japan, Cambridge, Mass.

Mowery, D.C. 1983. "Economic Theory and Government Technology Policy." *Policy Sciences.* 16: 27–43.

———. "The Impact of Federal Support of Transportation R&D: The Case of Aircraft." Paper presented at the National Academy of Sciences Symposium on the Impact of Federal Support of R&D, Washington, D.C.

Nabseth, L., and G.F. Ray. 1974. *The Diffusion of New Industrial Processes* (Cambridge: Cambridge University Press).

National Research Council, Committee on the Machine Tool Industry. 1983. *The Machine Tool Industry and the Defense Industrial Base* (Washington, D.C.: National Academy Press).

National Research Council, Committee on the Role of the Manufacturing Technology Program in the Defense Industrial Base. 1987. *Manufacturing Technology: Cornerstone of a Renewed Defense Industrial Base* (Washington, D.C.: National Academy Press).

Nelson, R.R., and E.S. Phelps. 1966. "Investment in Humans, Technological Diffusion and Economic Growth." *American Economic Review* 56: 69–75.

Nelson, R.R., M.J. Peck, and E.D. Kalachek. 1967. *Technology, Economic Growth, and Public Policy* (Washington, D.C.: Brookings Institution).

Nelson, R.R., and S.G. Winter. 1982. *An Evolutionary Theory of Economic Change* (Cambridge, Mass.: Harvard University Press).

New York Times. 1985. "Census, Counting Robots, Finds 5,535 Added in 1984" (September 3), p. 10.

Ray, G.F. 1984. *The Diffusion of Mature Technologies* (Cambridge: Cambridge University Press).

Rogers, E.M. 1983. *Diffusion of Innovations*, 3d ed. (New York: Free Press).

Rogers, E.M., and A.E. Havens. 1962. "Rejoinder to Griliches' 'Another False Dichotomy.'" *Rural Sociology* 27: 330–332.

Romeo, A.A. 1975. "Interindustry and Interfirm Differences in the Rate of Diffusion of an Innovation." *Review of Economics and Statistics* 57: 311–319.

———. "The Rate of Imitation of a Capital Embodied Process Innovation." *Economica* 44: 63–69.

Rosenberg, N. 1975. In *Perspectives on Technology* (New York: Cambridge University Press).

————. 1982. *Inside the Black Box: Technology in Economics*, (New York: Cambridge University Press).

Rosenberg, N., and C. Frischtak. 1984. "Technological Innovation and Long Waves." *Cambridge Journal of Economics* 8: 7–24.

Ryan, B., and N. C. Gross. 1943. "The Diffusion of Hybrid Seed Corn in Two Iowa Communities." *Rural Sociology* 8: 13–24.

Stigler, G. J. 1956. "Industrial Organization and Economic Progress." In *The State of the Social Sciences*, edited by L. D. White (Chicago: University of Chicago Press).

Stoneman, P.L. 1976. *Technological Diffusion and the Computer Revolution: The U.K. Experience* (Cambridge: Cambridge University Press).

————. 1983. *The Economics of Technological Change* (New York: Oxford University Press).

Stoneman, P.L., and N. Ireland. 1983. "The Role of Supply Factors in the Diffusion of New Process Technology." *Economic Journal* 93 (Supplement): 65–77.

Technology Management Center. 1985. "The Use of Advanced Manufacturing Technology in Industries Impacted by Import Competition: An Analysis of Three Pennsylvania Industries." unpublished report. Philadelphia, Penn.

Tierney, M.L. 1983. "Employer Provided Education and Training in 1981." In *Training's Benchmarks: A Statistical Sketch of Employer-Provided Training and Education, 1969–1981*, Task I report, *The Impact of Public Policy on Education and Training in the Private Sector*, edited by R. Zemesky (Philadelphia: Higher Education Finance Research Institute).

U.S. Department of Commerce. 1983. *The Robotics Industry* (Washington, D.C.: U.S.G.P.O.).

————. 1983. *U.S. Industrial Outlook: 1983* (Washington, D.C.: U.S.G.-P.O.).

————. 1984. *U.S. Industrial Outlook: 1984* (Washington, D.C.: U.S.G.-P.O.).

————. 1985. *A Competitive Assessment of the U.S. Flexible Manufacturing Systems Industry* (Washington, D.C.: U.S.G.P.O.).

————. 1987. *U.S. Industrial Outlook: 1987* (Washington, D.C.: U.S.G.-P.O.).

U.S. International Trade Commission. 1983. *The Competitive Position of U.S. Producers of Robotics in Domestic and World Markets* (Washington, D.C.: U.S.G.P.O.).

U.S. Office of Technology Assessment. 1984. *Computerized Manufacturing Automation: Employment, Education, and the Workplace* (Washington, D.C.: U.S.G.P.O.).

PAPERS COMMISSIONED BY THE PANEL ON TECHNOLOGY AND EMPLOYMENT

C. Michael Aho, Council on Foreign Relations, "Technology, Structural Change and Trade"

Martin Neil Baily, Brookings Institution, "An Analysis of the Productivity Growth Decline"*

Martin Binkin, Brookings Institution, "Technology and Skills: Lessons from the Military"

David E. Bloom, Columbia University, and McKinley L. Blackburn, University of South Carolina, "The Effects of Technological Change on Earnings and Income Inequality in the United States"

Joseph J. Cordes, George Washington University, "The Impact of Tax Policy on the Creation of New Technical Knowledge: An Assessment of the Evidence"

Robert M. Costrell, University of Massachusetts, Amherst, "The Impact of Technical Progress on Productivity, Wages, and the Distribution of Employment: Theory and Postwar Experience in the U.S."

Donald Critchlow, University of Notre Dame, "The Politics of Technology and Employment: The 1965 National Commission on Technology, Automation, and Economic Progress"*

*Available through Publication on Demand Program, National Academy Press, Washington, D.C. 20418 (Attention: Stephen Zubal).

Steven Deutsch, University of Oregon, "Technological Change, Worker Displacement and Readjustment, Employment and Job Training"*

Kenneth Flamm, Brookings Institution, "The Changing Pattern of Industrial Robot Use"

Jeffrey Hart, Indiana University, and Jeanne Schaaf, Telenet Corporation, "International Trade and U.S. Competitiveness in Services"*

Joseph Hight, U.S. Department of Labor, "Measuring Sources of Employment Change in U.S. Mining and Manufacturing, 1978–84: The Income Accounting Approach"*

Larry Hirschhorn, University of Pennsylvania, "Computers and Jobs: Services and the New Mode of Production"

Jonathan S. Leonard, University of California, Berkeley, "Technological Change and the Extent of Frictional and Structural Unemployment"

Alan Jay Marcus, "Military Training and the Civilian Economy"*

Michael Morgan, University of Washington, "Emerging Technologies: Implications for Worker Health and Safety"*

David C. Mowery, Carnegie-Mellon University, "The Diffusion of New Manufacturing Technologies"

Walter Y. Oi, University of Rochester, "The Indirect Impact of Technology on Retail Trade"

Michael Podgursky, University of Massachusetts, Amherst, "Job Displacement and Labor Market Adjustment: Evidence from the Displaced Worker Surveys"

Kenneth I. Spenner, Duke University, "Technological Change, Skill Requirements, and Education: The Case for Uncertainty"

Kan Young and Ann Lawson, U.S. Department of Commerce, "Effects of Changes in Technology and Demand on Employment Growth across Industries, 1972–84"*

SUBJECT INDEX

Accelerated Cost Recovery System (ACRS), 445, 450, 454
Adaptability advantage, 171
Administrative innovation, 503 n. 11
Aerospace industry, 187, 189, 283, 440 n. 4, 504 n. 18
Aegis missile, 214
Age factor in jobless duration, 25
Agricultural sector, 504 n. 18
Airbus, 440 n. 4
American Robot, 504 n. 14
Apparel industry, 64
"Arrow effect," 309
Asset Depreciation Range (ADR) system, 445
Austria, 488
Automated guided vehicles (AGVs), 270
Automation and job displacement, 20
Automobile industry, 473, 474; and robotics, 273–310 *passim*, 318, 323 ns. 26 and 27, 324 n. 31, 495
Average productivity of labor (APL), 118–119
Avionics systems, 207, 209–210
Armed services: aptitude test scores (AFQT), 199, 200; Armed Services Vocational Aptitude Battery, (ASVAB), 203; clerical positions, 188; computer-based command, control, and communications systems, 187; computer-based instruction, 205–206; computer models, 191; computer software maintenance, 213–214; educational attainment, 197–199, 200; electronics technology, 186–187, 188, 189, 207–208; electronics training eligibility, 201–203; enlistment standards/job performance, 203, 216–217, 219 n. 17; entry qualifications, 200–203; "ground combat" jobs, 186; high-performance weapons systems, 210–211, 214–215, 217; language skills training, 206; maintenance manpower requirements, 192–193; maintenance problems, 209–210, 211–214; occupational categories distribution, 187–188; organizational problems, 214–216; physical requirements/moral standards, 218 n. 14; recruitment

513

transfer machines, 270; sophisticated robots, 292–294; Unimate, 273; utilization level in other countries, 274–276, 491, 503 n. 8; VERSATRAN, 273; and welding, 277, 288–292, 294, 295, 303, 308, 313, 317, 318, 322 n. 25, 323 n. 27, 324 n. 31; workcells, 270
Rubber industry, 64

Safety standards, 98, 313
Sample, selection criteria, 9, 20–21, 37 n. 3
Sankyo Seiki, 311
Savings per head, 78, 79
Scale, returns to, 109, 429–431
Scale economies: external, 431, 435; and robotics, 299–301, 304, 326 n. 48
Scale effects, 229, 230
Scale monopolies, 430, 431
Scandinavia and robotics, 277
SCARA (Selective Compliance Assembly Robot Arm) robot, 311
Schmoller, Gustav, 434
Schumpeter, Joseph, 429
Secretaries, growth in number of, 387
Sectoral employment growth, 81
Sectoral productivity advance, 82
Sectoral shift, 99, 123 n. 13, 378–381; and earnings irregularity, 235–237, 242, 245, 246, 256, 259 n. 18; and labor's share, 85; and productivity growth, 80–83; and wage growth, 83–84, 115
Sectoral targeting, 442 n. 11
Sectors, growing/declining, 49, 86, 99
Self-employment in retail trade, 102, 123 n. 25
Semiconductors, 303; and Japan, 309–310
Seniority and earnings losses, 33
Service sector: capital intensity, 380; convergence with goods sector, 379, 380; and demand for services, 88, 108; "earnings ratio" for, 29; employment shares, 92, 93, 96, 102, 244; em-

ployment shift to, 5, 90–91, 102, 103, 109–110; investment per employee, 379–380; job creation in, 379; and job displacement, 14, 17; marginal productivity of labor, 115; output and productivity growth, 103, 108, 118, 121, 125 n. 36; relative wages, 97; reemployment in, 27, 33; R&D, 460; unemployment rate, 21; and wage loss for relative displacement, 99
Services to buildings, 97, 108, 124 n. 27
Shipments, real, 103
Shopping trip frequency, 339–340, 373 n. 49
Skill(s). See Job skill(s) entries
Skill-intensive employers, 69 n. 10
Skill mix, 193, 336
Skill shift hypothesis, contradictory, 157–158, 159
Skilled-unskilled workers ratio, 336
"Skillware" computer program, 402
Smith, Adam, 427
Society of Manufacturing Engineers (SME), 312
Socioeconomic environment, resource-rich/resource-lean, 165–166
Sony, 308
Staffing ratios, changes in, 387–388
Standard and Poor's Compustat, 463, 465
Structural change, 422–423
Subcontracting, 108, 297
Subsidies: for debt financing, 476–478; versus import duties, 435
Substitution, elasticities of, 77, 109
Supply-side theory, 4–5, 429
Sweden, 488, 499; R&D tax credits, 467–468; robotics, 274, 284, 287, 318, 489, 491

Target employment, 60
Targeted industrial policy, 440 n. 4

AUTHOR INDEX

LIST OF CONTRIBUTORS

C. Michael Aho, International Trade Project, Council on Foreign
   Relations
Martin Binkin, Brookings Institution
McKinley L. Blackburn, Department of Economics, University of
   South Carolina
David E. Bloom, Department of Economics, Columbia University
Joseph J. Cordes, Department of Economics, George Washington
   University
Robert M. Costrell, Department of Economics, University of Mas-
   sachusetts, Amherst
Kenneth S. Flamm, Brookings Institution
Larry Hirschhorn, Wharton School, University of Pennsylvania
Jonathan S. Leonard, School of Business, University of California,
   Berkeley
David C. Mowery, Department of Social and Decision Sciences,
   Carnegie-Mellon University
Walter Y. Oi, Department of Economics, University of Rochester
Michael Podgursky, Department of Economics, University of
   Massachusetts, Amherst
Kenneth I. Spenner, Department of Sociology, Duke University

535